Business Strategies and Approaches for Effective Engineering Management

Saqib Saeed
Bahria University Islamabad, Pakistan

Mohammad Ayoub Khan
C-DAC, Ministry of IT, India

Rizwan Ahmad
Qatar University, Qatar

BUSINESS SCIENCE
Reference

Managing Director:	Lindsay Johnston
Editorial Director:	Joel Gamon
Book Production Manager:	Jennifer Yoder
Publishing Systems Analyst:	Adrienne Freeland
Development Editor:	Myla Merkel
Assistant Acquisitions Editor:	Kayla Wolfe
Typesetter:	Christy Fic
Cover Design:	Jason Mull

Published in the United States of America by
    Business Science Reference (an imprint of IGI Global)
    701 E. Chocolate Avenue
    Hershey PA 17033
    Tel: 717-533-8845
    Fax: 717-533-8661
    E-mail: cust@igi-global.com
    Web site: http://www.igi-global.com

Library of Congress Cataloging-in-Publication Data

Business strategies and approaches for effective engineering management / Saqib Saeed, Mohammad Ayoub Khan, and Rizwan Ahmad, editor.
    p. cm.
  Includes bibliographical references and index.
  Summary: "This book brings together the latest methodologies, principles, practices and tools for engineering management, providing theoretical analysis and practical applications"--Provided by publisher.
  ISBN 978-1-4666-3658-3 (hardcover) -- ISBN 978-1-4666-3659-0 (ebook) -- ISBN 978-1-4666-3660-6
1. Industrial engineering--Management. 2. Production management. I. Saeed, Saqib, 1970- II. Khan, Mohammad Ayoub, 1980- III. Ahmad, Rizwan, 1978-
  T56.B87 2013
  620.0068'4--dc23
                          2012045774

British Cataloguing in Publication Data
A Cataloguing in Publication record for this book is available from the British Library.

Table of Contents

Preface xiii

Section 1
Process Management

Chapter 1
Order Release Strategies for Customer Order Scheduling Problems in Dynamic Environments 1
Paolo Renna, University of Basilicata, Italy

Chapter 2
Managing Organizational Change 22
Martin L. Bariff, Illinois Institute of Technology, USA

Chapter 3
Green Process Management Using Six Sigma Concepts 48
Seifedine Kadry, American University of the Middle East, Kuwait

Chapter 4
New Product Development: Challenges and Implications 62
Hassanali Rassouli, Aagahn Information Technology Consultants, Iran

Chapter 5
Supply Chain Analysis 84
Mohammad Anwar Rahman, University of Southern Mississippi, USA

Chapter 6
Intellectual Property Rights in Semi-Conductor Industries: An Indian Perspective............................ 97
Satish Chandra Tiwari, NSIT, India
Maneesha Gupta, NSIT, India
Mohammad Ayoub Khan, Government of India, India
A. Q. Ansari, Jamia Millia Islamia, India

Chapter 7
Defect Trend Analysis of MI-172 Helicopters Through Maintenance History 111
Mudassir Hussain, Centre for Advanced Studies in Engineering, Pakistan
Irfan Anjum Manarvi, HITEC University, Pakistan
Assad Iqbal, Bahria University Islamabad, Pakistan

Section 2
Project Management

Chapter 8
Component Failure Analysis of J69-T-25A Engine 128
Muhammad Asim Qazi, Center for Advanced Studies in Engineering, Pakistan
Irfan Manarvi, HITEC University Taxila, Pakistan
Assad Iqbal, Bahria University Islamabad, Pakistan

Chapter 9
Lean Development: A Tool for Knowledge Management in Software Development Process........... 142
Saqib Saeed, University of Siegen, Germany
Izzat Alsmadi, Yarmouk University, Jordan
Farrukh Masood Khawaja, Ericsson Telekommunikation GmbH & Co. KG, Germany

Chapter 10
Management Practices in Exploration and Production Industry 151
Kashif Saeed, Wintershall Holding GmbH – Kassel, Germany
Georg Ziegler, Wintershall Holding GmbH – Kassel, Germany
Muhammad Kashif Yaqoob, Mubadala Petroleum – Abu Dhabi, UAE

Chapter 11
Reasons Behind IT Project Failure: The Case of Jordan 188
Emad Abu-Shanab, Yarmouk University, Jordan
Ashraf Al-Saggar, Irbid Electricity Company, Jordan

Chapter 12
Issues and Technologies of Effective Energy Management 203
Edward T. Chen, University of Massachusetts – Lowell, USA

Chapter 13
An Examination of the Decision Making Styles of Egyptian Managers 219
Hisham M. Abdelsalam, Cairo University, Egypt
Reem H. Dawoud, Financial Consultant, Egypt
Hatem A. ElKadi, Cairo University, Egypt

Chapter 14
Identification of Major FOD Contributors in Aviation Industry 237
    Hammad Ahmed Rafiq, Centre for Advance Studies in Engineering (CASE), Pakistan
    Irfan Anjum Manarvi, HITECH University Taxila, Pakistan
    Assad Iqbal, Bahria University Islamabad, Pakistan

Section 3
Technology Management

Chapter 15
Media Management in Disaster Events: A Case Study of a Japanese Earthquake 252
    Eleonora Benecchi, University of Lugano, Switzerland
    Vincenzo De Masi, University of Zurich, Switzerland

Chapter 16
The Impact of Virtual Community (Web 2.0) in the Economic, Social, and Political Environment
of Traditional Society 262
    Irene Samanta, Technological Educational Institute of Piraues, Greece

Chapter 17
Website Performance Measurement: Process and Product Metrics... 275
    Izzat Alsmadi, Yarmouk University, Jordan

Chapter 18
Measuring the Conceptual Variables for E-Services Acceptance: A Descriptive Statistical
Analysis... 302
    Kamaljeet Sandhu, University of New England, Australia

Chapter 19
Empirical Analysis for E-Services Acceptance Model: Important Findings 310
    Kamaljeet Sandhu, University of New England, Australia

Compilation of References 323

About the Contributors 343

Index... 350

Detailed Table of Contents

Preface xiii

Section 1
Process Management

Chapter 1

Order Release Strategies for Customer Order Scheduling Problems in Dynamic Environments 1
Paolo Renna, University of Basilicata, Italy

The research proposes two strategies to release the parts in a job shop environment to handle the customer order-scheduling problem. The first strategy is based on the evaluation of a dynamic conwip level to take the decision. The second strategy tries to anticipate the production of components of a generic order when the utilization of the manufacturing system is low. In this strategy, a fuzzy approach is proposed to decide how many components to release. A simulation environment has been developed to test the proposed approaches. Two benchmark models are used to compare the performance measures: no order release strategy and classical conwip. Moreover, the simulations are conducted in a very dynamic environment. The simulation results show how the fuzzy approach leads to the better results in all conditions tested with a relevant reduction of the inventory level.

Chapter 2

Managing Organizational Change 22
Martin L. Bariff, Illinois Institute of Technology, USA

Many project deliverables extend beyond a product or a service for sale to customers. The deliverable may include a new or a revised process for internal workflow or relations with customers, suppliers, or partners. The success of these projects will depend upon adoption of the new or revised process in addition to typical metrics for cost, schedule, risk, and quality. The project manager and team will be responsible for "managing organizational change"—a skillset that is not addressed within the Project Management Institute Body of Knowledge. The purpose of this chapter is to provide sufficient knowledge about approaches and implementation for organizational change to achieve total project success. Case studies are included to illustrate best practices and lessons learned.

Chapter 3

Green Process Management Using Six Sigma Concepts 48
Seifedine Kadry, American University of the Middle East, Kuwait

The Six Sigma (6-σ) methodology, as it has evolved over the last two decades, provides a proven framework for problem solving and organizational leadership and enables leaders and practitioners to employ new ways of understanding and solving their sustainability problems. While business leaders now understand the importance of environmental sustainability to both profitability and customer satisfaction, few are able to translate good intentions into concrete, measurable improvement programs. Increasingly, these leaders are looking to their corps (six sigma experts) of six sigma "Master Black belts," "Black belts," and "Green belts" to lead and implement innovative programs that simultaneously reduce carbon emissions and provide large cost savings. Six sigma is a powerful execution engine and sustainability programs are in need of this operational approach and discipline. Six sigma rigors will help a business leader to design a sustainable program for both short- and long-term value creations. The aim of this chapter is to show the importance of applying six sigma methodologies to multidisciplinary sustainability-related projects and how to implement it.

Chapter 4

New Product Development: Challenges and Implications 62
Hassanali Rassouli, Aagahn Information Technology Consultants, Iran

Intensive global competition of 21st century calls for a systematically aligned set of approaches in all aspects of corporate management including new product development. Success of new products needs the strategic support of corporate executives to establish and deploy coherent policies that are compatible with approaching new paradigms in business and manufacturing. Lean thinking is the widely accepted paradigm in the beginning of 21st century, and this chapter defines challenges and implications that corporate management will face in attempting to adapt to new product development approaches in accordance with the lean thinking perspective. This attempt covers all aspects of the organization in respect to product lifecycle.

Chapter 5

Supply Chain Analysis 84
Mohammad Anwar Rahman, University of Southern Mississippi, USA

This chapter focuses on supply chain strategies that benefit organizations to meet the global challenges and business success. The chapter supports understanding of functional coordination in supply chain management, importance of supply chain partnership and performance measures, bullwhip cause-effect and consequences, business promotion strategies, and approach to synchronize operation schedule to improve supply chain operations. Bullwhip effect is often seen as a challenge to improve supply chain performances. It also discusses the strategies to reduce bullwhip effect, various well-known contract agreements that are currently practiced by many supply chain organizations, its impact and benefits on supply chain success.

Chapter 6

Intellectual Property Rights in Semi-Conductor Industries: An Indian Perspective............................97
Satish Chandra Tiwari, NSIT, India
Maneesha Gupta, NSIT, India
Mohammad Ayoub Khan, Government of India, India
A. Q. Ansari, Jamia Millia Islamia, India

Fundamentals of intellectual property rights are provided. In addition, the trends of patenting and patented technologies in India in different areas of semi-conductor technologies are analyzed. The authors discuss many aspects of the patents and patentability. They present patent practices in India covering required forms needed to be filled in order to file a patent. Finally, the importance of patenting and its growth is shown with few year wise statistics.

Chapter 7

Defect Trend Analysis of MI-172 Helicopters Through Maintenance History 111
Mudassir Hussain, Centre for Advanced Studies in Engineering, Pakistan
Irfan Anjum Manarvi, HITEC University, Pakistan
Assad Iqbal, Bahria University Islamabad, Pakistan

MI-172 helicopters are a variant of MI-17/171 helicopters. These helicopters were inducted in the organization, which was originally designed and tuned to the MI-17/171 series helicopters. Since the induction was not premeditated, it resulted in diverse problems that have accumulated over a period of 3 years. This chapter focuses on the defect trend analysis and identification of root causes of the accumulated problems. Technical and flying data is collected from the user Squadron and analyzed by statistical tools. The chapter is helpful in eradicating the existing problems and suggesting pragmatic solutions for an overall improvement of the maintenance setup.

Section 2
Project Management

Chapter 8

Component Failure Analysis of J69-T-25A Engine 128
Muhammad Asim Qazi, Center for Advanced Studies in Engineering, Pakistan
Irfan Manarvi, HITEC University Taxila, Pakistan
Assad Iqbal, Bahria University Islamabad, Pakistan

Reliability and serviceability of jet engines in the aviation industry is of paramount importance and is directly related to flight safety. Tight maintenance programs, including scheduled and preventive inspection are in place worldwide for jet engines to ensure air worthiness of aircraft. Old age provides maintenance maturity to the system, but on other hand, it requires focused efforts to ensure reliability due to aging factor. J69-T-25A falls in the same category, as it has been in service for the last six decades. Despite all maintenance efforts, a variety of defects are being faced on J69 engines. The major defects include RPM fluctuation, noise, oil gain, vibration, and smoke. The troubleshooting process identifies a number of components that cause these problems. this chapter is based on statistical analyses of component failure in terms of frequency and fault isolation. The top ten components were selected based upon failure rates and were compared against reported problems to establish a relationship between defects and failed components. Based upon the result, various remedial measures are suggested to reduce defects in the future and increase engine reliability.

Chapter 9
Lean Development: A Tool for Knowledge Management in Software Development Process........... 142
Saqib Saeed, University of Siegen, Germany
Izzat Alsmadi, Yarmouk University, Jordan
Farrukh Masood Khawaja, Ericsson Telekommunikation GmbH & Co. KG, Germany

Software development is a complex activity, which is human intensive in nature. In order to build quality software systems, organizations need to follow mature software development practices, which are continually improved. As a result, the concept of software development process emerged, which highlighted a systematic set of activities required to develop a software system. Recently, agile development methodologies have provided a rich set of innovative software development approaches, aiming to optimize the software process. In order to be successful in adopting these approaches, a thorough understanding of their implementation procedures is required. In this chapter, we took a look at the lean development approach to understand how its principles pave the way in fostering knowledge management initiatives in software process development.

Chapter 10
Management Practices in Exploration and Production Industry 151
Kashif Saeed, Wintershall Holding GmbH – Kassel, Germany
Georg Ziegler, Wintershall Holding GmbH – Kassel, Germany
Muhammad Kashif Yaqoob, Mubadala Petroleum – Abu Dhabi, UAE

This chapter is divided into three main sections; project management, HSE management, and quality management. A focus description of the different elements of exploration and production industry along with implementation of management practices on each of these elements including asset/portfolio, resources, time, project planning and scheduling, and proactive risk management are presented. Health safety and environment and quality management are dealt with as separate sections.

Chapter 11
Reasons Behind IT Project Failure: The Case of Jordan 188
Emad Abu-Shanab, Yarmouk University, Jordan
Ashraf Al-Saggar, Irbid Electricity Company, Jordan

Information Technology (IT) projects have high failure and escalation rates because of the nature of domain and the rapid technology changes. It is important to understand the factors causing IT project success or failure. This chapter reviews the literature related to project failure and escalation and concludes with 17 important factors that cause IT projects to fail and 10 factors that contribute to the escalation of projects in time, cost, or scope. The concluded factors are utilized in an empirical study to explore the Jordanian environment and check the rank of these factors as perceived by Jordanian specialists. Conclusions and future work are stated at the end of this chapter.

Chapter 12
Issues and Technologies of Effective Energy Management 203
Edward T. Chen, University of Massachusetts – Lowell, USA

The purpose of this chapter is to discuss critical issues and the role technology plays in today's energy sectors. Specific emphasis is placed on security, mobile dispatch solutions, and the so-called "Smart Grid." The industry continues to grow in both size and complexity, creating a multitude of challenges for companies as they struggle to keep the lights on. The utility business has traditionally lagged behind other sectors in the adoption and implementation of new technologies. However, mounting economic, environmental, social, and political pressures have thrust this once lumbering dinosaur out into the spotlight. Energy companies must look to innovative technology solutions to help them keep pace with our growing society. The chapter also touches upon how these issues create meaningful educational and employment opportunities.

Chapter 13
An Examination of the Decision Making Styles of Egyptian Managers 219

Hisham M. Abdelsalam, Cairo University, Egypt
Reem H. Dawoud, Financial Consultant, Egypt
Hatem A. ElKadi, Cairo University, Egypt

Many factors play roles in the success of managers. However, the manager's decision-making style is one factor that highly contributes to that success and, therefore, to the success of their organization. In this chapter, a survey that includes a sample of 138 Egyptian managers in different organizational levels (junior, middle, and senior) is conducted to explore their decision-making styles. The research, then, investigates the relation between the variety of managers' decision styles and seven variables: gender, age, ethnicity, educational level, educational major, administrative experience, and current position. Based on the findings, this research is able to provide baseline information to improve on the implications of decision-making styles on the selection and design of decision-support systems in Egypt.

Chapter 14
Identification of Major FOD Contributors in Aviation Industry 237

Hammad Ahmed Rafiq, Centre for Advance Studies in Engineering (CASE), Pakistan
Irfan Anjum Manarvi, HITECH University Taxila, Pakistan
Assad Iqbal, Bahria University Islamabad, Pakistan

Aviation safety is considered of paramount importance, and the Foreign Object Debris and the resulting Foreign Object Damage (FOD) is one of the major causes that put aviation safety at risk. FOD Prevention is thus a continual challenge for all aircraft operators and maintenance crew. It costs the aviation industry millions of dollars every year. This financial effect is a result of direct costs, such as harm to aircraft structures or damage of aircraft engines, as well as the indirect costs, which include flight schedule delays, cancellations, disruptions, and additional effort for the employees. In addition, on occasion, more critical than the financial impact, is the safety impact and potential loss of human life associated with occurrences caused by FOD. It is therefore ranked as the most likely potential ground-based cause that can lead to a catastrophic aviation event. The present chapter is based on statistical analysis of aircraft occurrences attributed to various types of FOD during the last ten years of operations in an aviation organization. Eight major cause factors contributing towards these cases have been identified. A broad FOD prevention and control plan is thus proposed to address the foremost cause factors and improve organizational response to FOD. The objective of the research is to promote ground and flight safety and the preservation of assets by reducing FOD.

Section 3
Technology Management

Chapter 15

Media Management in Disaster Events: A Case Study of a Japanese Earthquake 252
Eleonora Benecchi, University of Lugano, Switzerland
Vincenzo De Masi, University of Zurich, Switzerland

According to a survey by Goo Research (April 2011), the average Japanese person appears to have relied primarily on television news for gathering information in times of disaster, and as unlike a lot of overseas media, the public broadcaster NHK's news broadcasts were defined as very calm and measured. This chapter focuses on the NHK coverage of the earthquake and nuclear crisis in March 2011 compared with private channels' and specific websites' coverage with regard to specific events. The aim is to enlighten the ways and the tools through which Japanese Public Television played a double role: on one side it became a primary source of information for hard news and played a "service" role for the population in need; on the other side and with special regard to the coverage of the nuclear crisis, the duty to inform was balanced by the duty to reassure the public and promote harmony so that NHK privileged government and corporate statements about the Fukushima situation. The authors corroborate their study through an analysis of NHK's programming and private channels' changing schedules and advertising during the recent disaster. This chapter provides a concrete example of the potential television role in disaster mitigation, taking into account both the positive and critical aspects.

Chapter 16

The Impact of Virtual Community (Web 2.0) in the Economic, Social, and Political Environment
of Traditional Society 262
Irene Samanta, Technological Educational Institute of Piraues, Greece

The chapter enhances the scientific research in the area of the new digital era with a focus on diversity created in real society from the influence of social media. Specifically, it reveals the effects of social media on economic, political, and real society affairs. The latest riots in Middle East countries demonstrate that virtual social communities wield an influence on the citizens, and the changes they implemented show these countries will never be the same again. The effects of social media in real society are examined in highly developed countries such as the EU and North America (USA and Canada).

Chapter 17

Website Performance Measurement: Process and Product Metrics.. 275
Izzat Alsmadi, Yarmouk University, Jordan

Some tasks will be easier to implement and test, and others will either be un-applicable or difficult to test and implement in comparison with testing in traditional software development environments. For engineering management, product and process quality evaluation are important assessment tools by which managers can have significant indicators of the evaluated project or product. There are many ways and characteristics by which websites can be evaluated. Quality attributes can be external or internal. They can be measured based on the developed product (i.e. the website) or the developing process. In this chapter, the author describes in detail some of the product and process metrics by which websites can be evaluated. They are described based on the major classification: process and product metrics. In each one of those two major classes, the author describes possible measurements, how they can be evaluated, and examples of attributes and tools used in this measurement. Values of measurements can in combination provide useful information for project management and planning. Focusing on only one or two attributes can possibly be insufficient or misleading.

Chapter 18
Measuring the Conceptual Variables for E-Services Acceptance: A Descriptive Statistical
Analysis.. 302
Kamaljeet Sandhu, University of New England, Australia

This case study examines the Web Electronic Service framework for a University in Australia. The department is in the process of developing and implementing a Web-based e-service system. The user experience to use e-services requires insight into the attributes that shape the experience variable. The descriptive data about the attributes that form the experience variable is provided in this study.

Chapter 19
Empirical Analysis for E-Services Acceptance Model: Important Findings 310
Kamaljeet Sandhu, University of New England, Australia

This study investigates factors that influence the acceptance and use of e-Services. The research model includes factors such as user experience, user motivation, perceived usefulness, and perceived ease of use in explaining the process of e-Services acceptance, use, and continued use. The two core variables of the Technology Acceptance Model (TAM), perceived usefulness and perceived ease of use, are integrated into the Electronic Services Acceptance Model (E-SAM).

Compilation of References 323

About the Contributors 343

Index 350

Preface

Engineering projects are critical undertakings requiring serious commitment, clear vision, and robust long-term strategies to be successful. As a result, planning, monitoring, and evaluation activities are of utmost importance for successful execution. The projects may be carried out in technically sound ways, but failure can still be experienced due to weakness in management, such as improper budget and planning schedule. This book is aimed at highlighting this particular area and intended to give practical as well as conceptual knowledge of the latest methodologies, principles, practices, tools, and technologies used for engineering project management. This book is targeted for designers of Engineering Management (EM) processes, academicians, students, practitioners, professionals, and researchers working in the field of engineering management. The book is divided in three sections, which are Process Management, Project Management, and Technology Management.

The first section of the book is comprised of seven chapters and tries to discover new paradigms in various aspects of process management. The second section of the book focuses on issues, methods, theories, and research in the area of project management with specific reference to engineering projects. The latest practices, concepts, and theories of project management in various sectors are discussed in different chapters. Finally, the last section is comprised of five chapters and explores new paradigms in the area of technology management. Latest concepts, issues, methodologies, theories, and practices are delivered and discussed by considering various cases. This book attempts to establish new principles and suggestions or refine those existing in order for the maximum and safe use of technology for the best benefits of human kind. Transfer of technology, strategy, innovation, competitiveness, foresight, outsourcing, off-shoring, entrepreneurship in technology, etc. are the key areas of focus.

In the first chapter, Paolo Renna discusses two strategies to solve the customer order-scheduling problem. In order to understand the effectiveness, he has presented simulation results showing the performance of these approaches. Based on these results, he has also discussed some future work directions.

The second chapter by Martin Bariff focuses on organizational change as a project management agenda to achieve successful project outcomes. It discusses various models of organizational change management. The main objective of the study is to justify organizational change management as an important skill set for project management. The author reviews the strengths and limitations of existing organizational change management models and proposes a composite model of organizational change. An example case study is taken for the study of organizational change management to propose a composite model. At the end of the chapter, future areas for further research are recommended.

The third chapter focuses on six sigma, which provides a framework for real problem solving in almost any organizational context. It has the power to enable leaders and practitioners to employ new ways of understanding and solving sustainability problems. However, there is a dearth of practices and ability of translating good intentions into concrete profitability and customer satisfaction. Seifedine Kadry discusses the importance of applying six sigma methodologies to multidisciplinary sustainability programs and highlights ways to implement six-sigma in such projects.

In the next chapter, Hassanali Rassouli discusses challenges confronted by new product development. Lean thinking has been a widely accepted paradigm in order to cope with the intensive global competition of the 21st century. This chapter attempts to define challenges and implications faced by corporate management in an attempt to adapt to New Product Development approaches in line with lean thinking perspectives.

The next chapter focuses on supply chain strategies that help organizations to meet the global challenges and to achieve business success. The chapter supports understanding of functional coordination in supply chain management, importance of supply chain partnership, bullwhip cause-effect, and an approach to synchronize the operation schedule to improve supply chain operations. The bullwhip effect is often seen as a challenge to improve supply chain performances. It also discusses the strategies to reduce the bullwhip effect, various well-known contract agreements that are currently practiced by many supply chain organizations, its impact and benefits on supply chain success.

The sixth chapter highlights the Indian perspective of fundamental intellectual property rights. The focus of the chapter is to analyze the trends of patenting and patented technologies with specific reference to the Indian semi-conductor industry. In addition to discussing various aspects of patents and patentability, some year wise statistics on the importance and growth of patenting are depicted.

The last chapter of the Process Management section provides an analysis of defect trends in MI-172 Helicopters, which are a variant of the MI-17/171 helicopter series. The helicopters were brought to use in an organization, which was originally set up and tuned for the MI-17/171 series. This induction of the variant resulted in diverse problems. The chapter focuses to analyze trends in problems/defects accumulated over a period of 3 years to find the root cause. Real time data from technical and flying divisions is collected for the analysis. The chapter is expected to help eradicate the existing problems and suggest pragmatic solutions for an overall uplift of the maintenance system.

Muhammad Asim Qazi and his colleagues provide a valuable insight into the component failure analysis associated with J69-T25A engines in terms of frequency and fault isolation in the next chapter. The top ten components are selected for analysis based on failure rates, and their fault/defect history is studied to establish a relationship between defects and failed components. The comparison results are used to suggest remedial measures to reduce future problems and increase engine reliability.

The next chapter provides a view of the role of lean development approaches and its principles in fostering knowledge management initiatives in the software development process. A discussion on important principles of lean manufacturing is followed by an explanation of how these principles can help setting up a knowledge management culture in the software development process. This chapter will help practitioners and students better understand the knowledge management perspective of lean approach.

In the next chapter, Kashif Saeed and his colleagues discuss management practices in the exploration and production industry in three different sections; project management, health, safety, and environment management; and quality management. A focused description of the different elements of exploration and production industry including asset/portfolio, resources, time, planning and scheduling, risk management, etc. are discussed in these three sections.

Emad Abu-Shanab and Ashraf Al-Saggar discuss the reasons for IT project failures in the next chapter. This chapter focuses on the root causes of project failures as well as success factors. This literature review-based chapter highlights seventeen factors contributing to project failures and ten factors causing escalation of project cost, scope, and schedules. A ranking of these factors in the Jordanian business environment is explored based on an empirical study. Future areas for further research are suggested at the end of the chapter.

In the next chapter, Edward Chen focuses on critical issues and the role of technology in the energy sector with special emphasis on security, mobile dispatch solutions, and the so-called "Smart Grid." The utility business has traditionally lagged behind in the use of latest technologies to cope with the growth of other industries both in size and complexity. However, the mounting socio-economic and environmental pressures are now extensively demanding that the energy sector must look into innovative technology solutions. This chapter also discusses the various prospects of undertaking this initiative regarding meaningful educational and employment opportunities.

Hisham Abdelsalam and his colleagues draw conclusions on decision-making styles of Egyptian managers in the next chapter. They collected empirical data from 138 individuals working on management positions in various Egyptian organizations. Success of an organization is strongly linked with the personal attributes of managers working in that organization. One such attribute is the decision-making style of managers. Based on the findings of this survey, this chapter provides a baseline to select and design a decision support system in Egypt.

The next chapter focuses on identification of major Foreign Object Damage (FOD) contributors in aviation by studying the case of a particular aviation setup. FOD is one of the major challenges in aviation safety not only in terms of direct costs but also due to potential loss of human lives associated with such occurrences. This chapter provides a statistical analysis of aircraft accidents attributed to various types of FOD. Data for the last 10 years has been considered for analysis, which helped in identification of eight major factors. The chapter proposes a prevention and control plan to address the most critical cause factors and improve organizational response.

This chapter investigates the role of NHK (Japan's national broadcaster) in covering the earthquake and then the nuclear crisis in March 2011 in Japan. It explains how this Japanese public television played a double role during these crises and tries to establish a case to provide a concrete example of the potential television role in disaster mitigation.

In the next chapter, Irene Samanta studies the role of the new digital era in creating diversity in the real society and the influence of social media in it. It reveals the effects of social media on economic, political, and real society affairs. This is an attempt to promote innovation to the changes that take place in communities of particular participants in the global society.

Izzat Alsmadi describes various product and process metrics for evaluating websites in the next chapter. The description is based on two major classifications of metrics into product and process-related metrics. Empirical data in this chapter takes various values of measurements in combination to provide useful information for project management and planning.

In the next chapter, Kamaljeet Sandhu discusses the E-Service acceptance model. This is a case study-based research chapter that examines the Web Electronic Service framework for an Australian university. The case study examines the process of development and implementation of a Web-based e-service system in a department of the university. The user experience to use e-services requires insight into the attributes that shape the experience variable. The chapter includes descriptive data about the attributes that form the experience variable.

In the last chapter, the author investigates the acceptability and usability factors of e-Service systems. Factors like user experience, user motivation, perceived usefulness, and perceived ease of use are discussed in this research. The chapter makes an attempt to integrate the two core variables of the Technology Acceptance Model, i.e. perceived ease of use and perceived usefulness.

Saqib Saeed
Bahria University Islamabad, Pakistan

Section 1
Process Management

Chapter 1
Order Release Strategies for Customer Order Scheduling Problems in Dynamic Environments

Paolo Renna
University of Basilicata, Italy

ABSTRACT

The research proposes two strategies to release the parts in a job shop environment to handle the customer order-scheduling problem. The first strategy is based on the evaluation of a dynamic conwip level to take the decision. The second strategy tries to anticipate the production of components of a generic order when the utilization of the manufacturing system is low. In this strategy, a fuzzy approach is proposed to decide how many components to release. A simulation environment has been developed to test the proposed approaches. Two benchmark models are used to compare the performance measures: no order release strategy and classical conwip. Moreover, the simulations are conducted in a very dynamic environment. The simulation results show how the fuzzy approach leads to the better results in all conditions tested with a relevant reduction of the inventory level.

INTRODUCTION

The customer order-scheduling problem concerns the scheduling of an order composed by different parts typology (with different routings and volume) and nothing can be delivered to the customer until the order is complete. The aim of the problem is to optimize the performance measures as: the minimization between the release time of the order and the completion; the delay of the orders; the level of the work in process; the level of inventory; etc. Julien and Magazine (1990) introduced the customer order-scheduling problem analysing the problem with two product types and a given order processing sequence.

DOI: 10.4018/978-1-4666-3658-3.ch001

Several examples of real applications of customer order scheduling can be noticed. Yang (1998) describes some real applications; for example a car repair shop where several mechanics work on the different parts of a car at same time. The order concerns the complete reparation of the car. This maintenance example can be applied to other cases as airplane, ship, train, etc. Another example is the production of component for subsequent assembly as the electronics manufacturing facility (Yang, 2005). Other examples in manufacturing systems, computing system, and other industrial context are described in Li (2005).

The customer order problem concerns the order release of the parts that compose an order; in particular when the parts will be released and how many parts to release in the manufacturing system. These two decisions have an important impact on the delay of the order and the performance of the manufacturing system. Shapiro et al. (1992) reported an industrial example in which the 99% of all orders components are delivered on time, while only the 50% of the customer orders are delivered on time. Therefore, the performance of the manufacturing system can be considered elevated, but the customer perceived a lower performance.

Several studies showed that the customer order-scheduling problem is a *NP-hard*: minimizing the weighted sum of customer order delivery time (Yang, 2005); customer order scheduling problem on a single machine (Ghosh, 2007); minimizing the weighted sum of customer order delivery time (Ahmadi, et al., 2005).

Most existing research has focused on trying to optimize only one performance as the completion time, number of orders in delay and total weighted delay. These objectives can increase the stock of finished goods and reduce the logistic performance. Some examples of realistic customer order scheduling problem are the application of logos on shirts, jackets and other apparel, multiple-items orders arrived from a variety of retailers including stores such as Kmart (Blocher, et al., 2008). When the shipment process requires dock space, site preparation, the customer can require that all components of an order had to arrive at the same time (HBS, 1991). Another example is the production of personal computer systems; the finished parts must be bundled together before they can be delivered to the customers.

The customer order flow time characterized by the release of the first job of the order and the completion of the last job is a critical factor. In fact, the reduction of this flow time allows to reduce the work in process and the inventory level of finished goods. The decision model in a job shop related to the customer order scheduling problem concerns the following decisions: 1) when the raw parts are released in the manufacturing systems; 2) how many of the raw parts needs to be released; 3) the scheduling of the parts on the machines of the manufacturing systems.

The main problems of the customer order scheduling concerns the first two problems. Most of the research proposed in literature investigated manufacturing systems with limited number of machines (on a single machine or parallel machines) improving only one performance measure (as the order flow time, due date of the orders, etc.).

This chapter proposes two methodologies in order to optimize several performance measures. The first is based on the dynamic evaluation of the Work in Process (WIP) level of the manufacturing system in order to decide the release of the parts. The second methodology releases the production of some parts when the utilization of the manufacturing system is under certain level. Within this strategy, it is proposed an approach based on fuzzy logic to decide how many parts to release. The proposed strategies are tested in a simulation environment that emulates the dynamic market conditions in terms of volume, mix fluctuations and due date variations.

The performance measures evaluated are related to the customer order (order flow time and delay) and to the manufacturing system (work in process, stock level of finished goods). The proposed methodologies can be applied in a general job-shop manufacturing system.

Two benchmark models are used to compare the performance measures of the proposed approaches: one model with any policy and a model with a classical conwip methodology.

The rest of the chapter is organized as follows. A literature review on order releasing policy in job shop environments is presented in Section 2. A description of the order releasing policy and dispatching rule follows in Section 3. Then, in the fourth section, the simulation model and the experimental design used to study the performance of the scheduling system are discussed. Some general observations of the results in the fifth section are discussed. In the final section, we provide a summary of the results and make some suggestions for future work.

LITERATURE REVIEW AND MOTIVATION

Several approaches proposed in scientific literature are based on conwip policy (Spearman, et al., 1990). The main advantage of conwip policy is the capability to control Work in Process (WIP) using cards, and it can be used in a wider variety of manufacturing environments.

Several algorithms are proposed to solve the problem in a simplified manufacturing system composed by parallel machines; therefore, with one operation to perform by the parts. Leung et al. (2008) investigated scheduling orders on either dedicated or flexible machines in parallel to minimize the total weighted completion time. For due-date related objectives concerning the fully dedicated case, the reader is referred to Ng et al. (2003). For a more general description of order scheduling models and their many application examples, the reader is referred to Li (2005). Su et al. (2012) proposed three heuristics to minimize

the maximum lateness for the customer order problem where jobs are scheduled on a set of parallel machines and dispatched in batches.

Blocher et al. (1998) examined the performance of order-based dispatching rules in a general job shop, where the environmental factors are shop utilization and due date tightness. They compared dispatching rules from past job-based studies to some rules adapted to encompass order characteristics. Of the 16 dispatching rules tested, the results show that four of the simple rules dominate the others. Blocher and Chhajed (1998) investigated the parallel machines context minimizing the order flow time.

Yang and Posner (2005) considered scheduling problems where jobs are dispatched in batches. The objective is to minimize the sum of completion times of batches. A heuristic was presented for the parallel machine version of the problem. For large problems, the methods find solutions that are close to optimal.

Leung et al. (2006) considered m machines in parallel and n orders. They considered various due date related objectives such as the minimization of the maximum lateness and the total number of late orders. They proposed an exact algorithm based on Constraint Propagation and bounding strategy. The effectiveness of the algorithms is demonstrated through an empirical study.

Wang and Cheng (2007) considered the customer order scheduling case for m dedicated facilities and n orders. Each job only needs one operation at a dedicated facility. They showed that the problem is unary NP-hard. They proposed a heuristic method to minimize the total weighted order completion time and analyzed a worst-case scenario.

Li and Vairaktarakis (2007) considered an integrated scheduling and distribution model in which jobs completed by two different machines must be bundled together for delivery. The objective is to minimize the sum of delivery cost and customers' waiting cost. Such a model not only attempts to coordinate the job schedules on both machines, but it also aims to coordinate the machine schedules with the delivery plan. Polynomial-time heuristics and approximation schemes are developed for the model with only direct shipments as well as the general model with milk-run deliveries.

Liu and Ou (2007) proposed a model to coordinate the production and delivery schedules on the decentralized machines while taking into consideration the shipping cost as well as the waiting time of the customers. They developed polynomial-time heuristic algorithms for this problem and analyze their worst-case performance. Computational experiments are conducted to test the effectiveness of the heuristics and to evaluate the benefits obtained by coordinating the production and delivery of the two decentralized machines.

Hazir et al. (2008) investigated the problem of the customer order scheduling to satisfy the demand of customers who order several types of products produced on a single machine. A setup is required whenever a product type is launched. The objective of the scheduling problem is to minimize the average customer order flow time. Since the customer order scheduling problem is known to be strongly NP-hard, they solved it using four major metaheuristics and compare the performance of these heuristics, namely, simulated annealing, genetic algorithms, tabu search, and ant colony optimization.

Liu (2009) proposed the applicability of lot streaming to the customer order scheduling problem, in order to investigate whether the expected benefits of lot streaming can be observed in an order-based environment. The research proposed a genetic algorithm to determine the lot streaming conditions. The experiments led to the conclusion that their proposed algorithm is significantly superior over the other modes in terms of makespan, lateness and finished goods flow time.

Hsu and Liu (2009) proposed a new dispatching rule, referred to as Minimum Flow Time Variation for customer order scheduling in a normal job shop, in order to reduce the total time it takes to complete all jobs within the same order. The

simulation test showed that the dispatching rule proposed allows to reduce drastically the finished goods warehouse. The performance of their proposed method will become increasingly significant the more complex the system.

Liu (2010) proposed a coordinated scheduling of customer orders system that includes two main decisions: release the jobs and dispatch the jobs at station level. Extensive simulation experiments were performed to compare the proposed scheduling system with the benchmark mechanisms presented in previous studies. The simulation results show the benefits of the proposed system.

Based on the above literature review, the following limitations can be drawn:

1. Few studies concern the customer order scheduling problem in a general job shop environment. The research proposed concerns manufacturing systems with limitation on the number of machines (often, only one machine) or the number of tasks to perform. These restrictions reduce the possibility to introduce the approaches proposed in real industrial cases.

2. The approaches proposed in literature are tested in manufacturing system where the exceptions and rapidity of alterations were not investigated. The most tests are conducted in static conditions. Moreover, the performance measures investigated are often limited.

The research proposed in this chapter outcomes the above limitations in the following issues:

1. The proposed approaches regard a generic job shop environment with several machines and wider range of performance measures are analyzed.

2. A simulation environment is used to test the proposed approaches when some exceptions occur (for example, demand fluctuations) and the rapidity changes are investigated.

Moreover, the approaches proposed are characterized by a low computational complexity and simplicity in order to be appropriate in real case applications.

CONTROL POLICIES

The manufacturing system context consists of a given number of machines; each machine is able to perform a specific manufacturing operation. Each part type has a predefined routing depending on the manufacturing operations to perform. The customer orders arrive randomly with an inter-arrival time that is a normal distribution. A volume for each part type composes each customer order. The order can be delivered to the customer when all part types requested are manufactured. The queue of the customer orders that wait to be released is managed by Earliest Due Date rule.

The queues of the machines are managed by the First In First Out policy; each machine can breakdown randomly with an exponential distribution.

In this research, the transportation time of the material handling devices is included in the processing time, and the handling resources are always available.

The decision to release an order is made considering the Work In Process (WIP) of the manufacturing system. If the Expression (1) is verified then the customer order is released; otherwise the customer order waits in the queue:

$$WIP < \max wip \qquad (1)$$

The value of *maxwip* is the maximum level of WIP of the manufacturing system. This is the simpler policy to release the orders and keep the work in process lower than the maximum level defined. This policy is used as a benchmark to compare the performance measures of the approaches proposed. This is the control policy based on CONstant Work in Process (CONWIP) that allows to reduce the level of work in process (Spearman, et al., 1990).

Dynamic Conwip

The design of the maximum level of WIP in in pull systems can be obtained by two approaches (Framinan, et al., 2003):

- It is set the maximum level of WIP given the manufacturing conditions in order to obtain an adequate level of performance. The maximum level of WIP is fixed and can be re-designed when the manufacturing conditions change significantly.
- It is considered a set of rules that change or maintain the maximum level of WIP of the production system. It is obtained a variable value of the maximum level of WIP in order to adapt to the manufacturing conditions.

The methodology proposed can be ascribed to the second approach described above.

This control policy is performed evaluating the *pressure* of the customer orders in the queue that wait to release in the manufacturing system. The first value computed is the number of parts that compose the customer orders in the wait queue (see Expression 2):

$$\text{press} = \sum \text{orders in queue} \sum \text{parts typology} X_i \cdot V_i^o \qquad (2)$$

where, *orders in queue* is the number of customer orders arrived to the system but not released; *parts typology* is the number of possible typologies of part that can compose a customer order:

$$X_i = \begin{cases} 1, \text{if the part type } i \text{ is present in the customer order} \\ 0, \text{ otherwise} \end{cases}$$

and V_i^o is the volume that compose the customer order *o* for each part *i*.

Therefore, the Expression (2) evaluates the total parts that the manufacturing system will manufacture due to the orders in waiting state.

Then, it is computed the ratio between the *press* value and the *maxwip* of the classical conwip policy.

If (*press/maxvalue*) is lower than one, the policy used is the classical conwip. This means that the queue of customer orders is not relevant to change the threshold level of the conwip approach.

If the ratio (*press/maxvalue*) is greater than one, it is computed a new value of *maxwip (maxwip*)* as shown in Expression 3. In this case, the components to manufacture for the customer orders in the queue are relevant:

$$\max wip^* = \max wip \bullet \left\lceil \frac{press}{\max wip} \right\rceil \qquad (3)$$

In this way, it is obtained a dynamic value of *maxwip* to perform the conwip policy. The dynamic value depends on the volume of parts of the orders in waiting state. Then, if the wip is lower than *maxwip** the customer order can be released. Figure 1 shows the control level of the dynamic conwip approach proposed.

The main advantage of the approach proposed is the characterization of only one parameter "maxwip." The control system adapts dynamically to the inter-arrival of the customer orders and to the number of parts that compose the orders.

Manufacturing Utilization Policy

In this section, it is described the difference of the approaches based on manufacturing utilization compared to the dynamic conwip. These policies release the production of parts when the utilization of the manufacturing system is low. This policy leads to increase the component in the finished goods inventory, but can improve the performance in terms of due date and average flow time of the customer orders. The mean of the strategy is to release in advance some parts when the utilization of the manufacturing system is lower than a threshold level. This allows to keep a uniform level utilization of the manufacturing system reducing fluctuation of the utilization.

The policy is a periodic review control; the control of the manufacturing utilization is made at fixed periods. When the control is activated, the decision concerns if new parts can be released in the manufacturing system and how many parts release for each typology.

In the first case, the control verifies if the WIP is lower than the *maxwip*; in affirmative case, the control releases one part for each typology. This approach (*appr. 1*) is the simpler, because does not take into account the finished goods inventory level. Moreover, the parts are released equally for each typology.

The second approach (*appr. 2*) takes into account the finished goods inventory level. In particular, it is defined two levels (*S1* is the parameter

Figure 1. Control level of WIP

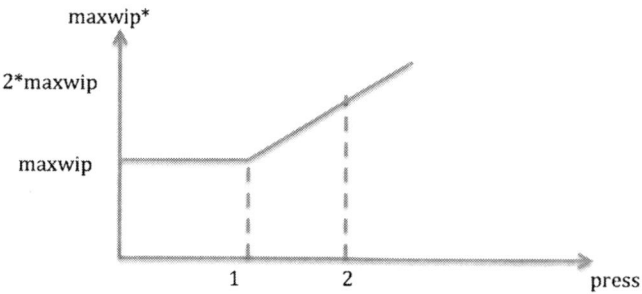

that defines the two levels) to decide the number of parts to release. Table 1 reports the approach used for each part typology.

If the finished goods of the generic part typology is lower than the parameter S1, then it can be released two parts. If the level of the finished goods of the generic part typology is greater than S1, it can be released one part. In the last case, the manufacturing system has a low utilization, but the finished goods level is high, therefore only one part is released. This strategy allows to uniform the parts in the finished goods.

The third case (*appr. 3*) concerns the decision based on three levels as shown in Table 2.

This approach is investigated to evaluate if a more complex strategy to release the number of parts can improve the performance measures.

The last approach is based on a fuzzy logic methodology. The steps to apply the methodology are the following:

- **Fuzzufication of the inputs:** In this research, the only input evaluated is the state of the finished goods inventory. Figure 2 shows the fuzzification of the input in terms of HIGH and LOW membership. The membership function is a graph that defines how each point in the input space is mapped to membership value between 0 and 1. The input space is defined by the values between 0 and the parameter "*a*" that is the maximum level of the finished goods. Triangular fuzzy numbers appear as useful means of quantifying the uncertain-

Table 1. Manufacturing utilization with two levels

Threshold	Number to release
Lower or equal than S1	2
Greater than S1	1

Table 2. Manufacturing utilization with three levels

Threshold	Number to release
Lower or equal than S1	3
Greater than S1 and lower or equal S2	2
Greater than S2	1

ty in decision-making due to their intuitive appeal and computationally efficient representation (Karsak & Tolga, 2001; Wang, 2009).

- The rules of the fuzzy engine are applied; The rules used are the following:
  - If finished goods inventory is high then the release of part is low.
  - If finished goods inventory is low then the release of part is high.
- The process of the inference concerns the result evaluation of each rule, these results should be combined to obtain a final result.
- The de-fuzzification to obtain the value of each part to release in the manufacturing system. The method used is the center of gravity of the fuzzy set.

Figure 2. Input fuzzy membership

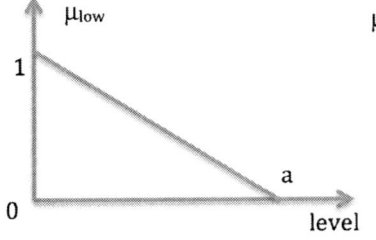

SIMULATION ENVIRONMENT

In order to test the proposed approaches, a simulation environment is developed. The author selected the Arena® discrete event simulation platform by Rockwell Software Inc. it was used to develop the simulation model of the presented approaches. Arena®– based on the known SIMAN simulation language—is well suited for modeling shop floors of production systems in which each entity (part) follows a manufacturing route through production resources (servers, material handling systems, buffers, and so forth) (Kelton, et al., 2007). Modern discrete-event simulation software has many modules to help analysts quickly construct simulation models of manufacturing systems. The survey in Smith (2003) deeply described the use of simulation in manufacturing system.

The manufacturing system consists of six machines type (*ws j*); the part typologies that can compose a customer order are three. Table 3 reports the routing and processing time for each part typology.

The processing times are all equal in order to investigate only the control policies and define the workload on the machines. In this way, the numerical results do not depend on the processing times of the parts. Each machine can breakdown randomly according to exponential distribution with Mean Time Between Failures (MTBF) and Mean Time To Repair (MTTR) as reported in Table 4.

The simulations are conducted in static and dynamic conditions. In static conditions, the parameters are reported in Table 5. The approaches are tested in two level of workload depending on the inter-arrival parameter. The customer orders can be four combinations of the part typologies (Mix) that compose an order. The volume of the order is the same, but the distribution among the part typologies change. The first mix (Mix 1) is balanced among the three typologies; the other three mixes are unbalance on one part typology.

Table 3. Routing and processing times

	ws 1	ws 2	ws 3	ws 4	ws 5	ws 6
Part type 1	5	5	5	5	5	5
Part type 2	5	-	5	-	5	-
Part type 3	-	5	-	5	-	5

Table 4. MTBF and MTTR

	ws 1	ws 2	ws 3	ws 4	ws 5	ws 6
MTBF	250	500	1000	250	500	1000
MTTR	50	50	50	50	50	50

The due date of a generic customer order *j* is assigned by the following expression:

$$duedate_j = 45 \cdot UNIFORM\left[3,5\right] \qquad (4)$$

The simulation length is 28800 unit times and in order to guarantee a statistical validity of the results, each experimental class is replicated to assure 95% of confidence level and the half width lower than 5% of the mean (Banks, 1998).

In order to emulate a dynamic environment the proposed approaches have been tested through a production run consisting of several alternating stages; each stage is characterized by different external attributes as: inter-arrival, mix and due date fluctuations.

Table 6 reports the value of inter-arrival, mix and due date fluctuations over the three alternating stages.

The lengths of the three alternating stages are:

- 2880 unit times; in this case the stages are alternated 10 times. This is the case with low dynamicity of the changes.
- 1440 unit times; in this case the stages are alternated 20 times. This is the case with medium dynamicity of the changes.

Table 5. Static parameters

	Low workload	High workload		
Inter-arrival Norm (mean, variance)	NORM(45,10%)	NORM(30,10%)		
	Mix 1	**Mix 2**	**Mix 3**	**Mix 4**
Mix customer orders	25%	25%	25%	25%
Part type 1	3	7	1	1
Part type 2	3	1	7	1
Part type 3	3	1	1	7
Total volume	9	9	9	9

- 720 unit times; in this case the stages are alternated 40 times. This is the case with higher dynamicity of the changes.

The above lengths are used to test the effect of the rapidity of the changes.

Table 7 reports the combination of the simulation conditions that lead to 11 experimental classes simulated for each models.

The performance measures investigated are the following:

- Throughput (*thr.*); it is the throughput of the manufacturing system.
- Average throughput time (av. thr); it is the average throughput time of the four typologies part.
- Work in process (*wip*); it is the average level of the WIP.

- Average utilization (*av. ut.*); it is the average utilization of the machines of the manufacturing system.
- Customer order flow time (*ord. thr.*); it is the average throughput time of the customer orders from the time of release to the time of delivery to the customer.
- Delay (*delay*); it is the sum of delay of the customer orders.
- Wait customer orders (*wait ord.*); it is the average number of orders released but in a wait state because the parts are in process.
- Finished goods inventory (*invent.*); it is the average value of the finished goods inventory for each typology part multiplied the average time of storage of these products. This is the index cost of inventory that takes into account the level and the time.

Table 6. Dynamic conditions

	Stage 1	Stage 2	Stage 3
Inter-arrival	NORM(45,10%)	NORM(30,10%)	NORM(45,10%)
Mix 1	20%	20%	20%
Mix 2	40%	20%	20%
Mix 3	20%	40%	20%
Mix 4	20%	20%	40%
Due date	UNIF(3,5)	UNIF(5,10)	UNIF(5,7)

Figure 3 shows the simulation model of the order arrival module. The activities simulated by the SIMAN blocks are the following:

- **"Create"** defines the parameter of the customer order arrival;
- **"Choose"** determines the mix type of the order using the probability of the mix;
- **"Assign"** assigns the composition of the part typology for each mix.

Figure 4 shows the simulation model of the order release module for the fuzzy approach. The activities simulated by the SIMAN blocks are the following:

- **"Queue"** contains the customer order that wait to satisfy because the parts that compose the order are not all produced.
- **"Scan"** The scan connected to the Queue checks if the parts of the customer order are available in the finished goods inventory. In affirmative case, the customer order is satisfied and goes on; otherwise the customer order waits in Queue.

The second part of Figure 4 is the simulation module that checks the release of raw parts in the manufacturing system:

Table 7. Experimental classes

Exp. No.	Inter-arrival	Mix	Due date
1	Static - low	static	Static
2	Static high	static	static
3	Dynamic (10)	static	static
4	Dynamic (20)	static	static
5	Dynamic (40)	static	static
6	Dynamic (10)	Dynamic (10)	static
7	Dynamic (20)	Dynamic (20)	static
8	Dynamic (40)	Dynamic (40)	static
9	Dynamic (10)	Dynamic (10)	Dynamic (10)
10	Dynamic (20)	Dynamic (20)	Dynamic (20)
11	Dynamic (40)	Dynamic (40)	Dynamic (40)

- **"Create"** activates the control mechanism to release raw parts.
- **"Scan"** checks the WIP of the manufacturing system. If the WIP is lower than threshold level the process goes on; otherwise the process waits until this condition is true.
- **"Duplicate"** applies the release policy described in the above sections in order to decide how many raw parts for each typology to release.
- **"Assign"** The assign blocks assigns all the information to the raw part (routing, typology, etc.).

Figure 3. Order arrival module

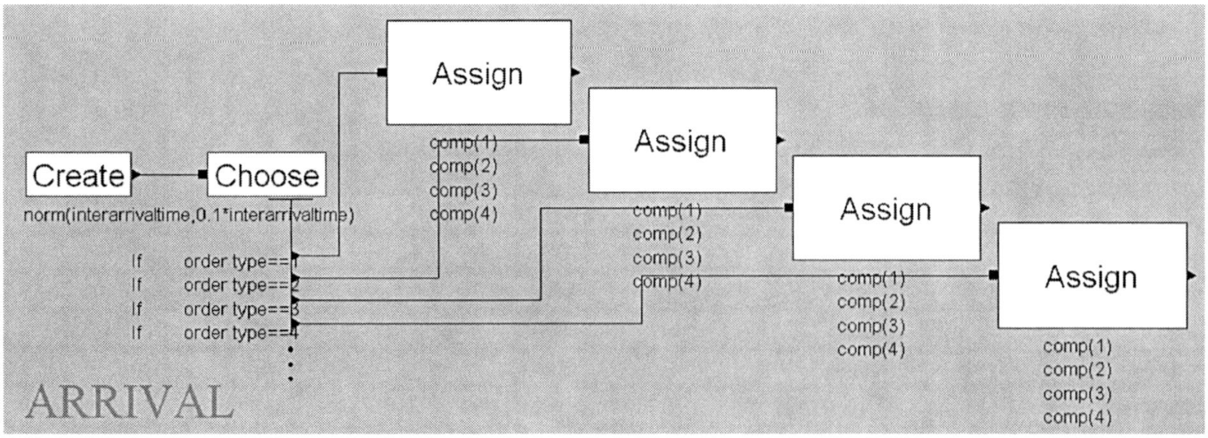

- **"Delay"** applies a delay in order to avoid the loop of the process.

Figure 5 shows the simulation model of the order release module for the others approach tested.

Figure 6 shows the simulation model of the finished goods inventory.

Figure 7 shows the simulation for the dynamicity of the manufacturing system conditions. The block 'Create' is activated at stage length and the 'Assign' block changes the conditions.

Figure 8 shows the simulation model of each manufacturing resources that produces the part typologies.

NUMERICAL RESULTS

The simulation results are reported in terms of percentage difference compared to the manufacturing system with no control policy.

The results concern the following models:

- Classical conwip with 18 maximum value of WIP (conwip 18);
- Classical conwip with 24 maximum value of WIP (conwip 24);
- Conwip with the dynamic evaluation of maximum value of wip (conwip dyn);
- Manufacturing utilization control policy simplified (appr. 1);
- Manufacturing utilization control with one threshold evaluation (appr. 2);
- Manufacturing utilization control with two threshold evaluation (appr. 3);
- Manufacturing utilization control with fuzzy methodology (fuzzy).

Manufacturing utilization with fuzzy methodology, but with the limitation of the finished goods inventory. In particular, the release of the parts can be activated if the inventory of the finished goods is lower than 5 parts for each typology (fuzzy max). This model is proposed to reduce the costs of inventory and evaluated how the performance measures change.

Figure 4. Order release

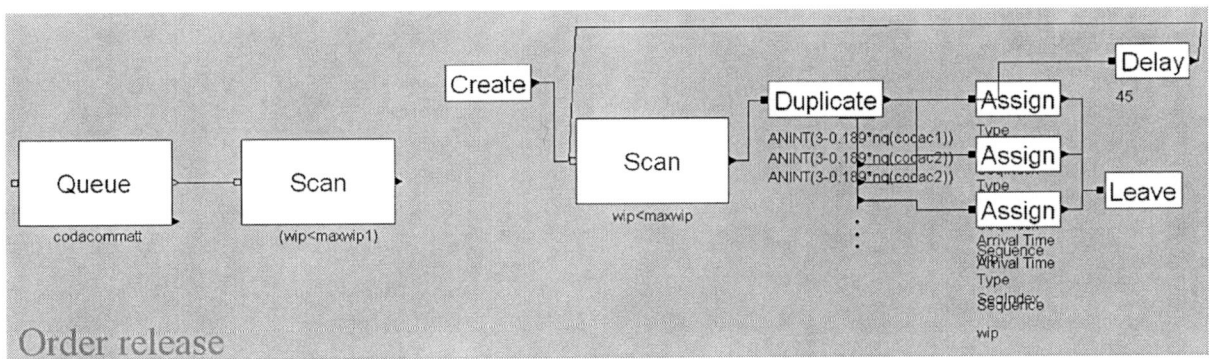

Figure 5. Order release

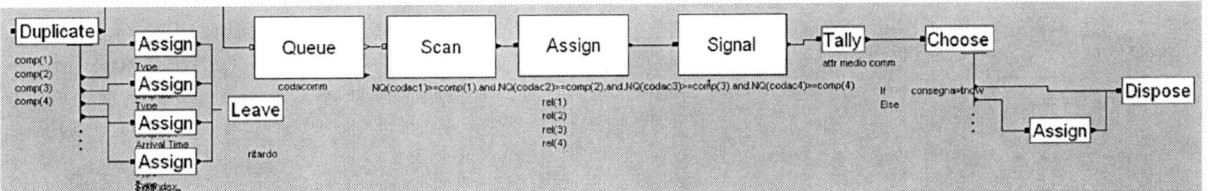

Figure 6. Finished goods inventory

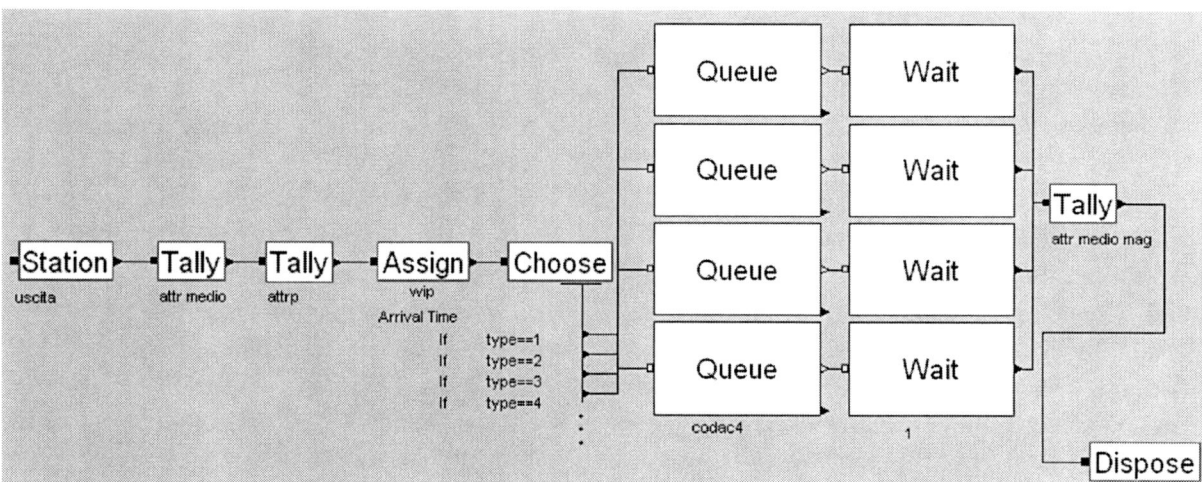

Figure 7. Dynamicity simulation

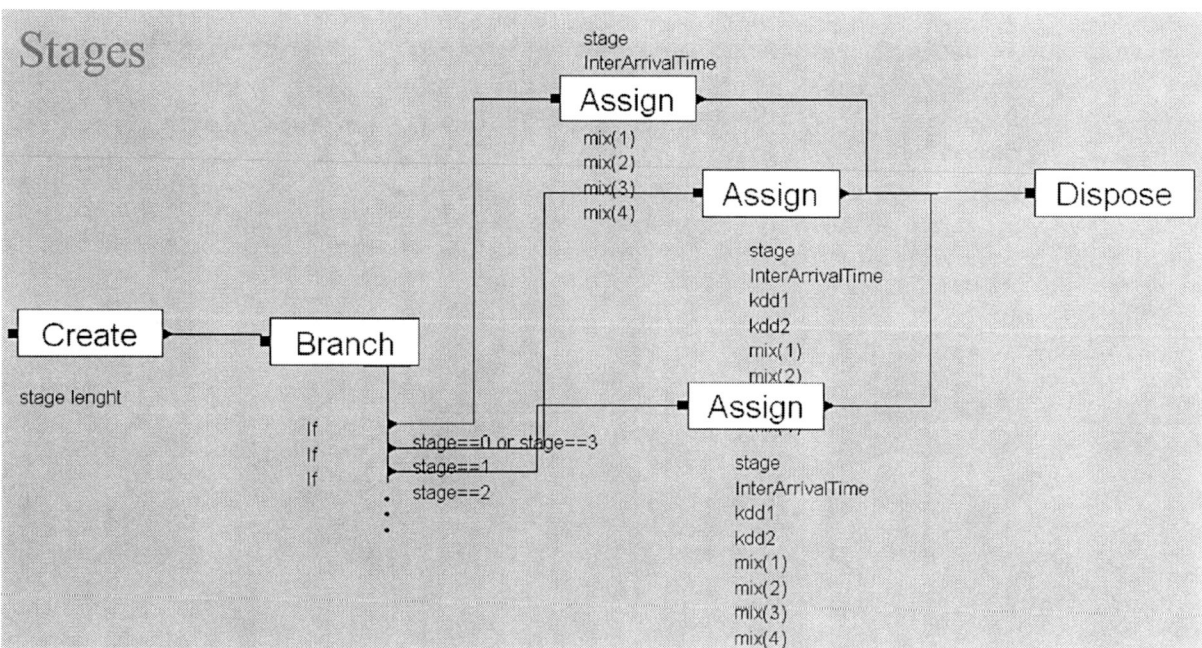

Figure 8. Manufacturing resources

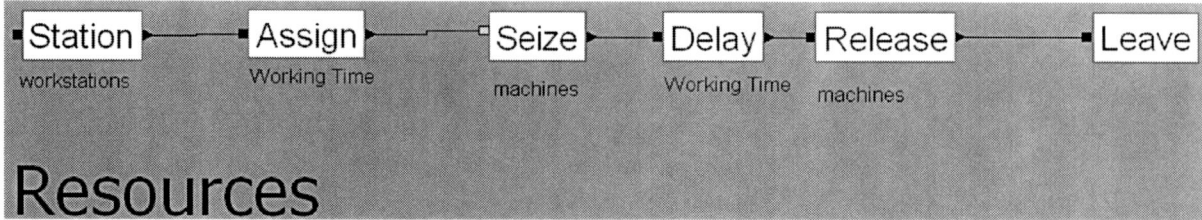

The manufacturing utilization controls *appr. 1, 2,* and *3* lead to simulation results very similar. This means that the threshold to define how many parts to release does not affect the performance. This result means that the simpler policy is the better among these approaches.

Because the results are very similar, it is reported the average between the *appr. 1, 2,* and *3*.

Table 8 reports the simulation results in static conditions with a low workload of the manufacturing system (inter-arrival 45).

The simulation results show that the dynamic evaluation of the maximum value (*maxwip*) for the conwip does not improve the performance of the system. The better result of the conwip is obtained for the maximum value of 24, but in this case the WIP of the manufacturing systems increases considerably. The approaches that anticipate the production of the parts (appr. 1, 2, and 3) reduce drastically the delay of the orders, but this benefit is obtained increasing the inventory costs. The better approach is the fuzzy approach that reduces the delay of the orders with a lower inventory costs than the appr. 1, 2, 3. The "*modulation*" of the fuzzy approach (fuzzy max) allows to define a better compromise between the delay reductions and inventory costs increment. In this case, the fuzzy approach methodology is performed by the evaluation of a maximum level of the inventory-finished goods. The parts are re-leased only if the finished good is lower a maximum level threshold. Therefore, the threshold of the finished inventory can be used to easily adapt to the performance more important between inventory costs and customer order delay.

Table 9 reports the simulation results in static conditions with a higher workload of the manufacturing system (inter-arrival 30).

The high workload leads to obtain similar results among the manufacturing utilization approaches (appr. 1, 2, 3, and fuzzy); in fact, the policy to anticipate the production is unimportant when the congestion of the manufacturing system is very high. This is caused by the absence of periods in which the manufacturing system has a low utilization. In these conditions, the conwip dynamic is a competitive approach with the manufacturing utilization approaches.

Table 10 reports the simulation results when the inter-arrival time changes with higher stage length (2880).

In these conditions, the dynamic conwip is the better approach among the conwip approaches. The fuzzy approach is the better model; the only undesirable effect is the increment of inventory costs. The limitation of the inventory of the fuzzy approach shows how the performance measures are improved with any increment of inventory costs.

Table 8. Static with low workload

	Conwip 18	Conwip 24	Conwip dyn	Appr 1, 2, 3	Fuzzy	Fuzzy max
Thr.	0,00%	13,86%	-0,08%	13,86%	1,42%	0,50%
Av. thr	-1,30%	38,53%	-1,40%	39,18%	2,80%	0,03%
Wip	-1,31%	57,72%	-1,51%	58,44%	4,25%	0,52%
Av. Ut.	-0,75%	0,00%	0,00%	14,00%	1,50%	0,50%
Ord. thr.	24,85%	0,46%	7,74%	-29,03%	-85,48%	-79,33%
Delay	299,61%	-3,76%	48,17%	-99,86%	-99,93%	-99,74%
Wait ord.	-5,86%	-1,56%	-1,56%	-97,27%	-99,61%	-89,45%
Invent.	-11,19%	-2,70%	-2,18%	7291,42%	771,24%	146,44%

Table 9. Static with high workload

	Conwip 18	Conwip 24	Conwip dyn	Appr 1, 2, 3	fuzzy	Fuzzy max
Thr.	-18,35%	-12,15%	-0,13%	-0,02%	0,00%	0,00%
Av. thr	-96,61%	-95,67%	-43,44%	-42,56%	-42,37%	-42,78%
Wip	-97,72%	-96,87%	-48,56%	-47,69%	-47,49%	-47,90%
Av. Ut.	-18,34%	-12,43%	1,78%	1,84%	1,97%	0,00%
Ord. thr.	29,58%	9,09%	-13,23%	-14,07%	-14,30%	-13,97%
Delay	15,32%	3,69%	-9,76%	-10,94%	-11,25%	-10,75%
Wait ord.	-97,68%	-96,83%	-47,86%	-49,21%	-49,56%	-49,01%
Invent.	-99,92%	-99,86%	-70,04%	-69,31%	-69,27%	-69,47%

Table 10. Static with dynamic workload: stage length high

	Conwip 18	Conwip 24	Conwip dyn	Appr 1, 2, 3	fuzzy	Fuzzy max
Thr.	0,00%	0,00%	0,00%	10,17%	1,11%	0,48%
Av. thr	-44,10%	-41,02%	-24,14%	-8,26%	-22,66%	-23,72%
Wip	-44,06%	-40,98%	-24,15%	1,12%	-21,78%	-23,32%
Av. Ut.	0,00%	0,00%	0,00%%	10,00%	1,43%	0,00%
Ord. thr.	48,45%	13,19%	1,68%	-48,55%	-56,01%	-44,69%
Delay	78,95%	20,48%	-4,99%	-66,45%	-52,35%	-35,75%
Wait ord.	-44,57%	-41,44%	-24,31%	-91,65%	-83,79%	-71,09%
Invent.	-37,10%	-32,87%	-42,63%	1351,57%	112,00%	-4,07%

Tables 11 and 12 report the simulation results when the inter-arrival time changes with medium and low stage length (1440 and 720).

The reduction of the stage length leads to reduce the benefits of the proposed approach. This because the peak of workload can be absorbed in a minor time available.

Tables 13, 14, and 15 report the simulation results when inter-arrival and mix fluctuations occur. The dynamic conwip and fuzzy approaches lead to the better results. The variations of the stage length have the following main effects:

- **Dynamic conwip:** The delay of the orders increases when the stage length reduces (this means that the conditions are more dynamic). Moreover, the benefit of reduc-

tion inventory is lower when the stage length reduces.

- **Fuzzy approaches:** The benefit in terms of delay reduction increases when the stage length reduces; however, this benefit is obtained by an increment of finished goods inventory.

Finally, Tables 16, 17, and 18 report the results when inter-arrival, mix, and due date changes occur. It can be noticed that the effects are similar to the above results.

Table 19 reports the results comparison between the dynamic conwip and fuzzy with finished goods inventory limitation. These are the better models in term of performance measures.

Table 11. Static with dynamic workload: stage length medium

	Conwip 18	Conwip 24	Conwip dyn	Appr 1, 2, 3	fuzzy	Fuzzy max
Thr.	-0,08%	-0,08%	-0,08%	11,72%	1,38%	0,49%
Av. thr	-20,10%	-17,02%	-12,55%	15,14%	-6,84%	-11,51%
Wip	27,79%	-17,01%	-12,60%	28,72%	-5,51%	-11,07%
Av. Ut.	0,00%	0,00%	0,00%	11,76%	1,47%	1,47%
Ord. thr.	14,93%	2,85%	5,82%	-35,98%	-71,63%	-61,64%
Delay	50,78%	4,77%	-2,73%	-75,92%	-76,75%	-54,48%
Wait ord.	-20,63%	-17,48%	-12,61%	-95,70%	-95,70%	-81,09%
Invent.	-48,56%	-45,64%	-23,71%	3418,67%	437,63%	57,59%

Table 12. Static with dynamic workload: stage length low

	Conwip 18	Conwip 24	Conwip dyn	Appr 1, 2, 3	fuzzy	Fuzzy max
Thr.	0,00%	0,00%	3,71%	12,72%	1,48%	0,58%
Av. thr	-7,46%	-5,02%	38,87%	29,44%	0,85%	-3,99%
Wip	-7,46%	-5,01%	43,88%	45,88%	2,27%	-3,55%
Av. Ut.	0,00%	0,00%	4,48%	13,43%	1,49%	1,49%
Ord. thr.	4,00%	0,58%	86,21%	-30,83%	-77,57%	-70,94%
Delay	27,98%	-1,22%	5,10%	-76,89%	-81,52%	-70,13%
Wait ord.	-7,34%	-4,90%	43,71%	-96,62%	-97,55%	-85,66%
Invent.	-27,64%	-24,42%	99,16%	5411,11%	632,04%	106,10%

Table 13. Static with dynamic workload and mix: stage length high

	Conwip 18	Conwip 24	Conwip dyn	Appr 1,2,3	Fuzzy	Fuzzy max
Thr.	0,00%	0,00%	-0,08%	9,53%	1,35%	0,48%
Av. thr	-42,59%	-38,65%	-19,04%	-2,48%	-14,61%	-17,87%
Wip	-42,56%	-38,63%	-19,06%	6,88%	-13,45%	-17,44%
Av. Ut.	0,00%	0,00%	0,00%	7,22%	1,46%	0,00%
Ord. thr.	48,60%	19,01%	8,81%	-34,29%	-51,88%	-36,63%
Delay	82,68%	31,88%	9,94%	-53,00%	-48,08%	-20,88%
Wait ord.	-43,48%	-39,70%	-19,66%	-87,84%	-84,88%	-65,97%
Invent.	-76,80%	-74,95%	-36,57%	1253,20%	193,57%	7,89%

Table 14. Static with dynamic workload and mix: stage length medium

	Conwip 18	Conwip 24	Conwip dyn	Appr 1, 2, 3	Fuzzy	Fuzzy max
Thr.	0,00%	0,00%	-0,08%	11,47%	1,38%	0,49%
Av. thr	-18,09%	-14,43%	-9,80%	19,82%	-2,87%	-8,23%
Wip	-18,09%	-14,42%	-9,81%	33,66%	-1,54%	-7,78%
Av. Ut.	0,00%	0,00%	0,00%	12,03%	1,50%	0,00%
Ord. thr.	17,63%	6,30%	8,71%	-30,62%	-67,54%	-58,33%
Delay	69,23%	21,42%	15,02%	-60,05%	-61,52%	-38,10%
Wait ord.	-18,64%	-15,09%	-10,06%	-92,60%	-92,90%	-78,40%
Invent.	-47,69%	-44,57%	-19,47%	3360,35%	444,32%	32,78%

Table 15. Static with dynamic workload and mix: stage length low

	Conwip 18	Conwip 24	Conwip dyn	Appr 1, 2, 3	Fuzzy	Fuzzy max
Thr.	0,00%	0,00%	-0,08%	12,14%	1,40%	0,49%
Av. thr	-6,06%	-3,42%	-4,85%	28,19%	2,23%	-2,77%
Wip	-6,05%	-3,42%	-4,85%	43,97%	3,89%	-2,28%
Av. Ut.	0,00%	0,00%	0,00%	12,86%	1,28%	1,02%
Ord. thr.	4,97%	1,92%	7,82%	-28,94%	-76,85%	-70,30%
Delay	42,90%	10,62%	21,25%	-63,77%	-71,42%	-59,83%
Wait ord.	-6,09%	-3,58%	-4,66%	-94,23%	-96,42%	-84,23%
Invent.	-29,13%	-25,59%	-12,06%	4497,66%	597,65%	107,41%

Table 16. Static with dynamic workload, mix and due date: stage length high

	Conwip 18	Conwip 24	Conwip dyn	Appr 1, 2, 3	Fuzzy	Fuzzy max
Thr.	0,00%	0,00%	-0,08%	9,56%	1,35%	0,48%
Av. thr	-42,42%	-38,66%	-19,08%	-2,58%	-14,65%	-18,12%
Wip	-42,40%	-38,63%	-19,09%	6,78%	-13,51%	-17,73%
Av. Ut.	0,00%	0,00%	0,00%	9,65%	1,46%	0,00%
Ord. thr.	50,29%	19,36%	8,86%	-34,41%	-51,88%	-36,74%
Delay	86,57%	26,08%	0,61%	-65,26%	-59,77%	-32,85%
Wait ord.	-74,48%	-39,70%	-19,66%	-87,90%	-84,88%	-66,16%
Invent.	-70,33%	-74,95%	-36,45%	1253,80%	194,32%	7,21%

Table 17. Static with dynamic workload, mix, and due date: stage length medium

	Conwip 18	Conwip 24	Conwip dyn	Appr 1, 2, 3	Fuzzy	Fuzzy max
Thr.	0,00%	0,00%	-0,08%	11,52%	1,38%	0,49%
Av. thr	-18,17%	-14,71%	-9,60%	19,55%	-3,36%	-8,42%
Wip	-18,19%	-14,72%	-9,61%	33,41%	-2,03%	-7,98%
Av. Ut.	0,00%	0,00%	0,00%	12,03%	1,50%	0,00%
Ord. thr.	16,48%	5,35%	9,24%	-30,86%	-67,97%	-58,30%
Delay	54,12%	-1,81%	-5,53%	-78,33%	-79,79%	-59,14%
Wait ord.	-62,43%	-15,38%	-9,76%	-92,70%	-92,90%	-78,40%
Invent.	-31,58%	-44,70%	-19,23%	3384,60%	441,86%	69,50%

Table 18. Static with dynamic workload, mix, and due date: stage length low

	Conwip 18	Conwip 24	Conwip dyn	Appr 1, 2, 3	Fuzzy	Fuzzy max
Thr.	0,00%	0,00%	-0,08%	12,57%	1,32%	0,49%
Av. thr	-6,07%	-3,73%	-5,04%	32,80%	2,06%	-2,74%
Wip	-6,05%	-3,71%	-5,09%	49,63%	3,48%	-2,28%
Av. Ut.	0,00%	0,00%	-0,26%	13,27%	1,02%	1,02%
Ord. thr.	4,55%	1,42%	7,69%	-28,46%	-77,22%	-70,11%
Delay	-4,07%	-39,47%	-35,75%	-90,59%	-92,01%	-86,37%
Wait ord.	-6,09%	-3,94%	-4,66%	-95,70%	-96,42%	-84,23%
Invent.	-2,57%	-23,48%	-9,29%	5301,13%	620,54%	115,32%

Table 19. Simulation results: summary

	Inter-arrival		Inter-arrival and mix		Inter-arrival, mix and due date		Average	
	Conwip dyn	Fuzzy max	Conwip dyn	Fuzzy max	Conwip dyn	Fuzzy max	Conwip dyn	Fuzzy max
Thr.	1,21%	0,52%	-0,08%	0,49%	-0,08%	0,49%	0,35%	0,52%
Av. thr	0,73%	-13,07%	-11,23%	-9,62%	-11,24%	-9,76%	-7,25%	-13,07%
Wip	2,38%	-12,65%	-11,24%	-9,17%	-11,26%	-9,33%	-6,71%	-12,65%
Av. Ut.	2,24%	0,99%	0,00%	0,34%	-0,09%	0,34%	0,72%	0,99%
Ord. thr.	31,24%	-59,09%	8,45%	-55,09%	8,60%	-55,05%	16,10%	-59,09%
Delay	-0,87%	-53,45%	15,40%	-39,60%	-13,56%	-59,45%	0,32%	-53,45%
Wait ord.	2,26%	-79,28%	-11,46%	-76,20%	-11,36%	-76,26%	-6,85%	-79,28%
Invent.	10,94%	53,21%	-22,70%	49,36%	-21,66%	64,01%	-11,14%	53,21%

From the analysis of Table 19 the following issues can be drawn:

- The average results show that the fuzzy approach allows to reduce significantly the customer order flow time and the delay of the orders. These benefits are obtained by an increment of the finished goods inventory of 53.21%.
- The dynamic conwip leads to reduce the finished goods inventory of 11.14% keeping the same value of delay of the model with no policy. Moreover, the customer order flow time increases and the WIP reduces but lower than the fuzzy approach.
- The results show that the fluctuations of inter-arrival and mix together are the most severe conditions for the approaches proposed.

The simulation results show that the fuzzy approach with the limitation of the inventory level lead to the better compromise among the performance of the manufacturing system and the customer satisfaction. Moreover, the values of the limitation of the inventory level can be used to control who get the main benefit: the manufacturing system performance (inventory costs) or the customer satisfaction (delay reduction).

CONCLUSION AND FUTURE DEVELOPMENT

The research presented several approaches based on conwip method to control a manufacturing system with customer order scheduling problem. The approaches proposed are characterized by a low computational complexity and simplicity in order to be appropriate in real case applications.

A simulation environment has been developed to test the proposed approaches in a very dynamic environment with the fluctuations of inter-arrival, mix and due date.

The results of this research can be summarized as follows:

1. The dynamic conwip approach proposed allows to improve the performance of the manufacturing system and the customer requirements. In particular, this approach reduces the finished goods inventory level.
2. The approaches that try to anticipate the production of the components when the manufacturing utilization is low improve significantly the customer order flow and the delay. These benefits are obtained with a very high increment of the finished goods inventory.
3. The anticipation of the production by the fuzzy method improves the customer order flow and delay, with a lower increment of the finished goods level. The simulations conduct show that the limitation of the inventory level in this approach can be used to obtain a compromise between the reduction of delay and customer order flow time and the increment of inventory level. Therefore, the fuzzy approach is a most promising approach to reduce the total cost components: delay, wip, and finished goods level.

Moreover, the simulation allows to evaluate the performance of the manufacturing systems in different conditions and choosing the most appropriate solution.

The main limitation of the research presented in this chapter regards the economic evaluation of the inventory level and the delay of the customer orders. The evaluation of these costs can be used to determine the better compromise between delay and inventory level for the fuzzy approach proposed.

Another limitations is the manufacturing system context analyzed.

Based on these limitations, further development concerns the study of different manufacturing system configuration to evaluate the benefits of the proposed approaches. The development of a learning method to forecast the market fluctuations and control the limit of the finished goods inventory of the fuzzy approach in order to obtain the better compromise between delay and inventory costs.

REFERENCES

Ahmadi, R. H., Bagchi, U., & Roemer, T. A. (2005). Coordinated scheduling of customer orders for quick response. *Naval Research Logistics, 52,* 493–512. doi:10.1002/nav.20092

Banks, J. (1998). *Handbook of simulation: Principles, methodology, advances, applications, and practice.* New York, NY: John Wiley & Sons Inc.

Blocher, J. D., & Chhajed, D. (1998). The customer order lead-time problem on parallel machines. *Naval Research Logistics, 43*(5), 629–654. doi:10.1002/ (SICI)1520-6750(199608)43:5<629::AID-NAV3>3.0.CO;2-7

Blocher, J. D., Chhajed, D., & Leung, M. (1998). Customer order scheduling in a general job shop environment. *Decision Sciences, 29*(4), 951–981. doi:10.1111/j.1540-5915.1998.tb00883.x

Case, H. B. S. (1991). *Digital equipment corporation: Complex order management.* Boston, MA: Harvard Business School.

Erel, E., & Ghosh, J. B. (2007). Customer order scheduling on a single machine with family setup times: Complexity and algorithms. *Applied Mathematics and Computation, 185,* 11–18. doi:10.1016/j.amc.2006.06.086

Framinan, J. M., Gonzales, P. L., & Ruiz-Usano, R. (2003). The conwip production control system: Review and research issues. *Production Planning and Control, 14*(3), 255–265. doi:10.1080/0953728031000102595

Hazır, Ö., Günalay, Y., & Erel, E. (2008). Customer order scheduling problem: A comparative metaheuristics study. *International Journal of Advanced Manufacturing Technology, 37,* 589–598. doi:10.1007/s00170-007-0998-8

Hsu, S.-Y., & Liu, C.-H. (2009). Improving the delivery efficiency of the customer order scheduling problem in a job shop. *Computers & Industrial Engineering, 57,* 856–866. doi:10.1016/j. cie.2009.02.015

Julien, F. M., & Magazine, M. J. (1990). Scheduling customer orders: An alternative production scheduling approach. *Journal of Manufacturing Operations, 3,* 177–199.

Karsak, E. E., & Tolga, E. (2001). Fuzzy multi-criteria decision-making procedure for evaluating advanced manufacturing system investments. *International Journal of Production Economics, 69*(1), 49–64. doi:10.1016/S0925-5273(00)00081-5

Kelton, W. D., Sadowski, R. P., & Sturrock, D. T. (2007). *Simulation with arena* (4th ed.). New York, NY: McGraw-Hill Publishing Co.

Leung, J. Y.-T., Li, H., & Pinedo, M. (2006). Scheduling orders for multiple product types with due date related objectives. *European Journal of Operational Research, 168,* 370–389. doi:10.1016/j.ejor.2004.03.030

Leung, J. Y.-T., Li, H., & Pinedo, M. (2008). Scheduling orders on either dedicated or flexible machines in parallel to minimize total weighted completion time. *Annals of Operations Research, 159,* 107–123. doi:10.1007/s10479-007-0270-5

Li, C., & Vairaktarakis, G. (2007). Coordinating production and distribution of jobs with bundling operations. *IIE Transactions, 39,* 203–215. doi:10.1080/07408170600735561

Li, C.-L., & Ou, J. (2007). Coordinated scheduling of customer orders with decentralized machine locations. *IIE Transactions*, *39*(9), 899–909. doi:10.1080/07408170600972990

Li, H. (2005). *Order scheduling in dedicated and flexible machine environments*. (Ph.D. Thesis). New Jersey Institute of Technology. Newark, NJ.

Liu, C.-H. (2009). Lot streaming for customer order scheduling problem in job shop environments. *International Journal of Computer Integrated Manufacturing*, *22*(9), 890–907. doi:10.1080/09511920902866104

Liu, C.-H. (2010). A coordinated scheduling system for customer orders scheduling problem in job shop environments. *Expert Systems with Applications*, *37*(12), 7831–7837. doi:10.1016/j.eswa.2010.04.055

Ng, C., Cheng, T., & Yuan, J. (2003). Concurrent open shop scheduling to minimize the weighted number of tardy jobs. *Journal of Scheduling*, *6*, 405–412. doi:10.1023/A:1024284828374

Shapiro, B. P., Rangan, V. K., & Sviokla, J. J. (1992). Staple yourself to an order. *Harvard Business Review*, *70*(4), 113–121.

Smith, J. S. (2003). Survey on the use of simulation for manufacturing system design and operation. *Journal of Manufacturing Systems*, *22*(2), 157. doi:10.1016/S0278-6125(03)90013-6

Spearman, M. L., Woodruff, D. L., & Hopp, W. J. (1990). CONWIP: A pull alternative to Kanban. *International Journal of Production Research*, *28*(5), 879–894. doi:10.1080/00207549008942761

Su, L. -.H., Ping-Shun Chen, P.-S., & Chen, S.-Y. (2012). Scheduling on parallel machines to minimise maximum lateness for the customer order problem. *International Journal of Systems Science*. doi:doi:10.1080/00207721.2011.649366

Wang, G., & Cheng, T. C. E. (2007). Customer order scheduling to minimizing total weighted completion time. *Omega*, *35*, 623–626. doi:10.1016/j.omega.2005.09.007

Wang, W. P. (2009). Toward developing agility evaluation of mass customization systems using 2-tuple linguistic computing. *Expert Systems with Applications*, *36*(2), 3439–3447. doi:10.1016/j.eswa.2008.02.015

Yang, J. (1998). *Scheduling with batch objectives*. (PhD Thesis). The Ohio State University. Columbus, OH.

Yang, J. (2005). The complexity of customer order scheduling problems on parallel machines. *Computers & Operations Research*, *32*, 1921–1939. doi:10.1016/j.cor.2003.12.008

Yang, J., & Posner, M. E. (2005). Scheduling parallel machines for the customer order problem. *Journal of Scheduling*, *8*, 49–74. doi:10.1007/s10951-005-5315-5

KEY TERMS AND DEFINITIONS

CONWIP: Is a pull-oriented control manufacturing systems. It uses a single card to control the Work In Process of the manufacturing system. The manufacturing systems are characterized by a maximum value of the Work in Process.

Customer Order Scheduling: The customer order consists of several jobs of different types, which are to be processed on m machines. Each machine is dedicated to the processing of only one type of jobs. The order can be delivered to the customer when all jobs are processed.

Discrete Event Simulation: It is a simulation in which the operation of a system is represented as a chronological sequence of events.

Finished Goods: It is the amount of manufactured product in inventory that waits to deliver to the customers.

Fuzzy Logic: Is a multivalued ogic that allows intermediate values to be defined between the conventional true/false. It is applied in expert system to develop methodology based on artificial intelligence.

Order Release: It is the policy used to authorize the production of materials previously identified as a planned order. It determines the timing of the job to release in the manufacturing system.

Scheduling: It is the activity in which the jobs are assigned to the resources at particular times n order to maximize some performance measures as makespan, due date, throughput time.

Chapter 2
Managing Organizational Change

Martin L. Bariff
Illinois Institute of Technology, USA

ABSTRACT

Many project deliverables extend beyond a product or a service for sale to customers. The deliverable may include a new or a revised process for internal workflow or relations with customers, suppliers, or partners. The success of these projects will depend upon adoption of the new or revised process in addition to typical metrics for cost, schedule, risk, and quality. The project manager and team will be responsible for "managing organizational change"—a skillset that is not addressed within the Project Management Institute Body of Knowledge. The purpose of this chapter is to provide sufficient knowledge about approaches and implementation for organizational change to achieve total project success. Case studies are included to illustrate best practices and lessons learned.

INTRODUCTION

Organizations operate in increasingly complex, dynamic, and uncertain environments. Shorter product life cycles, faster changes in customers' buying preferences and the globalization of markets are some of the drivers toward more frequent changes in products and services for sale, operating processes, and business strategy. Thus, organization change has become the norm rather than the exception.

DOI: 10.4018/978-1-4666-3658-3.ch002

Many organizations have shifted from a functional silo structure to a cross-functional, process-centric orientation. For example, an "order-to-fulfillment" process team in a manufacturing company would include participants from sales (order entry), production, possibly procurement, billing, warehouse, and shipping functional areas. These mostly digital processes extend outward to customers, distributors, suppliers, and partners. To improve the quality of the customer experience and employee productivity, business processes are reviewed frequently for opportunities to improve performance. These process improvement projects vary in scale from an incremental change to a transformational change. At all scales of improvement, some degree of organizational change will be included. Project managers must be prepared to address this challenge.

With the movement toward becoming a social business, collaboration among internal and external stakeholders has become more challenging for both business and information technology management. E-mail still has a role in communications; however, the growing portfolio of adopted social media tools has radically changed the nature of collaboration among an organization's internal and external stakeholders. Employees are able to utilize a variety of internal social media applications, e.g., blogs, Facebook-type, Twitter-type, and wiki's to share knowledge and improve coordination. Sandy Carter, an IBM Vice President, Social Business describes this phenomenon as:

At its core, a Social Business is a company that is engaged, transparent, and nimble. A Social Business is one that understands how to embrace social technology, use it, get value from it, and manage the risk around it. A Social Business embeds social tools in all its processes, and for both employees and clients—the entire ecosystem (Carter, 2012, p. 6).

Some employees are enthusiastic about these tools; others are resistant. In contrast to developing a new consumer durable goods with appropriate, antecedent market research, the outcome after implementation of a social media tool may be as intended or unintended. Many consultants use "ppt" (people, process, and technology) as one lens for planning an engagement. The people element is considered to be the greatest challenge. The project manager needs organizational change skills from within the team or an organizational development consultant to address this challenge.

Project management skills (Bolles, 2004) emphasize analytical and quantitative tools, e.g., CPM/PERT, earned value analysis; yet, behavioral skills also are included, e.g., leading a cohesive team, resolving conflicts, and communicating with all stakeholders. Managing organizational change, however, represents a new skill set that has become increasingly important in many projects to achieve a successful outcome. Further, project management skills can be applied beneficially to organizational change. "Dutch" Holland, an experienced change management consultant (Holland, 2000, p. xvi) shared "While I had been exposed to both project management and program management in the Air Force, my time at NASA helped me appreciate these two management disciplines and the roles they played in systematic organizational change." An examination of three case studies (Crawford & Nahmias, 2010) suggested that when the degree of planned behavior change is high, project managers with strong change management skills could lead the effort; else, a change management professional should co-lead the effort.

Thus, there are two perspectives on organizational change and project management. One view is that project management may include an organizational change agenda to achieve a successful project outcome. Another view is that an organizational change engagement using project management as its framework will improve its chance for success. Even an organization context that is changing dynamically could utilize an agile project management approach (Wysocki, 2012, Ch. 11). The focus of this chapter is the former view—including organizational change as a project management activity.

Managing organizational change may be viewed as a type of organizational design project itself. Galbraith's (1977, p. 31) Star Model includes these fully-connected elements:

- **Goals:** Choice of strategy.
- **Structure:** Mode of organization.
- **Information and decision processes:** Data and decision models.
- **Task:** Difficulty and inter-relatedness.

- **People:** Selection and training.
- **Reward systems:** Job design and compensation.

Given the total connectivity among these elements, a project manager must consider the direct and indirect effects from the change resulting from implementing the project deliverables. For example, the introduction of a new collaborative tool may not create the intended cooperation unless some portion of the team members' rewards is based upon team performance rather than totally dependent upon individual performance. Thus, the project manager must view project management as a socio-technical process when stakeholder behavioral changes are expected.

A successful change management effort is not limited to the closing phases of a project when training and cutover occur. Typically, a change in job task design and procedures increases employee anxiety for both future task performance and job security. Thus, obtaining and maintaining employee "buy-in" and participation (when appropriate) is essential. This challenge continues throughout the project life cycle.

Many approaches for Managing Organizational Change exist. Some models will be discussed in this chapter. The Lewin-Schein (Schein, 1987) model is chosen for more extensive discussion since there are three fundamental stages that exist in some form in the other models. Essentially, the approach includes:

- **Unfreezing:** Creating stakeholder readiness for change.
- **Changing:** Modify roles and beliefs (also other factors will promote change).
- **Refreezing:** Institutionalizing new roles and beliefs for sustaining change.

These stages will be discussed in detail in a later section. Indeed, a Composite Organizational Change model (see Figure 1) will be proposed as a comprehensive and feasible approach to managing organizational change.

The objectives of this chapter are:

1. To justify "Managing Organizational Change" as an important skill set for project management.
2. To review the major models of Organizational Change including strengths and limitations.
3. To propose a composite model of organizational change.
4. To discuss an exemplar case study of managing organizational change for illustrating best practices and lessons learned.
5. To recommend future research opportunities.

The content of this chapter should not be confused with Project Change Control—a well-structured and known set of procedures for evaluating, approving and documenting changes to project objectives, scope or procedures.

BACKGROUND

The section reviews selected models of change management and a sampling of research conducted on change management.

Change Management Models

Several approaches to managing organizational change have been proposed. A representative set of the more popular approaches will be reviewed in this chapter. The synthesis of each approach will describe: theme, change process and takeaways. Some proponents limit "organizational change" to enterprise-wide initiatives. Others include a full spectrum of change from incremental to enterprise-wide. The latter position is used in this chapter. Any elements of the organizational change process that are appropriate only to larger or smaller scale efforts will be noted.

Figure 1. Composite change management model

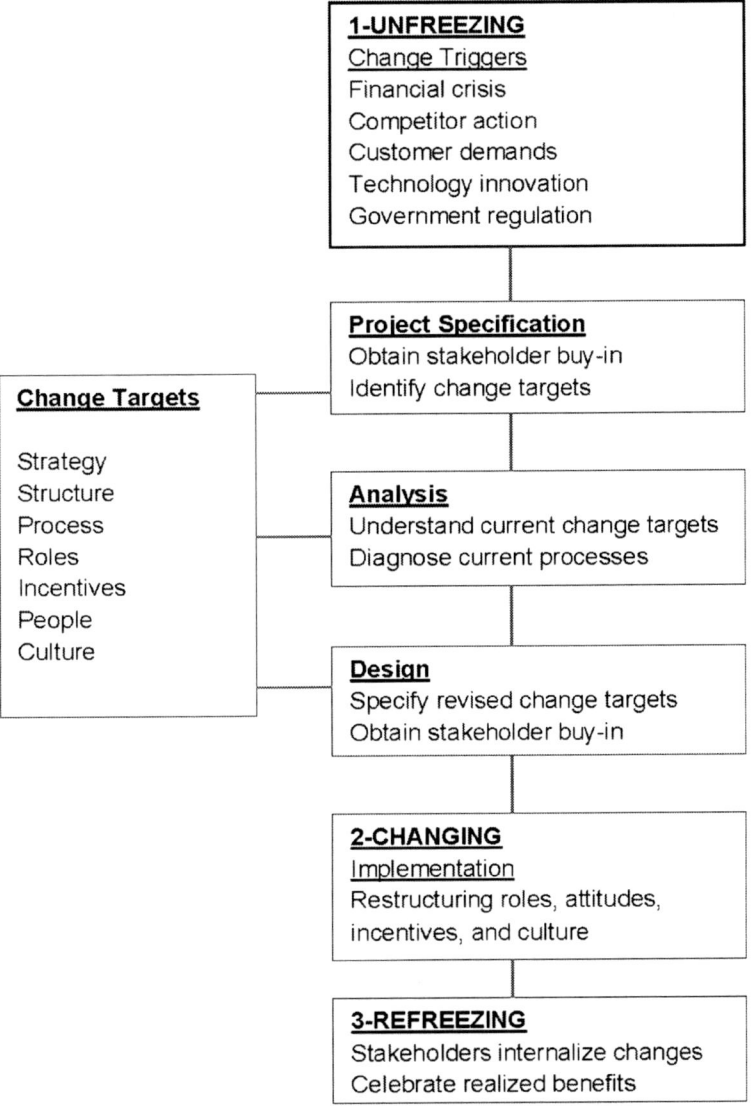

Lewin-Schein Three-Stage Model

This model (Schein, 1987) is an adaption from the original Lewin three-stage model (Lewin, 1951). The essential elements persist. More discussion will be provided in the next major section: Main Focus of the Chapter.

Theme

The objective is to persuade individuals that changes in role, beliefs, and behaviors are necessary either to avoid negative consequences from some event or harvest benefits from some opportunity. Typically, individuals are in a state of inertia or explicitly resist proposed changes. Thus, some intervention is necessary to achieve change.

Process

The first stage *Unfreezing* offers a rationale for buying into a change. Sometimes this trigger is known as the "burning bridge," e.g., sales decrease by a large percentage threatening financial viability. Further, the details of the change must be explained with an assurance that job security is not a risk and appropriate training will be provided, if necessary, for a continued ability to perform satisfactorily. Explicit support by senior management also is required.

The second stage *Executing Change* enacts the transformation of relevant cognitions and behaviors from the current state to the desired state. All the appropriate elements, i.e., change targets (see Figure 1) must be modified to provide a coherent and holistic change process. For example, training for a new role.

The third stage *Refreezing* provides continuing support to internalize and sustain the changes from the second stage. The new behaviors become routinized. Mentoring may be one lever to promote internalization of the new behaviors.

Takeaways

Changing beliefs and behaviors is a challenging task—especially within a dynamic, complex organization's environment. Sufficient motivation to change must be communicated effectively and transparently for individuals to be open to change. Sources of anxiety related to the change must be minimized, if not eliminated. Sustaining change requires some degree of post-change support. The model is simple, but the planning and execution are challenging.

Switch Model

Theme

The objective of this model (Heath & Heath, 2010) is to change [switch] both the cognitive and emotional elements of individuals' viewpoints on the current situation and the proposed changes. An analogy from Haidt (2006) presents two characters: a human rider on top of an elephant. The rider represents the cognitive or rational element. The elephant represents the emotional element.

The rider (planning resource) may overanalyze the situation, procrastinate, and be willing to accept longer-term benefits. The elephant [energy resource] is more impulsive and seeks short-term gratification. As much as the rider attempts to control the elephant, the sheer size of the elephant indicates that emotion eventually will overpower rationality.

Process

"*Directing* the Rider" includes three elements:

- If feasible, carry forward any aspects of the current situation that are working.
- Provide clear, digestible chunks of what the Switch process includes.
- Describe the desired state in detail and its related benefits.

"*Motivating* the elephant" includes these elements:

- The buy-in communication should be more than spoken words. Engage the individuals in some vivid activity that "activates their feelings." For example, to reinforce that the competitors are gaining market share, show a movie segment where a pirate is chasing the victim close to the edge of the plank over the side of the ship.
- Shrink the steps in the Switch process so that individuals feel that part of the switch already has been completed. For example, customers of a car wash were given a free wash stamps card (Heath & Heath, 2010, pp. 126-127). Some were given a blank card to take 8 stamps. Some were given a

card for 10 stamps on which 2 stamps were already attached. The latter group had a greater percentage of card completions in a shorter time period. In this scenario, the 10 stamp customers felt that the goal was more easily attainable because 20% already had been attained.

- Convey a positive set of messages for individuals to embrace a challenging change. Appeal to their core set of values to identify positively with the proposed change. In St. Lucia, a breed of parrots was becoming extinct (Heath & Heath, 2010, pp. 149-151). A proposed solution included passing a law with stiff penalties for killing the parrots and establishing a sanctuary for the parrots. The champion for this cause recognized that the St. Lucia citizens knew very little about the parrots and had no vested interest in saving this breed. Thus, the cause champion with limited funds from the Forestry Service launched multiple low-cost efforts to educate the residents about the beauty and importance of this breed to the country. Some initiatives included: t-shirts, church sermons, visiting schools, and distributing bumper stickers. The residents became aware of the situation and adopted the parrots as a national treasure worth preserving. The law was passed as well as the creation of the sanctuary. This approach successfully linked the preservation of the parrots to national pride. Modified attitudes and beliefs generated the necessary behavioral switch.

"Shape the Path" includes these elements:

- Behavioral changes follow from situation changes. Thus tweak the individuals' environment. Make it easier to adopt the proposed change. Students at a college were solicited to donate canned food (Heath & Heath, 2010, pp. 182-183). A

pre-experiment survey classified students as either more or less generous for making donations. Individuals in each group randomly received a brief request letter or a more detailed letter including a map to the donation location, suggesting that they think when they typically would be near the drop-off location, and a request for a specific type of food. The individuals receiving the detailed letter in both groups had a higher percentage making donations. Thus, "priming the pump" can increase the likelihood of switching behavior.

- Building habits should sustain a switch. Motivating children to brush their teeth even daily can be a challenge. If a child attended an overnight camp where brushing teeth was observed by the counselor, a good habit may form. Of course, rewards further can help reinforce the habit. Upon arriving back at home the child's tooth brushing should continue perhaps with some parental reinforcement.

- Rally the herd suggests that distributing good behavior switches through social media can promote mutual reinforcement once others "get on the bandwagon." Thus opinion leaders can influence adoption of a switch by planting a desired behavior in a blog, tweet, or Facebook wall.

Takeaways

Switch highlights the importance to obtain buy-in from both the rational and emotional elements of individuals targeted to change attitudes and behavior. A few differentiated approaches include: seed the propensity to switch (car wash example), present the switch as a series of attainable changes, and facilitate individuals to identify positively with a cause (St. Lucia parrots) in order to evoke the desired switch. Although the proposals are intriguing, there is no guidance if one approach from each of the three Process groups is required.

Which approaches complement each other or must be used concurrently. The Switch book (Heath & Heath, 2010) is a treasure chest of thoughtful stories that should provide valuable guidance.

Kotter 8-Step Model for Large-Scale Change

Theme

The objective of this model (Kotter & Cohen, 2002; Kotter & Schlesinger, 2008) is to change behaviors to the desired state using an incremental change approach. Leverage easy and early wins to build momentum for individuals to adopt the remaining changes. Techniques to achieve change emphasize a "see-feel-change" approach (emotional) with lessor attention to "analysis-think-change" approach (rational). The eight steps are described below.

Process

1. Increase a sense of urgency by communicating the need to change with strong, vivid statements and images. The tone should be positive, e.g., "we can work around this problem," or "we will take advantage of this opportunity." Negative statements may freeze individuals into their current behaviors.
2. Create a guiding team to champion the change cause. The members should be perceived as respected, trusting leaders.
3. Guiding team develops the vision and strategy that fits the planned changes.
4. Communicate an exciting vision of the planned outcome along with operational strategies to further obtain stakeholders' buy-in. Detailed procedures for change are insufficient to "rally the troops."
5. Empower the individuals to execute the requested changes in behavior. Remove any policies, procedures, and managers that are obstacles to adopting changes.
6. Partition the change journey into a set of smaller change projects. Position projects that can represent early, easy wins at the beginning of the journey to reinforce buy-in.
7. Stay the course. Maintain the strategy of completing smaller change projects successfully. Do not allow early wins to suggest combining remaining projects to leap forward.
8. Sustain the achieved changes with celebrations, possible selected promotions, and shaping first impressions for new employees. Consistent adoption of new behavior should generate new group norms and culture.

Takeaways

This eight-step model offers more detailed guidance. Although not stated by the authors, the steps would be considered as iterative if needed. For example, the communication of the vision to stakeholders may not be received as planned. Thus, the Guiding Team must rethink or reposition the vision to be communicated again. This model does include some new elements:

- Guiding team formally acts as the cause champions—similar to new product champions.
- Change process should be divided into mini-change projects with easy wins positioned at the beginning of implementation to generate enthusiasm and commitment to proceed through the entire change journey.

Although this model is positioned by the authors for large-scale change, the elements also apply to smaller-scale change projects. The book contains more detailed operational guidance and examples of change success stories.

Burke-Litwin Causal Model of Organizational Performance and Change

Theme

The objective of this model (Burke, 2008) is to achieve change through an assessment by all organization members of seven transformational factors (strategy and structure) and five transactional (operational) factors. The results from this assessment reveal elements of strength that will support the change process and elements of weaknesses that must be overcome to effect successful change. These twelve elements form a fully connected "causal" model of organizational performance (Burke, 2008, p. 187). The seven transformational factors are:

- External environment
- Leadership traits—all levels of leadership roles
- Mission and strategy
- Organization culture
- Organization structure
- Management practices
- Policies and procedures

The transactional (operational) factors are:

- **Work unit climate:** Collective perceptions by work unit members of leader behavior, work unit practices and norms.
- Motivation to perform job tasks.
- Job requirements.
- Personal needs and values.
- Individual and organizational outcomes.

This approach requires significantly more diagnostic effort to assess readiness for change.

Process

Pre-launch phase:

- Gather information from internal and external sources.
- Develop the business case for change.
- Create a clear statement of change objective and strategy.

Launch phase:

- Communicate need for change to stakeholders.
- Initiate change actions.
- Resolve any resistance to change.

Post-launch phase:

- Repeat the change need message.
- Provide continuing support for change.

Sustaining the change:

- Resolve unanticipated consequences.
- Choose leadership successors that support the completed change.

Takeaways

This model provides a comprehensive diagnosis of employee perceptions of the organization's strategic, structural, and job design factors. For most organizational change initiatives, the collected information may be more than necessary. The use of "causal" may be misleading since the causal inferences appear to be judgmental. If an appropriate technique, e.g., path analysis, were used, causal inferences could be stated.

Power of Habit Model

Theme

Organization change may be viewed as transforming existing habits into new habits. This model (Duhigg, 2012) describes a habit as a sequence of an experienced cue that triggers a familiar routine providing a desired reward. A change in habit occurs when a modified habit is adopted after identifying the related cue and reward. This process is somewhat similar to finding the root cause of a problem. Once this pattern of: cue→routine habit→reward is understood, then a plan can be developed to replace the existing routine with a better routine. The routine is the individual's behavior. This habit model is applicable to individual, organizational, and societal habits. The Appendix (Duhigg, 2012) presents a concise and interesting discussion how an individual's habit of taking a mid-afternoon break for eating a cookie with friends in the company cafeteria resulted in an unwelcome weight gain. A process for transforming this habit into taking a mid-afternoon break to chat with one friend in the office area resulted in weight loss and sustained the desired social break interaction. An example from the book (Duhigg, 2012, Ch. 4) describing organizational change is described below.

Process

The challenge for the new Alcoa CEO, Paul O'Neill was to find some means to improve mildly acceptable factory and administrative performance into a high performance company. The objective of zero injuries was presented at the company annual meeting and subsequent meetings with employees. Shareholders were puzzled why employee injury and not some financial performance metric was the primary target for the forthcoming year.

O'Neill searched for some objective that both management (no injury—no work absences) and

unions (better safety for workers) could embrace. Indeed, the clever agenda created a chain of positive outcomes:

- Analysis of work procedures to eliminate injuries.
- Identifying weaknesses in the manufacturing process.
- Quality control education provided to workers and 1st line management.
- A global e-mail system to report injuries also became a new means for improving overall collaboration.
- Improved factory efficiency.

A new injury reporting routine was adopted. When an injury occurred, a report to the CEO, O'Neill had to be sent with a plan to avoid this type of injury in the future. Only those employees adhering to this routine would be considered for promotion. O'Neill predicted if employees would adopt this workplace safety habit (representing one habit of excellence), other workplace habits would change for the better. At the end of the first fiscal year after the safety objective was introduced, both profits and market capitalization of Alcoa stock reached record highs. A routine habit that has the direct benefit of improved performance when changed and has an indirect benefit potentially to change other routines is known as a "keystone" habit. O'Neill was counting on this chain reaction to transform the company. Indeed, a well-respected senior executive did not report an injury to the CEO. The firing action did not cause an uproar because the company culture of "safety first" has become the norm.

Takeaways

The critical insight from this modeling approach is that careful analysis of the habits or behaviors desired to change can improve significantly the probability of success. Promotion was a strong incentive to change. There is some similarity with the Burke-Litwin causal model (2008) that evaluates individuals' extrinsic (e.g., compensation) and intrinsic (e.g., reputation among peers) motivators. Project management stresses the benefit from careful front-end activities (e.g., project definition and feasibility analysis). This Power of Habit approach also demonstrates the benefit from careful analysis for attaining the desired change.

Kolb-Frohman Consulting Model

Theme

The objective of this approach (Kolb & Frohman, 1970) is to conclude a successful consulting engagement that typically includes some degree of behavioral change. This seven-step model is equally applicable to external or internal consultants.

Process

1. **Scouting:** Consultant evaluates when is the best point to enter the system to begin the engagement, e.g., has senior management communicated to the stakeholders that a consultant will be assisting the proposed change process.
2. **Entry:** Once an Entry point is chosen, consultant formulates a "contract" with the client representing a mutual expectations about the objectives of the engagement. A definition of the problem or opportunity is determined with expected benefits. The "contract" may be renegotiated during the change process if circumstances merit a review.
3. **Diagnosis:** The general problem-opportunity statement is converted into specific measurable elements of the client's goals. Readiness to change also is evaluated by the consultant.
4. **Planning:** Define the change process stages, identify expected sources of resistance, involve the stakeholders to obtain their buy-in. If a disagreement between stakeholders and consultant occurs, the process iterates abck to the Entry stage for attempted resolution.
5. **Action:** Major targets include:
   a. **People:** Educate and train.
   b. **Authority structure:** Roles, positions and reporting may change.
   c. **Information systems:** Modification to assure the right information reaches the right people.
   d. **Task:** Technology support may change; job design and satisfaction should be considered.
   e. **Legal and culture:** Review organization policies for modification and attempt to reshape norms/values if appropriate.
   f. **Environment:** May reconfigure locations of jobs and assess regulatory and legislative bodies for recent and forthcoming changes.
6. **Evaluation:** Compare expected and actual performance metrics. If an unsatisfactory gap occurs, return to Action stage for further change initiatives once an explanation for the gap is determined.
7. **Termination:** Consultant withdraws when new behaviors have become the norm and coaching is not necessary.

Takeaways

This approach addresses both the cognitive and emotional elements to be changed. The seven-step model provides a logical progression for managing organizational change. Indeed two recursive loops from Planning and Evaluation are explicitly included to recognize the dynamic nature of a change process.

Change Management Tool Kit

Theme

This content is drawn from Holland (2000) consulting experience and Sharke (1999) 463-page guidance for business process re-engineering. The latter tool kit is applicable to changing any type of organization's processes and includes a section on change management. Holland is one of the four co-authors of the Change Management Tool Kit. Holland (2008) uniquely cautions that change management requires concurrently running the change program and the normal business operations. Most managers are trained for normal operations, not change management—an inherent risk.

Process

Holland (2000) proposes five steps for change management:

- Prepare a vision for change with a supporting business case.
- Inventory current people, equipment, and technology. What should be retained? What should be removed?
- Develop and negotiate new performance contracts with key stakeholders.
- Dismantle and remove elements not to be retained.
- Prepare and distribute the tactical action plan for changes.

Holland's viewpoints clearly had influenced the content of the Change Management Tool Kit (Sharke, 1999).

1. Position for Change Phase:
   a. Establish urgency and gain commitment including a business case and designated change champions.
   b. Create a map of current processes highlighting value-adding processes.
   c. Select processes for improvement and assign process owners if not existent.
   d. Establish a charter and project team(s).
2. Diagnose Existing Processes Phase:
   a. Define key process components.
   b. Identify potential organizations for benchmarking.
   c. Diagnose current process weaknesses.
   d. Establish new process performance targets.
3. Redesign Processes:
   a. Identify potential innovations.
   b. Develop initial design vision.
   c. Identify process improvements.
   d. Develop detail process redesign with customer input.
4. Transition to New Design:
   a. Begin change management with a set of plans differentiated for each customer segment.
   b. Develop a transition plan and teams.
   c. Prototype and test initial redesign changes with customers.
   d. Complete the transition, assess outcome, support continuous improvement.
5. A 23 question Change Readiness Assessment (Sharke, 1999, pp. 322-337) is included that represents 23 common sources of resistance to change with a scoring scale. The results are classified into two groups originating from the Chinese dual definition of a crisis:
   a. Individuals' perception that change represents danger.
   b. Individuals' perception that change provides opportunities.

Takeaways

The Change Management Tool Kit provides both detailed guidance and tools for use in process redesign. Holland (2008) however offers more detailed guidance specifically on change management. Jointly, both sources address most of the aspects addressed by the change management approaches reviewed above. Holland (2000) however offers unique guidance for managing both the change process and the normal operations of the organization.

Kaizen Model

Theme

Although the Kaizen approach (Colenso, 2000) is evolutionary, i.e., continuous improvement, some guidance for change management as a discrete process (revolutionary) is offered. Five stages for the change process are provided with related guidance.

Process

Stage 1: Negative Reactions

Many reactions to learning about a forthcoming change process surface, e.g., morale declines, sources for blame are sought, job insecurity and anxiety appear, groups justify why their operation should not be eliminated and uncertainty about ability to perform satisfactorily after the change is completed. Management should identify the root cause for the change and clearly communicate the urgency to the stakeholders. The planned changes, rationale, and timing should be clearly communicated to the impacted individuals. Change champions should be identified and supported. Reward and celebrate early change project successes.

Stage 2: Defining Event

The trigger for enacting a change process may be external, e.g., competitor or regulatory agency or internal, e.g., improve a broken business process. The "event" needs to be meaningful to all the impacted stakeholders; else, more than one trigger must be identified for each segment. Communication should be engaging, that is, hold forums with stakeholders to enable questions and answers.

Stage 3: Contracting – Making the Commitment

The setting and agreement of change process objectives and evaluation criteria with input from the stakeholders may be sufficient to form a psychological contract. All the senior managers must publicly support the change process and be among the first adopt the new behaviors as role models for the other employees.

Stage 4: Self-Discovery

Stakeholders will realize that the change will require some modified to new behaviors. To reduce related anxiety, job education, training, and coaching should be provided. Further, the planned changes may trigger unexpected beneficial changes in the stakeholders' other roles and personal values.

Stage 5: Internalization of Change

To anchor the new behaviors as the norm, management needs to recognize and reward adopters of the changes. Further coaching to help stakeholders master new behaviors should be provided until the desired behavior becomes the new routine.

Takeaways

These recommendations are offered if a radical change is required. The Kaizen approach builds continual change or improvement into standard operating practices, e.g., quality circles, morning meetings, and suggestion boxes. Thus, this set of recommendations is not as comprehensive or detailed as other change management approaches previously reviewed.

Organizational Transitions in Complex Change

A five-element model: (1) change motivation, (2) present state, (3) desired state, (4) change path, and (5) change actions (Beckhard & Harris, 1987) offers more detailed guidance for change management. The authors describe 29 potential actions to be taken, e.g., identifying relevant systems and subsystems, and information to be gathered, e.g., core mission. The expanded content is similar to the Kolb and Frohman (1970) and Lewin and Schein (Schein, 1987) models.

Leading Strategic Change Insights

Although the authors (Black & Gregerson, 2002) do not offer a specific change management model, a few useful insights are shared below:

1. Early adopters of desired change are needed to build traction for other stakeholders to adopt the changes. Early, easy wins facilitate this strategy.
2. The degree of inquisitiveness (IQ) by individuals may be a predictor of willingness to adopt new behaviors. A ten question instrument (Black & Gregerson, 2002, pp. 141-142) measures degree of inquisitiveness. A high IQ score suggests that the individual is a "master of change." A low IQ score suggests that the individual is a "creature of habit." This IQ score may be viewed as a proxy for an individual's "readiness for change."
3. Three types of change scenarios are described. These are ordered by decreasing degree of difficulty to motivate buy-in from stakeholders:
   a. **Anticipatory:** Recognition that a future event requires a present change to avoid a problem or take advantage of an opportunity.
   b. **Reactive:** Respond to a current event—either favorable or unfavorable.
   c. **Crisis:** Urgent need for immediate change.

A comparison (see Table 1) of these eight models for organizational change appears below. The Google Scholar Citation values with its strengths

and limitations only represents one proxy for assessing the impact of these organization change methodologies.

Summary of Best Practices from the Above Approaches

- Choose a point of entry that best advances the change management process, e.g., has senior management openly and enthusiastically endorsed the desired changes and presented the business case? Use positive language and images to reduce potential anxiety by stakeholders.
- Define mutually acceptable change project objectives and evaluation metrics with primary stakeholders. A crisis situation, however, may justify objectives mandated by senior management. Assess the feasibility of success, e.g., economic, adoption, resources, technology, and completion date.
- The stated objective might be a proxy objective for the intended objective to motivate buy-in by the stakeholders, e.g., no injury objective stated by the Alcoa CEO in Power of Habit model.
- If appropriate to obtain buy-in from stakeholders, select an intermediate project that could result in stakeholders positively identifying with the change cause. Then, the desired behaviors should follow, e.g., parrots of St Lucia in Switch model.
- Define change targets holistically for modification (see Figure 1). Secure change champions to help "sell" the change process.
- Identify expected resistance to change sources. Perhaps assess stakeholders "readiness for change." Develop a plan to reduce the stakeholders' anxiety and fear.
- Recruit interested stakeholders to become early adopters of changes to communicate the successes and influence others to adopt the desired changes.

Table 1. Comparison of change management models

METHOD NAME	AUTHORS	DATE	MAJOR FOCUS	Number Of Steps	Google Scholar Citations
3 Stages Model	Lewin-Schein	1951, 1987	Unfreeze-change-refreeze behaviors	3	Book 9416(a)
Switch	Heath & Heath	2010	Change both cognitive & emotional elements	3	Book 117
Large-scale Change	Kotter & Schlesinger	1979, 2008 reprint	Detailed guidance for sustaining incremental changes' journey	8	Article 1136
Causal Model Organization Performance & Change	Burke & Litwin	2008	12 transformational & transactional Factors analyzed to identify facilitators & inhibitors for change.	4	Book 474
Power of Habit	Duhigg	2012	Replace current cue→ routine→habit with new cue & reward to modify behavior with a new habit	4	Book (b)
Organizational Development	Kolb & Frohman	1970	Consultant's perspective with recursive elements to achieve change	7	Article 223
Change Management Toolkit	Sharke,et.al.; Holland	1999 2000	Sharke toolkit has extensive detail and forms-- includes Change Readiness diagnosis; Holland model guides what behaviors are to be deleted, continued and added.	5	Holland book 10; Sharke-n.a.
Kaizen	Colenso	2000	A continuous model for change is described; discrete steps for radical change are proposed	5	19

Notes:

1. Citation value overstated since book addresses many concepts beyond change management.

2. Book published March 2012 ranked in New York Times Best Seller list for 19 weeks since release with rank of 2 at 2nd week and rank of 15 at July 29, 2012.

- Plan change management initiatives to address both the cognitive and emotional (intrinsic and extrinsic motivators) concerns of stakeholders.
- Remove obstacles that could derail the change initiatives, e.g., company policies that limit the transfer of employees between locations or from acquiring new skills.
- Jumpstart the change process by "shrinking the steps," e.g., car wash stamps from Switch model.
- Partition the change management project into a series of smaller projects. Select easy changes first to establish early wins to create momentum for pursuing continued changes.
- Provide helpful guidance for behavior change by making it "easy" to perform a new behavior, e.g., canned food donation from Shape the Path in the Switch model.
- Provide sufficient post-change support, feedback, rewards, and celebration to internalize modified behaviors into new habits.

Review of Selected Research Studies on Change Management

A significant body of research on change management has been developed during the past fifty years. Indeed, there is a *Journal of Change Management* with a ten-year publication history. The following change management studies represent some of the research more closely related to project management. Reviews of change management research can be consulted (Burke, 2008; By, 2005; Weick & Quinn, 1999).

Factors Promoting Organizational Change Success

A ten-year study of forty international change situations in various stages of progress (Fendt, 2006) revealed four leadership factors that were associated with successful change:

- Top executives perceived as credible.
- Perceived sound strategic and financial decisions.
- Winning employee trust.
- Employees' belief that changes will be beneficial.

Three styles of CEO communication were discovered: (1) cartel—amass personal power, (2) aesthetic—likable, and (3) videogame—high energy with technical expertise. Fendt recommended a holistic communication style—innovative, inspirational, and consistent but flexible as most beneficial based upon the results. Further, a CEO should project determinism, relate to diverse stakeholders, and treat mistakes as opportunities for growth and learning. Communication credibility will be compromised by:

- Asking employees to change, but not changing oneself.
- Emphasize teamwork, but make decisions in isolation.
- Advocate feedback, but don't listen to suggestions.
- Advocate long-term change, but reward short-term fixes.

Vasella, CEO of Ciba-Geigy, lead the merger with Sandoz to create Novartis. Upon completion, the firm enjoyed nine consecutive years of record profits. Merger success was facilitated by:

- Goals were explained and negotiated; words transformed into actions quickly.
- Vasella was always available and projected a confident image.
- Merger updates were communicated frequently and honestly by Vasella.
- A 24 hour hotline and an intranet repository provided current merger information.

A study (Hoverstadt, 2004) of enterprise-wide introduction of TQM in six companies over a ten month period revealed that an organic approach to organizational change was successful after a planned change approach failed. The planned change approach failed due to mistakes with each of three leadership approaches:

- **Top-down:** Did not permit participation by workers.
- **Top-down and bottom up:** Workers were reluctant to make any decisions fearful that the any person above in the hierarchy could rescind the decision.
- **Attitude change:** Expected success with TQM training, but policies and procedures were not changed. Thus, no adoption.

A combination of economic and organization development approaches was found more successful for managing change than either approach alone (Beer & Nohria, 2000). At Scott Paper Company, the economic (top down) approach (remove hierarchical levels and employees) by CEO Dunlap resulted in the company sale to Kimberly Clark at a depressed price. Champion Paper CEO Norman adopted an organization development approach (top down plan, bottom-up) that improved company culture, but also resulted in a depressed value sale to PM-Kymme. In contrast, ASDA used economic and organizational development approaches in sequence with a subsequent sale to Walmart at a premium. Thus, a blended change strategy produced higher ultimate economic value to the shareholders.

A healthy organization change approach (Saksvik, 2007) was revealed from 180 interviews (57 interviews were with groups of 2-10 individuals) of managers and workers in ninety organizations. Change management process was rated higher if:

- Local norms were considered.
- Change management actions were tweaked for employee diversity in experience and maturity of the workforce.
- Future roles were clarified early.
- Managers were available for support.
- Discussions of concerns were welcomed.

These actions empowered workers to embrace change rather than feel insecure and defensive. Similar to the Switch approach (Heath & Heath, 2010) reviewed earlier, Smollen (2010) interviewed 24 individuals (each in a different organization) to investigate the role of emotions (elephant in Switch) in organizational change. When individual values were congruent with organization values, change was viewed more positively. Reactions to change were more negative if (1) process and outcomes judged unfair, (2) no organizational support, and voicing concerns was felt to be unsafe.

In a study of 395 U.S. Air Force officers (152), enlisted personnel (125), and civilians (118), higher scores on change readiness by officers and civilian resulted in greater involvement in the change process.

Initiation of a change effort should begin with modifying internal processes before introducing a new external technique, e.g., quality circles-a Japanese method (Abrahamson, 2000).

Rosenberg (2003) recommended that successful organizational change is dependent upon three factors:

1. Monitoring of external business, financial, political and regulatory information. Macdonald (1995) also emphasized monitoring the firm's environment to identify indicators of potential organizational change. Anticipatory change is the most difficult to influence employees that a sense of urgency exists. Yet, a valid indicator offers the best opportunity to avoid a future problem or pursue an opportunity.
2. Motivate employees to achieve high performance—characterized as passionate leaders, clear change focus, and a learning culture where mistakes are tolerated.
3. Vision, values, and expected results for change are communicated by leaders who model these characteristics

Factors Promoting Organizational Change Failure: Lessons Learned

A failure rate for organizational change initiatives often is stated (e.g., Balogun & Hope Hailey, 2004) to be 70 percent. Some of the pitfalls to avoid are discussed below.

Lucy (2008) interviewed nine thought leaders (6 academics, 3 practitioners) on why the failure rate for organizational change is high. The responses were categorized by the three stages of the Lewin-Schein (1987) model:

Unfreezing:

- Absence of clear executive vision.
- Absence of clear executive communication strategy.
- Inability to foster a sense of urgency.
- Poor consultation with stakeholders.

Changing:

- Lack of structured methodology (opportunity for project management?).
- No dedicated implementation team.
- Human resource policies not accommodating change.

Refreezing:

- No recognition or celebration of successes.
- Employees not engaged.
- No change champion.

Miles (2010) further offers these factors that can derail organizational change success:

- Culture of caution.
- Business as usual management processes (did not recognize parallel activities, i.e., normal business operations and change activities).
- Initiation gridlock—no consensus on mission, desired behaviors, and metrics.
- Tolerating recalcitrant executives.
- Disengaged employees.
- Loss of focus during execution.

Implementation of information systems may be viewed as a socio-political-technical process (Keen, 1981). A modified or new information system may change the distribution of power among users if access to information changes. Some users may deploy counter-implementation acts, e.g., delay change tasks or divert resources from the project to resist closure of the effort.

Kotter and Schleslinger (2008) offer these sources of resistance to change:

- Desire not to lose something of value.
- Misunderstanding of change objectives and implications for new behaviors.
- Lack of trust in leadership.
- Perception that change is not appropriate for the organization.
- Low tolerance for change.

The decision to adopt a slow or fast pace for organizational change is based on these criteria:

- Amount of expected resistance to change.
- Relative power of the initiator and resistor.

- Initiator's access to information on the design changes.
- Urgency for change.

Some recommendations to reduce resistance to change include:

- Education if employee claims lack of sufficient information.
- Encourage participation if commitment to change is low.
- Provide skills training and education of future job design is uncertain.
- Negotiate job responsibilities, performance targets and incentives if a power struggle exists.

If urgent completion is critical, coercion may be the only approach.

Another basis for resistance may be that organizational changes are perceived to be excessive (Stansakar, et al., 2001). Excessive is defined as either: (1) one change effort overlaps a following change effort or (2) changes are occurring concurrently. Thirty-seven interviews were conducted in three change projects (an acquisition and two administrative system strategic changes). Lower and middle managers stated that strategic changes were excessive due to:

- Insufficient implementation details.
- Partial understanding of total change management effort.
- Top management has more lead time to understand the scope and purpose of the entire effort.

Middle and lower managers coped by passively accepting requested changes, sabotage, or took control of the project. Performance impacts for the change projects included: (1) failed projects reported as successes and (2) less attention to regular responsibilities, e.g., customer support. To improve the chance for success, project could be partitioned into smaller projects and communications to stakeholders should become more effective.

Often, a recommendation is to select early change projects that are quick wins to gain buy-in and sustain momentum. Van Buren and Safferstone (2009) propose five traps that may limit or negate quick wins:

- Focus too heavily on project details.
- React to criticism negatively.
- Intimidate stakeholders.
- Jump to hasty conclusions from observations.
- Micromanage direct reports.

To harvest the benefit from quick wins, attempt to include all the stakeholders, even reluctant individuals to create a group quick win for moving forward.

Special Situations

One type of strategic change is a merger among organizations. Kavanaugh and Ashkansy (2004) studied three university mergers during a 3-year period including 5 data collection points for each merger. The outcome measured was the degree of satisfaction with the changes. For the university that continued to operate as separate units, stakeholders were indifferent. For the university that completed the change by closing buildings and transferring students, some were satisfied while others were not satisfied. For the university that adopted an incremental approach—some units combined, some remained autonomous; the greatest degree of satisfaction occurred. For all three scenarios, most (greater than 75%) stakeholders reported a significant change in institutional culture. These results should not be generalized to corporate or other not-for-profit organizations.

Many organizations have global operations. Change efforts may include expansion, contraction, or restructuring. Ericsson (Iveroth, 2010)

consolidated its decentralized data centers into 10 shared service centers with a single ERP system, IT governance, and a global network. Within 3 years, the change management objectives of lower costs, improved controls, and data consistency were achieved. A common architecture established the framework for the project. Further, the interviewed 12 change project managers were available for communications and coaching to both the business and IT employees. Open forums to discuss feelings about the project helped to reduce political obstacles and uncomfortable feelings about the project.

An earlier, but still interesting set of organizational change cases (Mirvis & Berg, 1977) may of interest to review.

MAIN FOCUS OF THE CHAPTER

Project management includes a complex set of both quantitative and behavioral skill set demands. The background of many project managers is engineering and technology. Thus, there is a comparative advantage for applying analytical skills to budgeting, scheduling (PERT/CPM), cost analysis, risk analysis, project leveling, and earned value analysis. The behavioral skills for project leadership, team building, conflict resolution, and stress reduction typically are not emphasized in an engineering curriculum. Formal project management courses may not provide significant attention to the behavioral issues. A MBA or Master's degree in Project Management should provide some exposure to these issues. Usually, an organizational behavior course includes Managing Organizational Change as one of many topics in a survey style experience.

A project typically includes identifiable stages, e.g., definition, feasibility, analysis, design, development, testing, training, implementation, operation, and evaluation. In parallel, project management stages include: charter, initial resource requirements, work breakdown structure, detailed resource requirements and budgeting, scheduling, team member acquisition (potential rescheduling), oversight of the project with milestone and other reviews, managing the team, managing procurement, quality assurance, communicating with all stakeholders, modifying the schedule and budget as needed and approved, approving project change requests, overseeing implementation, evaluating the project outcomes and project management performance. Typically, Managing Organization Change is not explicitly included as a major activity. Yet, the success of most projects significantly depends upon this challenging task. Further, Organizational Change activities permeate throughout the project life cycle—not just a task to address near the end of the project.

The proposed Composite Change Management Model (CCMM) (see Figure 1) is a fusion of the Lewin-Schein (1987) change management model, Galbraith (1977) organization design model, and the project management life cycle. These have been discussed previously; however a holistic analysis is presented below.

Unfreezing and Change Triggers

Representative, but not exhaustive triggers for change are suggested. These are relevant for profit, not-for-profit, and government organizations. The trigger may occur as an external force, e.g., competitor action or an internal need, e.g., restructuring. Whether the trigger is internal or external, it further may represent an anticipatory, responsive or crisis challenge. A well formulated statement of Project objectives including expected behavioral changes is needed to prepare stakeholders for the forthcoming challenges. The justification or "business case" also must be included even in preliminary form. Is the justification aligned with organization strategy? A difficult challenge is composing the language to capture stakeholder attention without creating mass panic, inertia and resistance. The project manager may need help from an organization development consultant to prepare these documents and communications.

Project Specification

The vision, objectives, deliverables, and outcome metrics are developed with representatives from all the stakeholder groups as one ingredient to obtain stakeholder buy-in. The organizational component of feasibility analysis addresses expected sources and types of resistance to change to prepare actions to reduce these threats. A change champion typically is needed to ignite buy-in from the stakeholders. A charismatic CEO could play this role. The extended digital enterprise potentially presents a greater challenge for obtaining buy-in. Customers, distributors, suppliers and partners may be included in the targeted stakeholder groups for change. Thus, there may be the need for multiple change champions for each stakeholder segment.

The change targets represent elements of organizational design (Galbraith, 1977). For project management, these targets may be characterized as organization elements for redesign unless a project might be creating a new or spinoff organization subunit. The important lesson is that all these change targets are interdependent. Thus all the elements must be formally considered for impacts from the proposed organizational changes. The stakeholders need to know exactly what changes are forthcoming to reduce anxiety and resistance. A plan for addressing the entire project or partitioning into segments needs to be decided in this stage.

Analysis

A project may be delivering externally a new or modified product or a service or a new or modified strategy, structure, role or process internally. Input from all the stakeholders regarding the current previously listed element—assessment of strength and limitations of current version if existent and expectations regarding the desired element. If feasible and appropriate, request stakeholders to interact with a prototype product or service to further cement buy-in and gain valuable feedback for easier changes during the early stages of a project. The engineering background of many project managers may focus the content of analysis on mechanistic aspects of the deliverables. In most cases however, the organizational context, e.g., social structure and culture of the organization must be evaluated. This type of "organizational diagnosis" is natural activity for a change management analyst (Alderfer, 2011).

The relevant change targets must be evaluated as part of the organizational diagnosis:

- Is the business model for converting (monetizing) the customer value proposition into cash flows changing due to a new business strategy? If yes, most likely all the other change targets will be impacted by this strategic change.
- Is the business structure changing by adding or deleting a business unit or how some business groups report to more senior management? Might the organization change from a hierarchical to a network, pure project, or matrix organization? This type of change also could have significant impacts on the other change targets.
- Are some business processes being improved incrementally or significantly? Is the change immediate or staged? Will job tasks also be modified? Are external stakeholders' means for interacting with the organization to be modified? Outsourcing business processes or functions also creates changes in roles and relationships.
- Will incentives be modified to align with new job roles and responsibilities?
- What education and training are required for stakeholder modified roles to be performed? What attitudes and beliefs of stakeholders need to change to be aligned with other planned changes?
- Will the organization culture and values change as other changes are implemented? Moreover, is a culture change a planned element of the organizational change project?

Most of the above changes related to organizational development are new to the typical project manager. Even with assistance from an internal or external organizational change expert, the project manager somewhat should be familiar with these challenges and issues to assure that a holistic plan for change is included to achieve the desired success for the planned project. To move stakeholders to desired behaviors and attitudes, the baseline assessment for these change targets must be evaluated. Communication from senior management supporting the change must continue enthusiastically.

Design

Alternative scenarios for the design of the deliverables, e.g. a product, a process, a subunit, jobs, business model, and other related revised change targets should be shared with the stakeholders to sustain buy-in. If the project is to be split into smaller projects, only the first or early projects should be shared with stakeholders to minimize information overload. Any opportunity for continued prototyping of the proposed alternatives by stakeholders would be invaluable.

Changing: Implementation

Assuming that stakeholder buy-in has been obtained and sustained, sufficient education and training for new internal stakeholder job roles and external stakeholder processes must be provided with coaching support. Retail customers might be offered videos of new procedures. Suppliers' order processors might need onsite training. Forums might be necessary to change internal stakeholders' views on organization's values and culture. In this stage, intended actions must translate into actual action. Do the stakeholders view the modified change targets as a holistic, consistent new world view?

Refreezing

The learned new behaviors and cognitions from the prior stage must be transformed into new habits. Additional coaching or mentoring may be required. Incentives may need tweaking. Formal recognition and rewards are needed to further cement the changes. If this success only relates to one segment of an organization, the journey continues to diffuse this adoption to other organization subunits or external stakeholders. A log of what worked and what failed should be entered and maintained for others to learn from this experience.

Another approach to combining organization design and change began with the McKinsey 7-S Model of Organization Effectiveness (Waterman, Peters, & Phillips, 1980). The seven factors: strategy, structure, systems, leadership style, staffing, skills, and shared values can identify candidate areas for organizational change. The Mckinsey model is comparable to the Galbarith model (1977) with leadership style and shared values included in the cultural context. Nohria and Khurana (1993) extended the McKinsey 7-S model by proposing a different set of 7-S factors for managing organizational change: strategic intent, substance of the change, scope of the change, scale of needed resources, speed of change, style of the leaders, and sequence of proposed actions. These 7-S factors have been addressed in the above Composite Change Management Model as well as the previously described Change Management models.

EXEMPLAR CASE STUDY

Texas Instruments was experiencing a strategic change in their customer demands as the market for computer chips shifted from a "build to stock" batch production process to a "build to order" custom chip process (Harvard Business School Business Process video, 1995). Order delays reached 180 days. Customers were threatening to move their business to Texas Instrument's competitors. This change somehow was not properly anticipated for an orderly change to this new production process. Urgent change was required.

A change management team was charged with the responsibility for redesigning the production processes to provide customers with exceptionally superior, not inferior order processing services. The current production configuration included chip assembly plants in Texas and a series of chip manufacturing plants in far east Asia. Customer orders were fulfilled by parts flown from Asia to Texas for final processing. The plants in Asia competed with each other for resources and performance rewards. The change management team correctly identified the root cause of the slow production as the poorly coordinated relationship between the Texas and Asia plants. There was no operative plan provided since the plants in each geographic location could not be relocated on short notice and thousands of workers' lives would be disrupted.

The team was instructed to create a feasible plan or would be disbanded. One of the team members suggested that virtual reality technology could provide a solution to this crisis. A set of virtual factories could be configured without changing the present physical location of Texas and Asian plants. Each facility in Texas would be paired with a facility in Asia to create a virtual factory. Operations would be tightly coordinated between the two facilities. New production processing and control software would facilitate the new arrangement. Further, 20% of the compensation would be based upon virtual team performance. The plan was welcomed by senior management, but how would implementation proceed successfully?

Employees required new education and training. To improve adoption of new knowledge, a group of senior production specialists in Texas Instruments first were trained in the new production methods. These "change champions" were called "Noahs." They returned to their home production facilities and trained fellow employees in the new production methodologies. The Noahs shared best practices through an electronic messaging network. The Noah knowledge transfer process was highly successful. Employees responded enthusiastically to learning from respected peers. An external training approach probably would not have achieved this degree of success. This strategic change at Texas Instruments retained their customer base and earned recognition as the best performing company in the chip manufacturing industry.

This organizational change process included all the change targets in the Composite Change Management model (see Figure 1). Manufacturing strategy changed to a successful "build to order" process. Organization structure was virtually modified. Business processes were redesigned. New technology for manufacturing and information processing was introduced. The reward system was modified to encourage improved team performance. Innovative production methods were introduced and diffused throughout Texas Instruments by the network of Noahs. Employees were retained and retrained. A new culture of worker sharing rather than competing emerged. New behavior, beliefs and values were adopted. This was a total or strategic change experience. Behavioral change management skills were the critical ingredient for success. The best practices and lessons learned from this case generalize to many projects requiring organization change.

FUTURE RESEARCH DIRECTIONS

This area of managing organizational change requires education and training in social science concepts and research methodology. Concepts from psychology, sociology and anthropology are relevant. Research design approaches include laboratory experiments, field experiments, and survey data collection. This portfolio of knowledge and tools are not included as part of the typical project management education curriculum. Thus, col-

laboration between project management domain experts and social science researchers represent a most productive approach.

From the research described in this chapter, an absence of well planned, longitudinal research on the comparative benefits from alternative change management approaches under what situational characteristics drive the most successful outcomes. The conducted research is highly anecdotal. Large-scale field research is difficult to plan and execute. However, a planned set of comparative case studies (Yin, 2008) could surface a set of research questions and hypotheses that could be investigated with less global research methods with experiments in the laboratory or field to tease out causal, rather than correlational relationships.

The project management body of knowledge and certification process are well defined and have been adopted globally. This era of "change is normal" however presents some new challenges. How can projects in the anticipatory change category be accepted by stakeholders when the triggering event is a probability rather than a reality? How do project management methods change to accommodate a context of an agile organization quickly responding to external change threats and opportunities? Should the Agile (e.g., SCRUM) project management or Extreme project management approaches be compared in the field to a traditional project management approach to record actual relative performance rather than professional recommendations (Wysocki, 2012)?

Which existing project management methods and techniques are appropriate for adoption is a Change Management effort? Is a well-structured work breakdown process and associated PERT-CPM scheduling too rigid for the dynamics of change management? Would agile or extreme project management better match the needs of managing organizational change? A features analysis followed by field research can address this opportunity.

If organizational development experts are participating in a project team, how do the two world views of the engineer and the social scientist become complementary rather than combative? Mutual understanding and appreciation of each other's discipline is necessary if a productive working relationship is to be promoted in a medium to long-term project. Thus, attitudinal adjustments may be necessary to facilitate cooperation. Laboratory and field research can address this issue. In general, how can recent project management methods be adapted to continue to provide historical value from more stable operating environments to the more dynamic environment of organizational change?

CONCLUSION

Project management has a successful record of contributing to project success using quantitative methods and behavioral approaches for project team development and management. Organizations have become more process-centric internally as well as externally tightly linked with customers, suppliers, and partners. Thus, both product-oriented and process-oriented projects are conducted in a much more complex network of stakeholder relationships. Intuitive judgments by project managers with mostly an engineering background on how to orchestrate successful organizational change for their projects represents a greater risk to project success. Formal expertise for managing organizational change is necessary. This chapter reviewed a number of change management models with their strengths and limitations. Some examples of their application were included. Projects of any scale consume valuable organizational resources. The challenge is how to incorporate more formal change management expertise to promote project success in this era of continuing change. Collaboration by academics and practitioners can provide research guidance to help resolve this challenge.

REFERENCES

Abrahamson, E. (2000). Change without pain. *Harvard Business Review*, 78(4), 75–79.

Alderfer, C. (2011). *The practice of organizational diagnosis: Theory and methods*. Oxford, UK: Oxford University Press.

Balogun, J., & Hope Hailey, V. (2004). *Exploring strategic change* (2nd ed.). London, UK: Prentice-Hall.

Beckhard, R., & Harris, R. (1987). *Organizational transitions: Managing complex change* (2nd ed.). Reading, MA: Addison-Wesley.

Beer, M., & Noria, N. (2000). Cracking the code of change. *Harvard Business Review*, 78(3), 133–141.

Black, J., & Gregersen, H. (2002). *Leading strategic change: Breaking through the brain barrier*. Upper Saddle River, NJ: Prentice-Hall.

Bolles, D. (2004). *A guide to the project management body of knowledge* (3rd ed.). Newton Square, PA: Project Management Institute.

Burke, W. (2008). *Organizational change: theory and practice* (2nd ed.). Thousand Oaks, CA: Sage Publications.

By, R. (2005). Organizational change management: A critical review. *Journal of Change Management*, 5(4), 369–380. doi:10.1080/14697010500359250

Carter, S. (2012). *Get bold: Using social media to create a new type of social business*. Boston, MA: Pearson Education, Inc.

Colenso, M. (2000). *Kaizen strategies for successful organizational change: Enabling evolution and revolution within the organization*. London, UK: Pearson Education Ltd.

Crawford, L., & Nahmias, A. (2010). Competencies for managing change. *International Journal of Project Management, 28*(4), 405–412. doi:10.1016/j.ijproman.2010.01.015

Duhigg, C. (2012). *The power of habit: Why we do what we do in life and business.* New York, NY: Random House, Inc.

Fendt, J. (2006, Winter). Are you promoting change—Or hindering it?. *Harvard Management Communication Newsletter*, 1-6.

Galbraith, J. (1977). *Organization design.* Reading, MA: Addison-Wesley Publishing Inc.

Haidt, J. (2006). *The happiness hypothesis: Finding modern truth in ancient wisdom.* New York, NY: Basic Books.

Heath, C., & Heath, D. (2010). *Switch: How to change things when change is hard.* New York, NY: Broadway Books.

Holland, W. (2000). *Change is the rule: Practical actions for change: On target, on time, on budget.* Chicago, IL: Dearborn Financial Publishing, Inc.

Hoverstadt, P. (2004). Mosaic transformations in organizations. *Journal of Organizational Transformation and Social Change, 1*(2-3), 163–177. doi:10.1386/jots.1.2.163/0

Iveroth, E. (2010). Inside ericsson: A framework for the practice of leading global IT-enabled change. *California Management Review, 53*(1), 136–153. doi:10.1525/cmr.2010.53.1.136

Kavanagh, M., & Ashkanasy, N. (2006). The impact of leadership and change management strategy on organizational culture and individual acceptance of change during a merger. *British Journal of Management, 17*, 81–103. doi:10.1111/j.1467-8551.2006.00480.x

Keen, P. (1981). Information systems and organizational change. *Communications of the ACM, 24*(1), 24–33. doi:10.1145/358527.358543

Kolb, D., & Frohman, A. (1970). An organizational development approach to consulting. *Sloan Management Review, 12*, 51–65.

Kotter, J., & Cohen, C. (2002). *The heart of change: Real-life stories of how people change their organizations.* Boston, MA: Harvard Business School Press.

Kotter, J., & Schlesinger, L. (2008, July-August). Choosing strategies for change. *Harvard Business Review*, 130–139.

Lewin, K. (1951). *Field theory in social science: Selected theoretical papers.* New York, NY: Harper and Brothers.

Lucy, J. (2008). Why is the failure rate for organizational change so high? *Management Services, 52*(4), 10–18.

Lyons, J. (2009). The impact of leadership on change readiness in the U.S. military. *Journal of Change Management, 9*(4), 459–475. doi:10.1080/14697010903360665

Macdonald, S. (1995). Learning to change: An information perspective on learning in the organization. *Organization Science, 6*(5), 557–568. doi:10.1287/orsc.6.5.557

Miles, R. (2010). Accelerating corporate transformations (don't lose your nerve!): Six mistakes that can derail your company's attempts to change. *Harvard Business Review, 88*(1/2), 68–75.

Mirvis, P., & Berg, D. (1977). *Failures in organization development and change: Cases and essays for learning.* New York, NY: John Wiley & Sons.

Nohria, N., & Khurana, R. (1993). *Executing change: Seven key considerations.* Harvard Business School Note 9-494-038. Boston, MA: Harvard Business School Press.

Rosenberg, R. (2003, Summer). The eight rings of organizational influence: How to structure your organization for successful organizational change. *Journal for Quality and Participation*, 30–34.

Saksvik, P. (2007). Developing criteria for healthy organizational change. *Work and Stress*, *21*(3), 243–263. doi:10.1080/02678370701685707

Schein, E. (1987). Process consultation: *Vol. II. Lessons for managers and consultants*. Reading, MA: Addison-Wesley Publishing Company, Inc.

Skarke, G. (1999). *The change management toolkit: A step-by-step methodology for successfully implementing dramatic organizational change* (2nd ed.). Houston, TX: Winhope Press.

Smollan, J., & Sayers, J. (2009). Organizational culture, change and emotions: A qualitative study. *Journal of Change Management*, *9*(4), 435–457. doi:10.1080/14697010903360632

Stensakar, I. (2003). *Excessive change: Unintended consequences of strategic change*. Briarcliff Manor, NY: Academy of Management Press.

Van Buren, M., & Safferstone, T. (2009). The quick wins paradox. *Harvard Business Review*, *87*(1), 54–61.

Waterman, R. Jr, Peters, T., & Phillips, J. (1980). Structure is not organization. *Business Horizons*, *23*(3), 14–26. doi:10.1016/0007-6813(80)90027-0

Weick, K., & Quinn, R. (1999). Organizational change and development. *Annual Review of Psychology*, *50*, 361–386. doi:10.1146/annurev.psych.50.1.361

Wysocki, R. (2012). *Effective project management: Traditional, agile, extreme* (6th ed.). Indianapolis, IN: John Wiley & Sons.

Yin, R. (2008). *Case study research: Design and methods* (4th ed.). Thousand Oaks, CA: Sage Publications, Inc.

ADDITIONAL READING

Battilana, J., & Casciaro, T. (2012). Change agents, networks and institutions: A contingency theory of organizational change. *Academy of Management Journal*, *55*(2), 381–398. doi:10.5465/amj.2009.0891

Beer, M. (2009). Why change programs don't produce change. *Harvard Business Review*, *68*(6), 158–166.

Bennet, A., & Bennet, D. (2004). *Organizational survival in the new world, the intelligent complex adaptive system: A new theory of the firm*. New York, NY: Elsevier Publishing, Inc.

Bigelow, J. (1982). A catastrophe model of organizational change. *Behavioral Science*, *27*(1), 26–42. doi:10.1002/bs.3830270104

Doktor, R., Schultz, R., & Slevin, D. (Eds.). (1979). The implementation of management science. In R. Machol (Ed.), *Studies in the Management Sciences*. New York, NY: North-Holland Publishing Company.

Dove, R. (2001). *Response ability: The language, structure, and culture of the agile enterprise*. New York, NY: John Wiley & Sons.

Kerzner, H. (2009). *Project management case studies* (3rd ed.). Hoboken, NJ: John Wiley & Sons.

Ostroff, F. (2006). Change management in government. *Harvard Business Review*, *84*(5), 141–147.

Ottaway, R. (1983). The change agent: A taxonomy in relation to the change process. *Human Relations*, *36*(4), 361–392. doi:10.1177/001872678303600403

Padar, K. (2011). A comparative analysis of stakeholder and role theories in project management and change management. *International Journal of Management Cases*, *13*(4), 252–260.

Pennings, J. (1985). *Organizational strategy and change*. San Francisco, CA: Jossey Bass Publishers.

Prokesch, S. (2009). How GE teaches teams to lead change. *Harvard Business Review*, *87*(1), 99–106.

Sashkin, M., Morris, W., & Horst, L. (1973). A comparison of social and organizational change models: Information flow and data use processes. *Psychological Review*, *80*(6), 510–526. doi:10.1037/h0035568

Senge, P. (1999). *The dance of change: The challenges to sustaining momentum in learning organizations*. New York, NY: Doubleday Publishing Company. doi:10.1002/pfi.4140380511

Senge, P. (2008). *The necessary revolution: How individuals and organizations are working together to create a sustainable world*. New York, NY: Doubleday Publishing Group.

KEY TERMS AND DEFINITIONS

Business Process: A set of cross-functional activities the generate business value for the customers.

Managing Change: A set of activities to introduce and achieve successful adoption of some or all of revised objectives, structures, processes, roles, and culture.

Chapter 3
Green Process Management Using Six Sigma Concepts

Seifedine Kadry
American University of the Middle East, Kuwait

ABSTRACT

The Six Sigma ($6 - \sigma$) methodology, as it has evolved over the last two decades, provides a proven framework for problem solving and organizational leadership and enables leaders and practitioners to employ new ways of understanding and solving their sustainability problems. While business leaders now understand the importance of environmental sustainability to both profitability and customer satisfaction, few are able to translate good intentions into concrete, measurable improvement programs. Increasingly, these leaders are looking to their corps (six sigma experts) of six sigma "Master Black belts," "Black belts," and "Green belts" to lead and implement innovative programs that simultaneously reduce carbon emissions and provide large cost savings. Six sigma is a powerful execution engine and sustainability programs are in need of this operational approach and discipline. Six sigma rigors will help a business leader to design a sustainable program for both short- and long-term value creations. The aim of this chapter is to show the importance of applying six sigma methodologies to multidisciplinary sustainability-related projects and how to implement it.

INTRODUCTION

In 2000, the carbon disclosure project (Baietti, et al., 2012) was launched as a centrally organized effort to get companies to be transparent about carbon emissions, and by the end of 2009, almost 2500 companies were participating. In 2010, the U.S. Securities and Exchange Commission issued guidance (2010) to public companies saying that they should explain the impacts of climate change and climate regulation on their financial disclosure forms. Whether the initial triggers are intrinsic or extrinsic, there are a multitude of triggers that compel a company dialog to consider launching a formal environmental sustainability program.

DOI: 10.4018/978-1-4666-3658-3.ch003

The aim of this chapter is to show the power of Six Sigma to solve the current global challenge of environmental sustainability. One of the most complex problems that organizations face today is achieving success through strategies that are compatible with and supportive of environmental sustainability. The goal is to show how typical Six Sigma Define, Measure, Analyze, Improve, and Control (DMAIC) structures, such as program governance, transfer functions, measurement systems, risk assessment, and process design, lend themselves to environmental sustainability. In this chapter, a case study of sustainability problems, such as excess oxygen reduction, is analyzed using Six Sigma tools.

WHAT IS SIX SIGMA?

The use of Total Quality Management (TQM) as an overall quality program is still prevalent in modern industry, but many companies are extending this kind of initiative to incorporate strategic and financial issues (Puksic & Goricanec, 2005). After the TQM hype of the early 1980s, Six Sigma, building on well-proven elements of TQM, can be seen as the current stage of the evolution (Harry, 2000): although some conceptual differences exist between TQM activities and Six Sigma systems, the shift from the firsts to a Six Sigma program is a key to successfully implement a quality management system (Wessel & Burcher, 2004).

Six sigma methodology was originally developed by Motorola in 1987 and it targeted a difficult goal of 3.4 parts per million (ppm) defects (Barney, 2002). At that time, Motorola was facing the threat of Japanese competition in the electronics industry and needed to carry out drastic improvements in their quality levels (Harry & Schroeder, 2002). In 1994, Six Sigma was introduced as a business initiative to 'produce high-level results, improve work processes, and expand all employees' skills and change the culture. This introduction was followed by the well-revealed implementation of six sigma at General Electric beginning in 1995 (Slater, et al., 1999). Sigma is the Greek letter that is a statistical unit of measurement used to define the standard deviation of a population. Therefore, Six Sigma refers to six standard deviations. Likewise, Three Sigma refers to three standard deviations. In probability and statistics, the standard deviation is the most commonly used measure of statistical dispersion; i.e., it measures the degree to which values in a data set are spread. The standard deviation is defined as the square root of the variance, i.e., the root mean square (rms) deviation from the average. It is defined in this way to give us a measure of dispersion. Assuming that defects occur according to a standard normal distribution, this corresponds to approximately 2 quality failures per million parts manufactured. In practical application of the six sigma methodology, however, the rate is taken to be 3.4 per million.

Initially, many believed that such high process reliability was impossible, and three sigma (67,000 Defects Per Million Opportunities, or DPMO) was considered acceptable. However, market leaders have measurably reached six sigma in numerous processes.

According to the Six Sigma methodology a 6 σ process yields fewer defects than a 3, 4, or 5 σ processes. It is a name given to indicate how much of the data falls within the customers' requirements. The higher the process sigma, the more of the process outputs, products and services, meet customers' requirements—or, the fewer the defects. Table 1 and Figure 1 provide further resolution of the riddle involving the relationship between σ value and process performance and capability. The associated assumed process distributions in Table 1 are used to construct Figure 1.

The challenge of the Six Sigma methodology is to utilize a set of quality and management tools, through a systematic process, to improve key operational and business processes so they achieve 6 σ performances for key process indicators/metrics. Table 2 provides examples of 6 σ perfor-

Table 1. The relationship between σ,

process performance and process capability

Sigma value	Process performance	Process capability	Process distribution
σ =1.67	3 σ	1.00	Normal ($\bar{x}$ =10, σ =1.67)
σ =1.25	4 σ	1.33	Normal ($\bar{x}$ =10, σ =1.25)
σ =1.00	5 σ	1.67	Normal ($\bar{x}$ =10, σ =1.00)
σ =0.83	6 σ	2.00	Normal ($\bar{x}$ =10, σ =0.83)

Figure 1. Three, four, five, and six sigma processes for our laboratory example

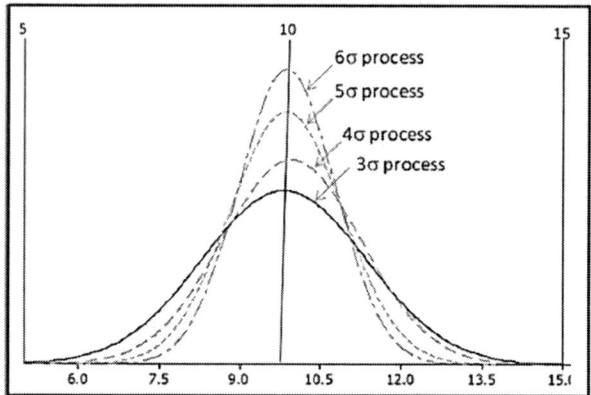

mances for selected processes. Table 2: Examples of 6σ performances for selected key processes indicators/metrics.

According to Mikel Harry and Richard Schroeder, each sigma improvement in a business process (e.g. moving from a 5 σ to 6 σ) translates into about "10% net income improvement, a 20% margin improvement, and a 10 to 30% capital reduction" (Harry & Schroeder, 2000) This is supported with several success stories such as:

- By 1998, AlliedSignal saved $1.5 Billion from implement-ing its 6σ program in 1994.
- By 1998, GE realized from initiating 6σ programs in 1996 the following gains:
  - Revenues rose 11%,
  - Earnings rose 13%, and
  - Working capital turns rose to 9.2% from 7.2% in 1997.

DMAIC CYCLE

Six sigma methodology is basically including 5 steps. They are Definition, Measure, Analysis, Improve, and Control (DMAIC). The systematic

improvement methodology has been successfully approved in solution of forging defects, achieved lower costs and met customer requirements.

The DMAIC problem-solving methodology and the associated tools and training to support the methodology have evolved over the past 20 years to become a set of powerful, robust, and widely adopted practices. The methodology was specifically developed to help teams get root-cause problem solving more efficiently. The DMAIC (McCarty, et al., 2011) problem-solving methodology (see Figure 2) was developed to help teams answer five key questions with regard to any problem:

Define

The purpose of the define phase is to identify and/or validate the project opportunity, develop the process that will drive the green initiative, define critical stakeholder requirements, and prepare team members to act as an effective project team. This focused session has the effect of pulling the team together around a common understanding of the green problem that they are trying to solve and the goals and objectives that they share. Key activities of the define phase (see Table 3) include the following:

Table 2. Examples of 6 σ performances

Sector	Key process indicator	6 σ performance
Manufacturing	Outer diameter of a shaft produced on a lathe	3.4 defects out of 1 million shafts are produced
Healthcare	Waiting time of patients receiving primary healthcare service at a clinic	3.4 out of 1 million patients wait excessively
Higher education	Publications from funded research projects by the research administration	3.4 out of 1 million funded projects fail to produce publications
Telecommunication	Interruptions in mobile calls made by customers of a local service provider	3.4 interruptions out of 1 million calls

Table 3. Define phase

Objectives	Activities	Tools
• Identify the improvement opportunity • Develop the current state process • Define critical shareholder requirement • Prepare to be an effective project team	• Create team • Develop team charter • Perform stakeholder analysis • Document process map • Identify barriers within process • Perform value stream analysis	• Team charter • Stakeholder analysis • Flowchart • Value analysis
		Deliverables
		• Prioritized shareholder requirements • Current state process map • Clear team charter • Quick wins

- Validate/identify the green improvement opportunity.
- Validate/develop the team charter.
- Identify and map processes.
- Identify quick wins, and refine the work process.
- Gather expectations of various stakeholders and convert those expectations into critical project requirements.
- Develop team guidelines and ground rules.

This activity helps to get the team excited about the potential for the project and motivated team members to set an aggressive work plan and agree on team norms. With its define workshop completed, the team was ready to move into the measure phase.

Measure

In the measure phase, teams determine what they should measure and what techniques and tools they can use to conduct the measurement and data collection, and then they review methods for ensuring that their measurement process is valid and accurate. Once the measurement plan is in place, the measure phase continues as the measurement and data collection take place. Data collection continues until the team finds that it has a statistically valid sample size from which to conduct valid data analysis. Typical activities during the measure phase (see Table 4) include the following:

- Determine process performance.
- Identify input, process, and output indicators.

Figure 2. DMAIC processes

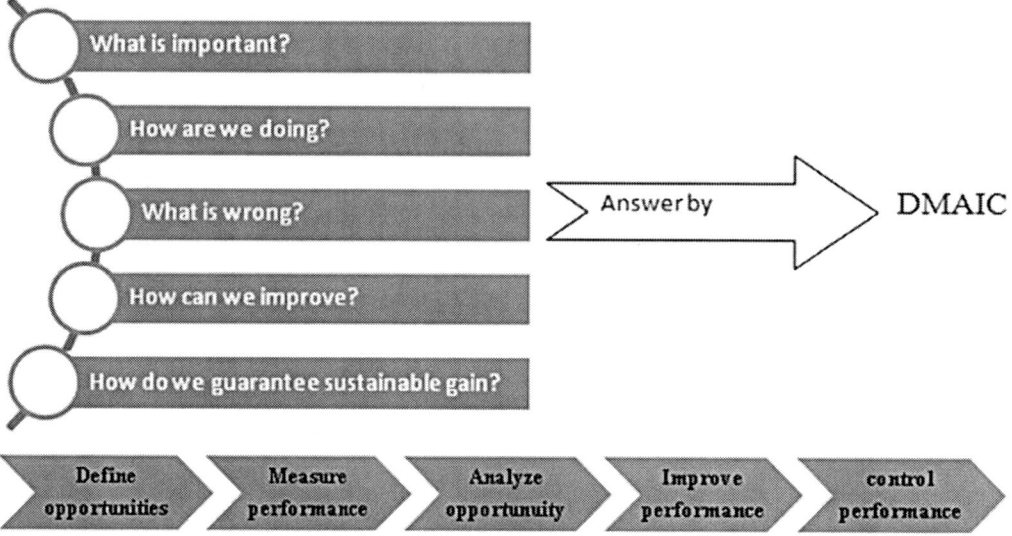

- Develop operational definitions and a measurement plan.
- Plot and analyze data.
- Determine if special causes exist.
- Collect other baseline performance data.

Analyze

The purpose of the analyze phase is to provide teams with the techniques and tools they need to stratify and analyze the data collected during the measure phase in order to identify a specific problem (root cause) and create an easily understood problem statement. When teams reach a point in which they want to analyze available data, they are confronted with two potential failure modes. These failure modes are either a lack of relevant data or too much data and an inability to determine how to analyze those data in ways that will lead to relevant conclusions aligned with the problem the team is trying to solve. Teams typically follow a process of first creating a problem statement or hypothesis of what the problem is (e.g., "Lighting is the number one source of energy loss in this data center"). Then teams use data-stratification

techniques, comparative analysis, and regression analysis to either prove or disprove the hypothesis. Teams will run through a number of hypothesis statements and the associated analysis until they can statistically prove that they have identified the sources of variation that are the most valid root causes of the problem. The list of activities and techniques employed by teams in the analyze phase (see Table 5) typically could include the following:

- Development of the problem statement.
- Stratification of the data.
- Comparative analysis of multiple data sets.
- Performing sources-of-variation studies.
- Analysis of failure modes and effects.
- Regression analysis to determine the strongest correlations with the problem statement.
- Identification of root causes.
- Design of root-cause verification analysis.
- Validation of root causes.
- Design of experimental studies to statistically prove the root cause.

Table 4. The measure phase

Objectives	Activities	Tools
• Identify key measure to evaluate the success • Establish baseline performance for the processes the team is about to analyze	• Identify input process, and output indicators • Develop operational definition and measurement plan • Plot and evaluate data • Determine if special cause exist • Determine performance level • Collect benchmark data	• Flowchart • Data check sheet • Benchmark data collection • Surveillance • Graph and charting
		Deliverables
		• Data collection plan • Baseline data set

Table 5. The analyze phase

Objectives	Activities	Tools
• Analyze the opportunity to identify a specific problem defined by an easily understood problem statement • Determine true sources of variation and potential failure modes that lead to shareholder dissatisfaction	• Analyze current state • Develop problem statement • Identify cost causes • Validate root causes • Perform statistical analysis • Identify performance gaps	• Run charts • Control charts • Cause and effect diagrams • Statistical tools
		Deliverables
		• Source of variation study • Validated root causes • Problem statement • Potential solutions

Improve

The purpose of the improve phase is to enable teams to identify, evaluate, and select the right improvement solutions and then to develop a change-management approach to assist the organization in adapting to the changes introduced through solution implementation The typical sequence of activities during the improve phase (see Table 6) is as follows:

- Generate solution ideas.
- Determine solution impacts and benefits.
- Evaluate and select solutions.
- Develop the process map and high-level plan.
- Develop financial analysis and the business case.
- Develop and present the solution storyboard.
- Develop the change-management plan.

- Communicate the solution to all stakeholders.

Control

The purpose of the control phase is to help teams understand the importance of planning and executing against the plan and to determine the approach to be taken to ensure achievement of the targeted results. The control phase also helps teams to understand how to disseminate lessons learned, to identify replication and standardization opportunities processes, and to develop related plans. Most important, the control phase forces teams to think through strategies so that identified benefits and financial impacts actually will be realized when the solution is fully implemented and institutionalized. It also will ensure that the solution will deliver results over a long period of time.

Typical activities that occur during the control phase (see Table 7) are as follows:

- Develop the pilot plan.
- Conduct and monitor the pilot.
- Verify reduction in root causes resulting from the solution.
- Identify whether additional solutions are necessary to achieve goal.
- Identify and develop replication and standardization opportunities.
- Integrate and manage solutions into the daily work processes.
- Integrate lessons learned.
- Identify the team's next steps and plans for remaining opportunities

In summary, the DMAIC problem-solving methodology, as well as the associated tools and training to support the methodology, is a powerful, robust, and widely adopted set of practices designed to improve the success rate of problem-solving teams. The methodology was developed

specifically to help teams get to root-cause problem solving more efficiently and with greater consistency and repeatability across teams. This overview was developed to help the reader gain an appreciation for how the methodology can he applied in the green project team arena and encourage team members to learn the methodology and supporting tools.

While the DMAIC methodology provides teams with the process and tools required, that methodology is not sufficient to ensure that the solutions developed will achieve any level of organizational acceptance and adoption. Throughout a sustainability initiative, the leadership team must implement solid change-management strategies to ensure that the team remains committed, the overall organization understands and supports the sustainability objectives, and the organization therefore is ready to support adoption of the green project team's solutions.

Table 6. The improve phase

Objectives	Activities	Tools
• Identify, evaluate, and select the right improvement solutions • Assist the organization in adapting to the changes introduced through solution implementation	• Brainstorm possible solutions • Perform cost/benefit analysis • Design and execute implementation plan	• Brainstorming • Process simulation • Staff feedback • Implementation planning
		Deliverables
		• Ideal process design • Business case approved • Implementation plan

Table 7. The control phase

Objectives	Activities	Tools
• Understand the importance of execution against the plan • Assure targeted results • Disseminate lessons learned • Prevent reversion to current state	• Determine approach to assure targeted results • Track metrics that will show if ideal process is in control • Review progress reports regularly and adjust as needed to support adoption of new process	• Control charts • Statistical process control • Leadership and change management
		Deliverables
		• Process control plan • Ongoing monitor and reporting plan • Replication opportunities

CASE STUDY 1: REDUCE EXCESS OXYGEN IN PLANT X

In this section the application of the DMAIC cycle to reduce the excess in plant X (see Figure 3) is explained.

Define Phase

In this phase the problems of excess oxygen of six boilers in plant X is examined. It is observed that there are some essential problems of the current system; the percentage of excess oxygen which leads to high cost and indirect pollution. The system structure is believed to be convenient for six sigma approach and DMAIC cycle. Additionally in this phase, we must define the defect, opportunity, expected annual savings, the objective, and the project plan:

- **Defect:** Any day for any boiler (B1, B2, B3, and B4) average Excess O_2 > 4% and B5 and B6 O_2 > 4.5%.

- **Opportunity:** Average reading of 66% of excess O_2 reading > 4.0% for the 4 boilers & 4.5% for the remaining 2.
- **Objective:** Reducing 70% of existing defect, i.e. Reduce excess O2% for (B1, B2, B3, and B4) $\leq$ 4.0% and B5 and B6 O2 $\leq$ 4.5%.
- **Annual savings:** 148.300 \$/Year.

Project plan is shown in Figure 4.

Measure Phase

For measure phase, one has to measure the right process and in the right time. It is so important for latter phases of the project. So the oxygen excess percentage in the boilers has been analyzed and relevant times are measured.

The current measure of the oxygen average excess for last three years (2008-2011) is given in the following chart (see Figure 5), and the current 6-sigma calculation is given in Figure 6.

DPMO: In process improvement efforts, a defect per million opportunities or DPMO is a measure of process performance.

Figure 3. Plant X, 6 boilers

Figure 4. Project plan

Key Deliverables		Aug, 10				Sep, 10				Oct, 10				Nov, 10				Dec, 10				Jan, 11			
		W1	W2	W3	W4	W1	W2	W3	W4	W1	W2	W3	W4	W1	W2	W3	W4	W1	W2	W3	W4	W1	W2	W3	W4
Currunt Process Map	P	X																							
	A	X																							
Stakeholders/Communication plan	P		X																						
	A		X																						
Basline - measure	P			X																					
	A			X																					
Current Six sigma calculation	P																								
	A																								
Update the charter	P					X																			
	A					X																			
Possible root causes brainstorm/5 Why's	P						X																		
	A						X																		
Root causes validation chart	P							X																	
	A							X																	
Data regression	P								X																
	A								X																
Fix obvious -sustain it	P									X															
	A									X															
Finalyze project/update storyboard	P										X														
	A										X														
Possible solution brainstorm	P												X		X										
	A												X		X										
Evaluating solution	P																X	X							
	A																X	X							
Risks/benefits	P																		X						
	A																		X						
FMEA	P																			X					
	A																			X					
Implementation plan	P																				X				
	A																				X				
Control Plan	P																					X			
	A																					X			
Control Procedure	P																						X		
	A																						X		
Update storyboard	P																							X	
	A																							X	
Documentation	P																								X
	A																								X
Transfer/Audit Questinnare	P																								
	A																								X

Planned ▒   Actual ▓

Analyze Phase

After it is decided that correct and enough data is collected the analyze phase has begun. During the analysis of the data it is determined that there are five main Root-Causes (RC) were validated that directly affect the oxygen excess problem:

1. Control parameter not connected to APC (Air Pollution Control) system and manual most of the time (67%) (see Figure 7).
2. **No close follow-up and supervision:** Based on the survey results, 50% of surveyed operators confirmed lack of adequate follow-up.
3. O_2 analyzer reading not matching with Lab analysis:
   a. Operator leaves O_2 in excess.
   b. Operator does not take action to reduce O_2.
   c. Operator does not refer to Analyzer.
   d. Operator does not trust Analyzer reading.
   e. Lab analysis does not match Analyzer reading.
4. **Operators not aware of excess O_2 operating limits:** Based on the survey results, 40% of surveyed operators answered correctly.
5. **B2 Working below the Low air pressure alarm:** Combustion air pressure Low Alarm was set at 100 mm-W.G. Most of the time, operations were done while the alarm was on.

Improve Phase

In improve phase, relevant solutions are investigated. While searching for solutions, their applicability is also taken into account. Additionally, its cost should be low (see Figures 8, 9, 10, and 11).

Improvement result (see Figure 12) and Six-Sigma before and after (see Figure 13).

Figure 5. Oxygen average excess

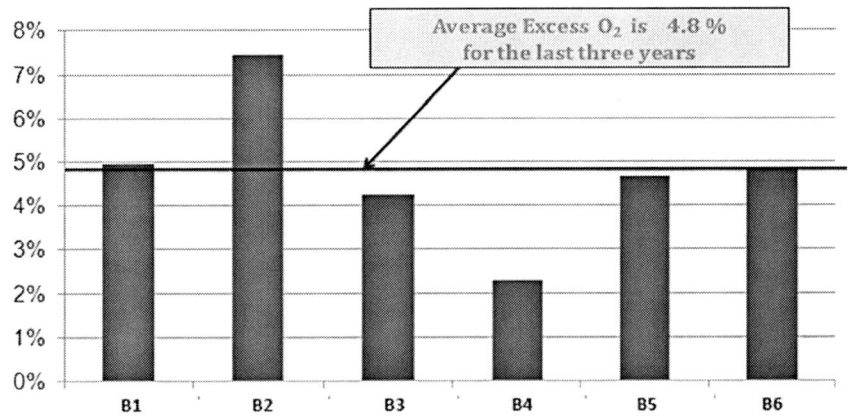

Figure 6. Current 6-sigma calculation

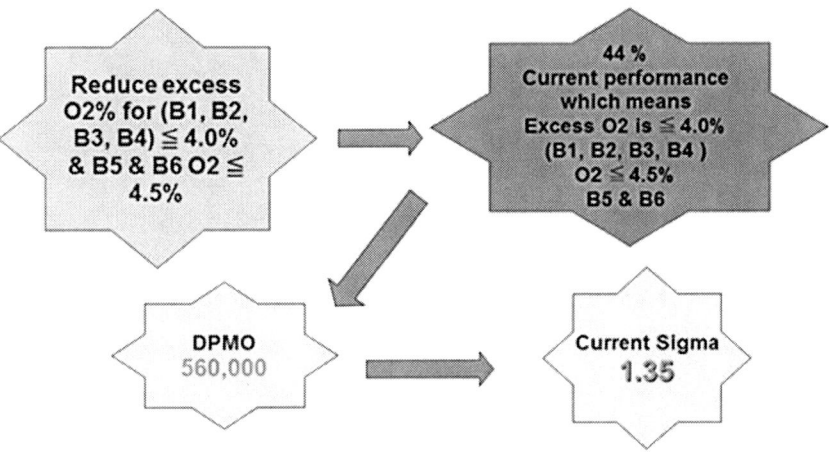

Figure 7. APC manual most time

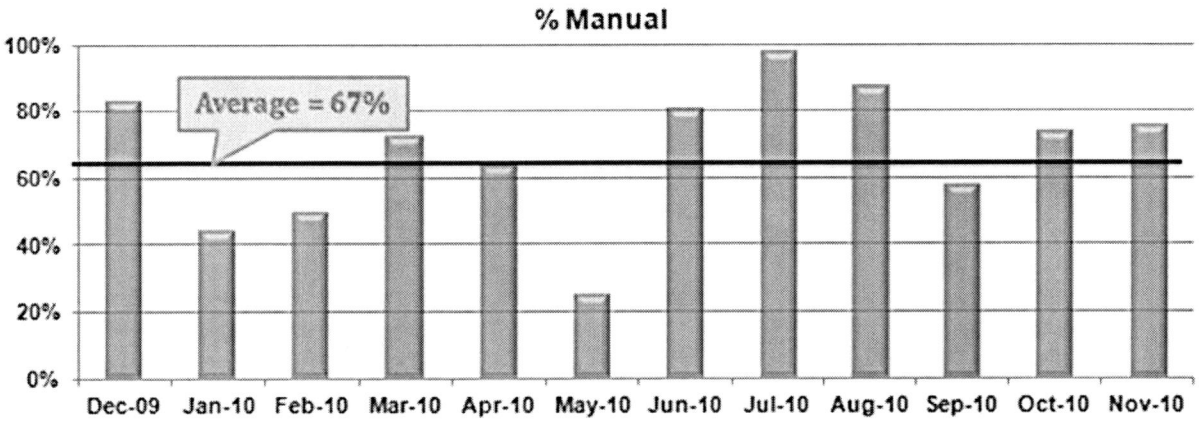

Control Phase

The control phase is applied such that the changes are indeed is valid in the reducing of oxygen excess. Therefore, the O_2 excess percentage is being examined continually. In the phase, we should propose a control plan (see Figure 14).

CONCLUSION

Consumers, regulators, and shareholders are all clamoring for sustainability. With the public's growing environmental awareness, consumers are actively seeking "greener" options. Regulators and legislators are changing the landscape for environmental reporting, compliance, and

Figure 8. 1ˢᵗ and 2ⁿᵈ root-cause solutions

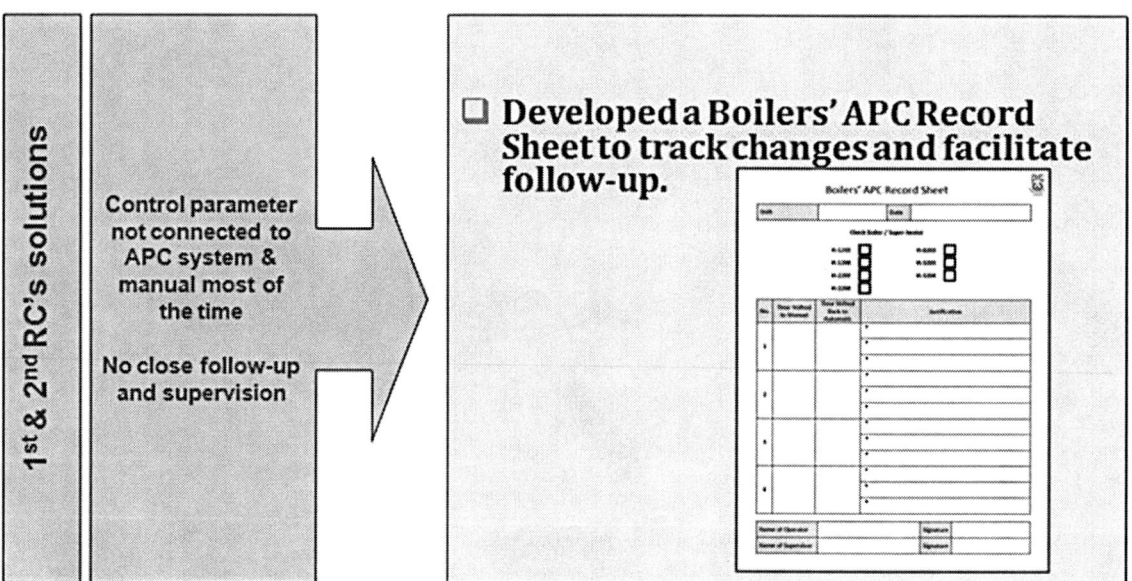

Figure 9. 3ʳᵈ root-cause solution

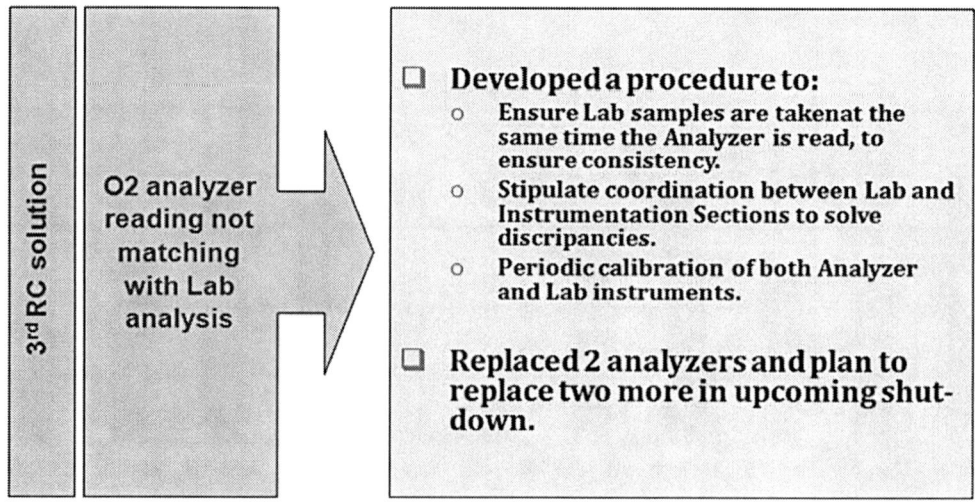

Figure 10. 4th root-cause solution

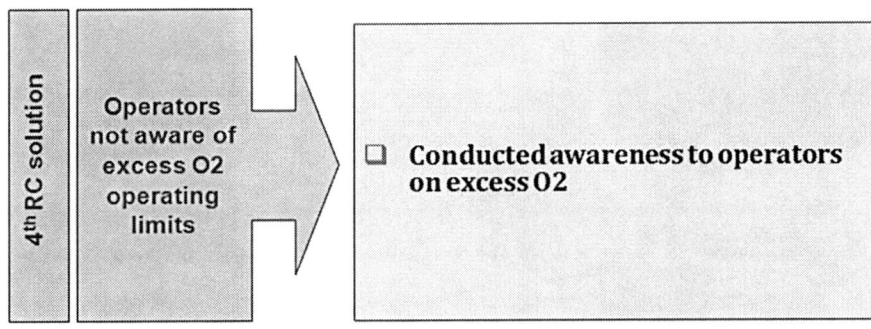

Figure 11. 5th root-cause solution

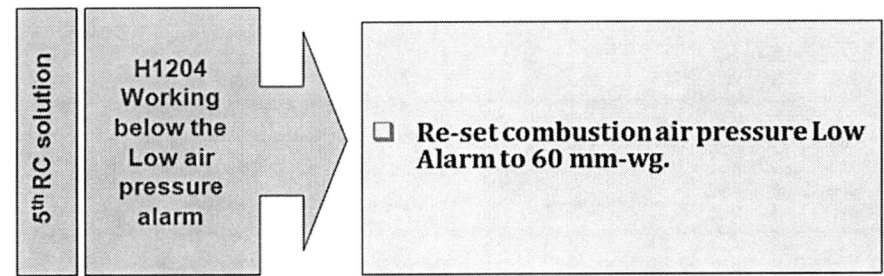

Figure 12. Result of the improvement

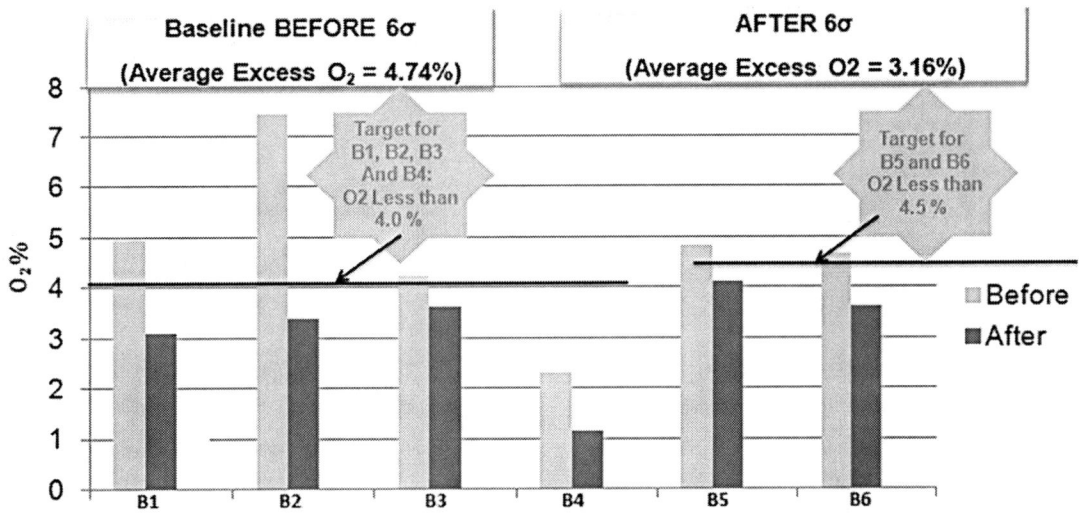

Figure 13. Six-sigma calculation before and after

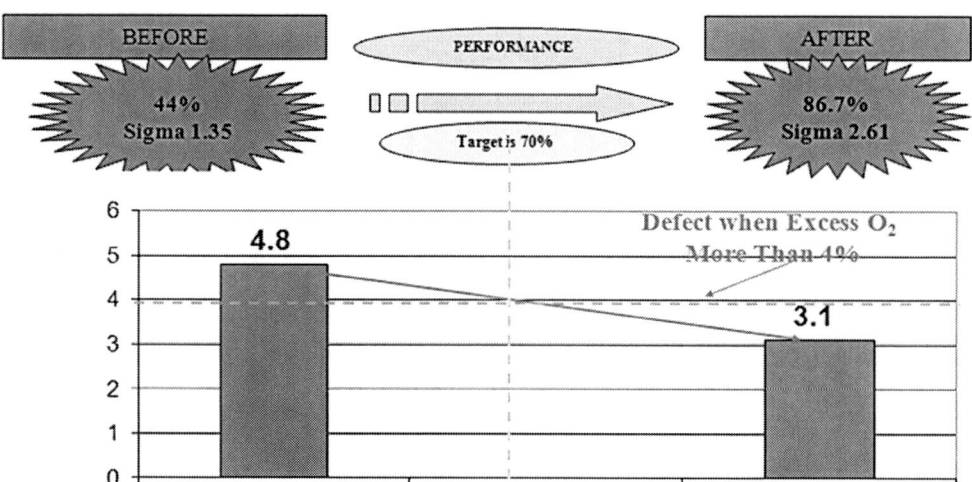

Figure 14. Control plan

Indicators	Performance Standards	Item	Frequency	Contingency Plans	Procedures
					Standards
	Specs, targets, control limits	What to check	When to check	Corrective actions	
Excess O₂ : B1, B2, B3 B4, B5, B6	Not more than 4.0% and 4.5%	Item 1 Item 2	Monthly (Include Daily Average)	Reduce air flow to bring O2 reading back to less than 4.0% and 4.5%	Follow the excess O2 instruction
Combustion Air Pressure	MMWG 80-120 450-500	Item3 Item 4	Monthly (Include Daily Average	Re-conduct awareness to the Control Room operators and supervisors	Follow the excess O2 instruction
Boilers on APC mode	on Automatic mode all the time	APC	Monthly report	Reconduct awareness to the Control Room operators and supervisors	Follow the excess O2 instruction
Monthly report	Less than 30% of the defect	Excess O₂	Monthly	Reconduct awareness to the Control Room operators and supervisors	Follow the excess O₂ instruction

transparency. Shareholders and investors have made environmental and social performance a top consideration. At the same time, many environmentalists claim that cutting greenhouse gases, reducing waste, increasing recycling, and broadly shrinking a company's "impact footprint" will reduce costs.

The sustainability imperative is growing, but along with it comes the recognition that improving sustainability is more difficult than some companies hoped – and many environmentalists would admit. However, by broadening Lean Six Sigma to include sustainability goals, companies can leverage a powerful and well-established performance improvement methodology to jump-start new sustainability programs or substantially boost existing ones. In this way, companies may well be able to marry together the critical goals of being good corporate citizens while improving their bottom line.

In this chapter, we study the applicability of Six Sigma concept to sustainable project. The studied case study shows a remarkable improvement to sustainability.

REFERENCES

Baietti, A., Shlyakhtenko, A., La Rocca, R., & Patel, U. (2012). *Green infrastructure finance: Leading initiatives and research.* Washington, DC: World Bank Publications.

Barney, M. (2002). Motorola's second generation. *Six Sigma Forum Magazine, 1*(3), 13–16.

Harry, M., & Schroeder, R. (2000). Six sigma – The break-through management strategy revolutionizing the world's top corporations. *Soundview Executive Book Summaries, 22*(11).

Harry, M. J. (2000). A new definition aims to connect quality with financial performance. *Quality Progress, 33*(1), 64–66.

Harry, M. J., & Schroeder, R. (2002). *Six sigma: The breakthrough management strategy revolutionizing the world's top corporations.* New York, NY: Doubleday.

McCarty, T., Jordan, M., & Probst, D. (2011). *Six sigma for sustainability.* New York, NY: McGraw-Hill Professional.

Puksic, M., & Goricanec, D. (2005). Increasing quality and economic efficacy of health institutions in public and private sectors in Slovenia. In *Proceedings of the 5th WSEAS International Conference on Distance Learning and Web Engineering,* (pp. 59-64). Corfu, Greece: WSEAS.

Securities and Exchange Commission. (2010). *Website.* Retrieved from www.sec.gov/news/press/2010/2010-15.htm

Slater, R. (1999). *Jack Welch & the GE way: Management insights and leadership secrets of the legendary CEO.* New York, NY: McGraw-Hill.

Wessel, G., & Burcher, P. (2004). Six sigma for small and medium-sized enterprises. *The TQM Magazine, 16*(4), 264–272. doi:10.1108/09544780410541918

Chapter 4
New Product Development:
Challenges and Implications

Hassanali Rassouli
Aagahn Information Technology Consultants, Iran

ABSTRACT

Intensive global competition of 21ˢᵗ century calls for a systematically aligned set of approaches in all aspects of corporate management including new product development. Success of new products needs the strategic support of corporate executives to establish and deploy coherent policies that are compatible with approaching new paradigms in business and manufacturing. Lean thinking is the widely accepted paradigm in the beginning of 21ˢᵗ century, and this chapter defines challenges and implications that corporate management will face in attempting to adapt to new product development approaches in accordance with the lean thinking perspective. This attempt covers all aspects of the organization in respect to product lifecycle.

NEW PRODUCT DEVELOPMENT: CHALLENGES AND IMPLICATIONS

In the last decades of 20th century, Western industries faced intensive competition with Japanese industries. Automotive industry was one of the industries that were seriously affected by this competition. Attempts to find Japanese advantages led to different results. Some believed that not having warehouse is the main advantage, some others believed Quality Circles is the advantage and some others saw some other aspects of Japanese industries as the main advantage. In the NBC documentary (1980, June 24) *If Japan Can... Why Can't We?* Americans found out about Dr. Deming. Chrysler Corporation, Ford Motor Company, and General Motors Corporation (1995) developed *Advanced Product Quality Planning (APQP) and Control Plan: Reference Manual*, a

DOI: 10.4018/978-1-4666-3658-3.ch004

standard procedure for simultaneous engineering as a part of QS-9000 which was a standard based on ISO 9000. Finally Massachusetts Institute of Technology (MIT) (1980) executed the International Motor Vehicle Program (IMVP) to find out what Japanese were actually doing. IMVP project leaders Womak, Jones, and Rose (1991) reported IMVP findings in the book *The Machine that Changed the World* and consequently Womak and Jones (1996) theorized on these findings in the book *Lean Thinking*.

Lean Thinking is the paradigm shift to design, manufacturing, and business in the 21st century. Womak and Jones defined Lean Thinking principles as:

- Specify value
- Identify the value stream
- Make value flow continuously
- Let customers pull value
- Pursue perfection

In addition, basic value streams are identified as:

- **Product design and development value stream:** From design concept to product launch.
- **Production and delivery value stream:** Raw materials transformation into deliverable products and services to customers.
- **Information value stream:** From order to delivery.

Thus, product design and development is a basic value stream in any business. Keeping this in mind, New Product Development chapter covers the following subtopics:

- Business and product strategy
- Product development strategy
- Product planning
- Product portfolio management

- Technology planning
- **Product design process:** Integrated Product and Process Development, Advanced Product Quality Planning (APQP), Lean Product Development
- **Design for X:** Design For Manufacturing (DFM), Design For Assembly (DFA), design for reliability, design for six-sigma
- **Product documentation and change management:** Configuration Management (CM)
- Product Lifecycle Management (PLM)
- **Product development tools and methods:** Quality Function Deployment (QFD), Failure Modes and Effect Analysis (FMEA), Geometric Dimensioning and Tolerancing (GD&T), Design of Experiments (DOE) conventional and Taguchi methods . . .

BUSINESS AND PRODUCT STRATEGY

Product is the resultant of business and product strategies of the organization that produces and delivers the product. Product is the mean by which organization creates and delivers value to its customers and enhances customer satisfaction. Quality, cost, and delivery of the product are main elements of customer satisfaction and strongly affected by strategic alignment of organization's policies. Business leaders have explicit or implicit policies for whatever concerns them in relations with their customers, shareholders, employees, suppliers, and other stakeholders, in dealing with economic, political, social, cultural, and environmental issues, in coming up with financial, budget, and other resource limitations, and for any other issue which affect and threaten their business and organization. Misalignments and fluctuations in these policies may deteriorate product quality, increase product cost and production lead time, and delay product delivery. In order to avoid misalignments and fluctuations in these policies strategic alignment of them is a must. According to Jackson (1995) major business concerns might be defined as customer satisfaction, corporate alignment, bureaucracy elimination, stakeholders' satisfaction, on time availability of information, business improvement, waste free production and productivity, failure free and defect free ongoing production, and timely usage of market opportunities.

Concerning customer satisfaction necessitates establishing policies and approaches to define customer requirements from product quality, cost and delivery points of view, manage customer relations and collect information about customer feedback and satisfaction, and encourage and enable all related functions and disciplines in the organization to participate in designing, establishment and improvement of product and order to delivery process.

Corporate alignment needs shared vision, communicated mission statement, and deployed strategies among all involved parties. Corporate strategies should be based on its core capabilities and careful consideration of competitive, demographic, social, political, and environmental factors, and regulatory and statutory obligations. Strategic management based on cascading policy deployment, Hoshin Kanri, is a common approach for this purpose.

Team working and process approach is the way to eliminate bureaucracy, increase effectiveness and efficiency of the organization, enhance product quality, decrease product cost, and shorten lead time of product development and release.

Win-win partnership approach is the key to stakeholders' satisfaction and delight. Corporate should involve and/or take into consideration stakeholders including corporate staff, suppliers, shareholders, and society in corporate strategy and policy development and deployment.

On time availability of information is a key factor to ensure timely and cost effective delivery of the product. Information system and architecture should enable visual control of work place, ensure fast feedback of issues and problems happening anywhere in the product lifecycle, from product development to production and product usage, and facilitate performance measurement and reporting.

Involving people and utilizing their creativity is the key to create a culture of improvement within the organization. Corporate should involve people in any related aspect from strategy definition and strategic goal setting to small daily improvements. Corporate should establish policies and approaches to encourage and enable people in their involvements. These approaches may include standardization, technology utilization, and training towards a learning organization.

Lean thinking and lean production are means to waste free production and productivity. Waste is whatever absorbs resources and does not add value to product. Toyota's Taiichi Ohno (1988) identified and classified wastes as:

- Waste of overproduction
- Waste of time on hand (waiting)
- Waste in transportation
- Waste of processing itself
- Waste of stock on hand (inventory)
- Waste of movement
- Waste of making defective products

New product development is the common approach to use market opportunities. Product development should be aligned with business strategy. In fact business strategy is directly reflected in corporate product portfolio. All the strategies and policies of the organization should support new product development and release, ongoing production of demanded products, product support and customer services during product life cycle, product improvements based on customer and market feedback, and customer satisfaction and delight based on quality, cost, and delivery requirements of customer.

PRODUCT DEVELOPMENT STRATEGY

Product development strategy is a directive policy on corporate product development approach based on internal strengths and weaknesses, and external opportunities and threats, i.e. SWOT study, and competitive market analysis and studies. Crow (2012) believes that companies must determine their primary product development strategic orientation and defines six primary product development strategic orientations as:

1. Time-to-Market
2. Low Product Cost
3. Low Development Cost
4. Product Performance, Technology, and Innovation
5. Quality, Reliability, and Robustness
6. Service, Responsibility, and Flexibility

Strategic orientation of product development would affect product development approach, product technology, and innovation to enhance corporate competitive dimensions such as: product performance and quality requirements, product development and production costs, and on time product release and delivery.

There are two major strategic approaches to product development, known as functional and Integrated Product Development (IPD). Functional, also known as over the wall, approach is the old way of developing new products in which each individual discipline-based function does its job and transmits it to next function for proper action, either continuing the design process or commenting to and/or requesting changes from previous functions, which is a time-consuming and costly process. In the globally competitive environment of 21st century, this approach is neither feasible nor competitive any more. On the other hand, IPD enhances collaborative team based product development which leads to decreased product and process development lead time, and cost effective production and timely delivery of the developed quality product.

PRODUCT PLANNING

Product planning is the decision making process to turn proposed new product ideas into marketable product specifications. Any new product idea by itself is not suitable for a commercially feasible product and should go through several steps to be recognized as new product. New product ideas may come from different sources like sales, marketing, customer services and engineering. New product ideas, from whatever source they come, have something in common; they are a response to a need or a proposed solution of a problem. Whatever makes an idea suitable for being a successful marketable product is not only the nature of the idea itself and generality of the need or problem plays a significant role in product success because more people, in fact more potential customers, are involved with more general needs or problems.

Product planning cycle typically consists of the steps such as:

- Making a list of potential or actual problems or needs that may lead to a product idea based on generated information and gathered data from sources that have direct or indirect interaction with potential and actual customers.
- Summarizing listed problems and needs into groups and categories based on demographic factors of potential customers.
- Prioritizing grouped and categorized problems and needs based on their generality.
- Selecting prioritized groups based on compliance with corporate policies.
- Defining new product concepts and proposals for selected groups.

- Approving new product proposals based on product development strategies and marketing justifications to be added to product portfolio as well as rejecting them, asking for more information, or holding them for future.
- Defining features and functional specifications for approved products.
- Initiating product and production process development.

Above steps may be reduced into combined steps.

PRODUCT PORTFOLIO MANAGEMENT

Every company may have a range of products or services, so called product portfolio, for different applications to satisfy various needs and requirements of different groups of customers. Allocating resources to and investments in new product development projects is a critical challenge for corporate executive managers. Executive mangers face the problem of what combination of potential and actual products creates maximum profit for the company and maximum value for its customers, and how should limited resources of the company be allocated to them. Providing solutions for this problem is the subject of product portfolio management.

Product portfolio management aligns new product development projects with corporate strategy and increases return of investment in these projects, keeps the company competitive, enable productive, i.e. effective and efficient, allocation of corporate resources, and establish balance between different types of product development projects. Cooper, Edgett, and Kleinschmidt (2012) defined four goals for product portfolio management: maximizing value of product portfolio, achieving a desired balance of projects in term of a number of parameters, aligning product portfolio with corporate strategy, and right number of projects; and explained the potential of conflicts between these four goals and dependence of the right portfolio approach selection to highlighted goal of corporate leadership.

Portfolio management approaches include mathematical models, scoring methods, and graphical techniques. Mathematical models such as Net Present Value (NPV), Expected Commercial Value (ECV), and Productivity Index (PI) are more concerned with financial factors rather than balance or strategic alignment. Scoring methods deal more with financial and strategic issues by scoring criteria like strategic alignment, feasibility, attractiveness, etc. Graphical techniques such as Bubble charts show the balance of two factors like profitability versus risk, costs versus benefits, costs versus time and so forth. Since these approaches do not cover all issues, it is recommended that a combination of them be applied.

Portfolio management is selecting and executing the right project first rather than just executing projects right and leads to a balanced strategic aligned product portfolio.

TECHNOLOGY PLANNING

The purpose of new product development is satisfying potential customers' recent, and changed or unsatisfied expectations and needs. This implies application of newer and better technologies in new products. As an instance, ever-increasing fuel price motivates buyers to purchase vehicles with lower consumption and this directs automotive industry to new car models with lower fuel and energy consumption. In order to meet this goal, car manufacturer have to look for and apply newer technologies in engine, power train and other systems of the automobile. Thus companies, in order to stay competitive, should have a technology plan to find new and most of the time innovative technologies on a timely manner and apply these new technologies through new product development projects. New technologies may affect new products in adapting to new customer requirements and improving product specification, decreasing product development and production costs, decreasing product time-to-market and delivery time, and facilitating development and production processes.

Mohan (1992) based on case study findings on the technology planning process for individual New Product Development (NPD) projects subdivided technology management process for NPD into technology planning and technology implementation activities and subsequently defined interactive relation of technology selection process with earlier activities of stating product requirements, determining technology options, and formulating technology implementation plans as "the descriptive framework model of technology planning."

PRODUCT DESIGN PROCESS

Turning ideas that are sparked in response to a customer requirement or as a solution of a problem into a marketable product takes several steps that make the design process. Pugh (1996) defined design process stages as: market, specification, concept design, detail design, manufacture, and sell. Earle (2008) described problem identification, preliminary ideas, refinement, analysis, decision, and implementation steps for design process. Khandani (2005) defined defining the problem, gathering information, generating solutions, analyzing and selecting a solution, and testing and implementing the selected solution as five steps for design process. Similar definitions for design process steps may be found in any design engineering textbook. Thus, common sense consideration of design engineering process as defined above reveals that:

- There should be a problem or requirement to initiate the design process.

- The problem or requirement should be clarified and clearly defined.
- More information should be gathered about the problem to ensure the problem is real and defined accurately, there is not any available solution or why available solutions are not applicable and a new solution is needed, to realize generality extent of the problem, and to know important (economic, social, environmental, aesthetic, ethical, safety, etc.) factors affecting the problem.
- Ideas for probable solution should be generated by means like brainstorming and searching technical literature and categorized into applicable design concepts.
- Design concepts should be refined, analyzed, and prioritized.
- Higher priority design concepts should be adequately detailed to appropriate extent to enable selection and decision-making.
- A design concept with optimum specifications should be selected for full detail design.
- Detailed product design should be accomplished while considering important factors like manufacturability, assembly, quality, reliability, cost, marketability, maintainability, and product retirement and disposal.
- Production processes, equipment, and tooling should be planned, designed, and provided as well.
- At certain stages of design process, product verification should be accomplished via appropriate analytic and testing techniques.
- Production launch and market release should be done after validating product design by prototype testing and successful pilot production.

It is necessary to make some explanatory notes here. Certain combinations of above activities may be summarized into fewer design steps. Design verification is the set of activities to ensure that product is designed as per purposed product specification. Product validation is the set of activities to ensure that designed product satisfy customer requirements.

Early stages of product design process might have been initiated and accomplished in product planning process and new product development projects should comply with product development strategies and product portfolio management policies and decisions.

As mentioned earlier, there are two major strategic approaches to product development, functional and integrated product and process development (IPD). These approaches as well as lean product development, which is an advanced version of IPD, are described in this section.

Over the Wall Process

Functional approach to product development leads to sequential, also known as over the wall, design process. In this approach, different steps of design process are assigned to different departments of the organization based on functional duties of each department. Each department receives step inputs from the department that is responsible for pervious step of the process and delivers step outputs to the department responsible for the next step of the process. Sometimes it might be necessary for a department to communicate with other departments, which are responsible for pervious steps of the process to discuss a problem or ask for a change. Best optimistic result might be pervious department agreeing and applying the requested change and worst pessimistic case might happen if the pervious department disagree the change which sometimes might lead to disputes between departments and higher managerial efforts might be necessary to solve it. This communication process even for the best optimistic case is costly and time consuming. Worst pessimistic cases are much more costly and time consuming. That is why this approach is not feasible for intensive competitive conditions of the 21st century and has been replaced with concurrent engineering

approach. Simultaneous engineering is another name for concurrent engineering.

Integrated Product and Process Development

Integrated Product and Process Development (IPPD) is a systematic approach to concurrent engineering. Concurrent engineering is not a new concept since early pioneers and entrepreneur of manufacturing workshops concerned production process while designing new products, in fact product and process designed by the same person and this person sometimes involved in other related functions like purchasing raw materials, providing equipment and tools, and sale of products. As workshops grew into large companies and it was not possible for one person to undertake and control all these duties, functional and disciplinary assignment of duties and, as a result, sequential design process replaced this approach which by that time was done instinctively in an unsystematic way and not known as simultaneous engineering, concurrent engineering or any similar terms. However, this concept was not vanished entirely and gurus like Deming transferred it to Japan. Japanese adapted their cultural heritage to this concept and developed it into an advanced approach, which was later named Total Quality Management. As evidenced in the introduction section of this chapter, intensive competition with Japanese firms brought this concept back to United State and European countries up to the point that the Institute for Defense Analysis (1986) in report R-338 developed the term "concurrent engineering" as:

Concurrent engineering is a systematic approach to the integrated, concurrent design of products and their related processes, including manufacture and support. This approach is intended to cause the developers, from the outset, to consider all elements of the product life cycle from concept through disposal, including quality, cost, schedule, and user requirements.

DoD, United States Department of Defense (1998), referring to DoD Regulation 5000.2-R, Mandatory Procedures for Major Defense Acquisition Programs (MDAPs) and Major Automated Information System (MAIS) Acquisition Programs defined IPPD as:

A management technique that simultaneously integrates all essential acquisition activities through the use of multidisciplinary teams to optimize the design, manufacturing and supportability processes. IPPD facilitates meeting cost and performance objectives from product concept through production, including field support. One of the key IPPD tenets is multidisciplinary teamwork through Integrated Product Teams (IPTs).

DoD defined Integrated Product Team as a multidisciplinary group of people who are collectively responsible for delivering a defined product or process and grouped IPPD tenets into the following main principles:

- Customer Focus
- Concurrent Development of Products and Processes
- Early and Continuous Life-Cycle Planning
- Proactive Identification and Management of Risk
- Maximum Flexibility for Optimization and Use of Contractor Approaches

DoD divided the acquisition process into five stages:

- **Phase 0:** Concept Exploration (CE)
- **Phase 1:** Program Definition and Risk Reduction (PDRR)
- **Phase 2:** Engineering and Manufacturing Development (EMD)
- **Phase 3:** Production, Fielding/Deployment, and Operational Support (PFDOS)
- Demilitarization and Disposal (DD)

US Air force (1994) faced the following challenges in applying Integrated Product Development (IPD) to The F-22 Integrated Product Development (IPD) Implementation Improvement Plan:

1. Trust
2. Leadership
3. Resistance
4. Resources
5. Contracting
6. Measurement
7. Understanding By a long term commitment to making and keeping agreements, maintaining credibility, and maintaining openness trust was established between the F-22 team members. Leaders were empowered at the lowest levels possible to make the right decisions to provide leadership at all levels of acquisition to support IPD. Teamwork and communication through IPTs was the approach to ease resistance by functional groups (i.e., engineering, finance and contracting). Early resource commitment, (i.e., time, people, and funding) at adequate levels was performed by identifying the requirement for IPD at the beginning of the Engineering and Manufacturing Development phase (EMD). Contracts were changed to support IPD and CE concepts. The F-22 team developed IPD Implementation Improvement Plan to have its own understanding and measurement of its IPD implementation progress.

DoD presents a high end approach to IPPD which may not be applicable to SMEs and organizations with more limited resources. Chrysler Corporation, Ford Motor Company, and General Motors Corporation's *Advanced Product Quality Planning (APQP)* might be a more suitable approach for lower end applications.

Chrysler Corporation, Ford Motor Company, and General Motor Corporation defined APQP as a structured method of defining and establishing necessary steps to assure customer satisfaction and timely completion of these steps by facilitating communication between involved parties. According to Chrysler Corporation, Ford Motor Company, and General Motor Corporation benefits of Product Quality Planning are directing resources to satisfy customer, promoting early identification of required changes, avoiding late changes, and timely provision of a quality product.

An organization that decides to apply APQC should organize a cross functional team, define the scope of development project, communicate with other teams of customer and suppliers, train resources appropriately, involve customer and suppliers in development project, apply simultaneous engineering in design process, and prepare proper control plans for making prototype, running pilot production, and regular production after launching products. The team should document encountered issues as well as resolutions found for them. The team should also develop a well-organized timing plan and adhere to it.

Five steps of APQP cycle are: Plan and Define Program, Product Design and Development, Process Design and Development, Product and Process Validation, and Feedback Assessment and Corrective Action. Plan and Define Program begins with concept initiation and/or approval and leads to Program Approval. Product Design and Development and Process Design and Development steps start simultaneously well enough before Program Approval but do not finish at the same time. Product Design and Development leads to prototype build. In addition, Process Design and Development results in pilot production. Product and Process Validation starts well enough before prototype build and ends well enough after pilot production. Feedback Assessment and Corrective Action is an ongoing process to enable continual improvement of product and while production is in action.

APQP is a concurrent engineering approach to new product development by a cross- functional team that apply tools, techniques, and approaches such as: Quality Function Deployment (QFD), Failure Mode and Effects Analysis (FMEA), benchmarking, cause and effect diagram, characteristic matrix, project management knowledge, Design Of Experiments (DOE) and Taguchi robust design, Design For Manufacturing (DFM), Design For Assembly (DFA), design for reliability, design for six-sigma, mistake proofing (POKA-YOKE), etc., as appropriate.

Lean Product Development

As competition with Japanese car manufacturers became more and more intense for US automakers in 1970s and 80s, Massachusetts Institute of Technology (MIT) initiated International Motor Vehicle Project (IMVP) to understand challenges facing the global automotive industry. Womak, Jones and Rose (1991) based on Clark, Chew, Fujimoto (1987) research work showed that Japanese automakers, in fact Toyota, develop new cars with much less efforts, at much shorter time, and with lower production cost and less defects than American carmakers. Fifteen year later Morgan and Liker (2006) based on Sociotechnical System (STS) thinking that says an organization must find appropriate fit between social and technical system that fits the organizatinal purpose and the external environment to be successful, used a sociotechnical model that has three mutually aligned process, people, and tools and technology subsystems to describe Toyota product development system and defined 13 principles that correspond to each of these subsystems.

Origin of Lean Product Development is Toyota Product Development System up to the point that they can be referred to interchangeably. According to detailed description of above subsystems and principles by Morgan and Liker (2006) most important features of Lean Product Development are:

- Waste free design process and prevention of wastes, that are generally due to poor design, in manufacturing, assembly, production, product delivery, and customer services.

- Simultaneous screening of alternative solutions and concepts using set-based concurrent engineering, verses traditional point-based concurrent engineering.
- Tailoring matured product platforms into new applications to satisfy evolving needs and requirements of potential and actual customers.
- Embedding new and innovative technologies into present and future platforms.
- Streamlining process flow in product development.
- Embedding engineering knowledge in standard design, process, and engineering resource competencies.
- Enabling and empowering engineering resources as well as suppliers to act in design process.
- Adapt technology to serve to the product development process instead of product development system serving to the technology.
- Aligning organization via policy deployment
- Continuous improvement and organizational learning using appropriate powerful tools.

Morgan and Liker suggested a five level approach for transforming from traditional design process to lean product development as: *Initial Preparation, Pilot Lean Processes, Lean Organization, Lean Tools and Technology, and Lean enterprise.*

DESIGN FOR X (DFX)

The term DFX stands for Design for X which sometimes is interpreted as Design for eXcellence perhaps because in the actual concept X is a variable which can be replaced with a terms from a set of terms that their consideration in design may lead to excellence in engineering, and busi-

ness. This set of terms include but is not limited to the words like quality, cost, modularity, interchangeability, reliability, logistics, procurement, six sigma, manufacturing, fabrication, assembly, test, serviceability, maintainability, environment, disposal, or obsolescence. Something that is common to these terms is that they may refer or belong to any of the phases of the product lifecycle. Product lifecycle in broadest terms can be divided into development, production, utilization, and disposal phases.

Commitment of top corporate executives to DFX up to the point that they devise it as a strategic approach and corporate policy is the key to success of this approach. Because DFX may need break through changes in organization management. For instance, DFX calls for a cross-functional approach and without top management support it is not possible to break structural barriers between functions in the organization to involve every interested party from different disciplines in the approach.

DFX is a good mean to shorten development lead-time, enhance product quality, decrease development and production costs, and ensure timely delivery of products and services to customers. The following subsections are introductions to some of the most common instances of DFX:

Design for Manufacturability (DFM)

Design for Manufacturability (DFM) is an instance of Design for X as well as a pioneering driver of simultaneous engineering. DFM facilitate early consideration of manufacturing issues in initiating phases of product development. DFM shortens product development lead time by facilitating communication between design engineers, process planners, quality experts, tool designers, purchasing agents, suppliers, and any other party involved in product development and reduces unnecessary iterative correspondence between these parties. DFM reduces product development costs as well as production costs by reducing design changes that will otherwise be initiated to remedy design shortcomings due to disregarding process capabilities, tooling provision, test conditions, quality requirements, interchangeability of parts, and availability of materials and standard components. DFM will result in application of fewer parts, and more standard components in the designed product. Adapting to modular design concepts can be another consequence of DFM approach.

Design for Assembly (DFA)

Design for Assembly (DFA) is a subset of Design for Manufacturability as well as an instance of DFX. DFA aims at ensuring easier and faster assembly of products at lower costs. Easier assembly means less effort to handle the parts and add them to other parts. Handling is picking the part and orienting it into a position that is suitable for adding to other parts. There are certain design concepts like symmetry, or an unsymmetrical feature in the part that can facilitate orientation. Self-lucking or self-locating features can make adding parts together easier. Easier orientation and insertion of a part result in shorter assembly time and lower assembly cost. The most common approach in DFA is reducing number of parts in the assembly that leads to reduction of provision activities, i.e. manufacturing or purchasing activities, shorter assembly and production time, and lower production costs.

Design for Reliability

Definition of reliability in the simplest form is trustworthiness. However, this definition is not adequate for engineering applications and these applications require a broader measurable definition. Night, Jervis, and Herd (1955) presented a definition for reliability that was widely accepted and cited in several texts concerning reliability. As an instance, Bozovsky (1961) cited this definition in his book *Reliability Theory and Practice*. In addition, Billinton, and Allan (1992) referring to

Bozovsky book broke down this definition into four parts as: Probability, adequate performance, time, and operating conditions. Levinston (1964) presented a similar definition and described four basic factors associated with reliability as: operating cycle or function, tolerance or accuracy of the output function, operating time, and ambient environmental conditions. Welker (1966) also presented a definition for reliability that is very similar to above definitions. From common aspects of above definitions, it can be concluded that reliability of a product, whether a component or a system, is probability of repetitive performance of the product without failure for a rather lengthy defined period of time under specified working condition. In addition, Design for Reliability (DOR) is designing a product that can withstand specified working conditions over a specified period of time.

Withstanding specified working condition means reducing the probability of failure of the product during its lifecycle down to a point that is satisfactory for customer or user.

Thus, Design for Reliability is designing a product and its production processes in such a way that minimizes actual failure occurrences down to a desired point. Zero defects that will be discussed in the next section might be the ultimate of goal of a Design for Reliability program.

Design for Six-Sigma

Sigma, denoted by the Greek letter σ, is the symbol of standard deviation of normal distribution in statistics. In normal distribution, probability of a measured quantity to remain within $\pm 1\sigma$ deviation limit from the nominal value is %31. This means that %31 of measured samples is acceptable and %69 of measured samples should be considered defective if acceptable tolerance limits of a measured quantity equals $\pm 1\sigma$. In traditional Statistical Process Control (SPC) $\pm 3\sigma$ is an acceptable tolerance limit of measured quantities. Probability of a measured quantity to remain within $\pm 3\sigma$ is %99.73. At first sight, this figure appears to be quiet satisfactory but a deeper look at the probability of the measured quantity to be out of acceptable range, that is %0.27, may lead to a quiet different conclusion. Defining %0.27 more precisely will show 2700 DPMO (Defective Parts per Million Opportunities). Harry and Lawson (1992) based on years of process and data collection in Motorola believed that processes vary and drift over time and defined Long-Term Dynamic Mean Variation as a figure between 1.4 to 1.6. As a result, they shifted mean of normal distribution 1.5σ. This shift determined actual deviation limits as 0.5σ, 1.5σ, 2.5σ, 3.5σ, and 4.5σ correspondingly for 2, 3, 4, 5, 6 Six Sigma level. According to this definition, actual probability of measured quantities to be out of acceptable range in level 3 sigma is 66808 DPMO, which is a quiet large figure. Although, this 1.5σ shift has provoked some criticism by statisticians like Wheeler (2004) but has been a successful approach in action and saved billions of dollars for companies like General Electric as declared by Welsh (2003), ex. CEO of General Electric. Six sigma implementation steps are Define, Measure, Analyze, Improve, and Control (DMAIC).

Appling Six Sigma concept in new product development leads to Design for Six Sigma (DFSS). DFSS is an approach that enhances product design process to design quality products that meet and exceed customer requirements. DFSS is an extension of Design for Reliability that aims to reduce failure modes of engineered system to 3.4 DPMO that is the practical example of zero defects.

DFSS starts with design team hearing the Voice of Customer as well as Voice of Process via tools like Quality Function Deployment to define need and requirements of external and internal customers and proceeds with assessing customer needs and specifications and analyzing process capabilities. Then, design team develops products and processes that meet customer needs and requirements. Finally, design team verifies that the design meets customer requirements. These steps that are named Define, Measure, Analyze, Design, and Verify (DMADV) will usually be embedded in a concurrent product development process.

PRODUCT DOCUMENTATION AND CHANGE MANAGEMENT

Outcomes of product and process development are sets of documents that define which raw materials by what processes should turn into specified forms and fit together to do proposed functions and construct the required product. Product documentation and process documentation as well, may include but not limited to items like bill of materials, material specification, product specification, drawings, control plans, manufacturing and assembly instructions, documents to support field application like product operation and maintenance manuals, documents generated during design process like QFD and FMEA spreadsheets, change control documents like Engineering Change Requests (ECR) and Engineering Change Orders (ECO), paper works like minutes of meetings, and reports, engineering management documents like standards, design manuals, and engineering procedures manuals. Recording, filing, distributing, storing, and retrieving these documents especially in complex design cases is a very complicated job and needs a sophisticated system. Since design engineering work has an iterative nature, several versions of these documents might be generated and this adds to the complexity of the job. Availability of the latest version of documents and obsolescence of old versions is very important for engineering staff and other users. Nowadays Product Data Management (PDM) and Engineering Data Management System (EDMS) software solutions are available to handle this work via computers. However, managing engineering documents is far beyond just purchasing and installing a software solution and needs some managerial effort to define required features, acquire optimum solution, train operators, and users, and manage supporting functions. PDM can be a component of a more sophisticated solution for Product Lifecycle Management (PLM). A structured approach to documentation control and change management is Configuration Management (CM).

Configuration Management

According to Oxford Advanced Learner Dictionary (2007), formal or technical meaning of configuration is an arrangement of the parts of something or a group of things or shape that this group of things produces. ANSI/EIA-649 (2011) defined configuration management as a process for establishing and maintaining consistency of a product's performance, functional and physical attributes with its requirements, design and operational information throughout its life. According to ISO10007:2003 configuration management can be defined as coordinated activities to direct and control interrelated functional and physical characteristics of a product defined in requirements for product design, realization, verification, operation and support. Military handbook MIL-HDBK-61B (2002) defined functions of configuration management as: CM Lifecycle Management and Planning, Configuration Identification, Configuration Control, Configuration Status Accounting, and Configuration Verification and Audit. ISO 10007:2003 defined configuration processes as: Configuration Management Planning, Configuration Identification, Change control, Configuration Status Accounting, and Configuration Audits. NASA (2008) defined five elements for CM systems, which have to be created by its functions and suppliers as: CM Planning, Configuration Identification, Configuration Control, Configuration Status Accounting, and Configuration Verification and Audit. Configuration management functions should be active in the entire lifecycle of the product.

According to NASA (2008), Configuration Planning is the process to create a Configuration Management Plan (CMP). Configuration Management Plan should be a controlled document, which defines what configuration items should be under control and how to implement CM functions and covers the life cycle of a product to ensure effective, predictable, and repeatable CM processes throughout the engineering, manufacturing, and utilization of the product. CMP may include computer-aided tools and methodologies if applicable. CMP should define training requirements for effective implementation of CM processes and provision of a consistent basis for understanding the CM functions and procedures for trainees. CMP is a good mean for explaining the process to trainees, as well as customers, quality assessor, and auditors. CM Training is the continuing process that addresses both performance of assigned CM tasks and cross training to provide awareness of relationships and interactions with others having CM-related responsibilities.

Configuration Identification is the process to select configuration items based on product structure, define relevant and traceable configuration information, establish configuration baselines, and assigns unique identifiers to each product and product configuration information Establishes a structure for products and product configuration. As defined in MIL-HDBK-61B (2002) a Configuration Item is any hardware, software or combination of both that satisfies an end use function and is designated for separate configuration management. In addition, Configuration Baseline is an agreed-to description of the attributes of a product, at a point in time, which serves as a basis for defining change. Each Stage of the product lifecycle may have its own Configuration Baseline as appropriate.

Configuration Control as defined in MIL-HDBK-61B (2002) is a systematic process that ensures that changes to released configuration documentation are properly identified, documented, evaluated for impact, approved by an appropriate level of authority, incorporated, and verified. An organization function, a customer, a supplier, or any other involved party may initiated a typical change via a change proposal document, sometimes named Engineering Change Proposal (ECP) or Engineering Change Request (ECR), that defines configuration item and its related information, change description, other affected configuration items, change justification, change classification, identity of change originator, and date of proposal. Change proposal goes for approval after an evaluation that defines technical aspects, potential risks, and impacts of change on cost, quality, and delivery due to consequent effects on interchangeability, manufacturing process and tooling, other configuration items to have proper interface and adaptation, and usability or disposal of current items on the stocks, and customer service. Evaluated change proposal when approved by an appropriate authority will be released for proper action sometimes via a document named Engineering Change Order (ECO) or Engineering Change Notice (ECN). Change verification might be done by auditing, testing, or inspecting.

Configuration Status Accounting is the process to record and report information about configuration item and their change status. As defined in MIL-HDBK-61B (2002) Configuration Status Accounting (CSA) is the process of creating and organizing the knowledge base necessary for the performance of configuration management. In addition to facilitating CM, the purpose of CSA is to provide a highly reliable source of configuration information to support all program/ project activities including program management, systems engineering, manufacturing, software development and maintenance, logistic support, modification, and maintenance. NASA defines that the CSA process manages the capture and maintenance of product configuration information necessary to account for the configuration of a product throughout the life cycle. It provides a historical record and provides for engineering release of design documentation for manufacturing

and other stakeholders in the design and manufacturing process. CSA is a by-product of all the other CM processes to ensure that information is systematically recorded, safeguarded, validated, and disseminated. Decisions on the information to be captured are based on judgment of what knowledge will be needed in the future for operations, sustainment, and CM audit requirements. Baseline and release records with appropriate history and metadata are examples of required information. The CSA should provide a knowledge base for information on Critical Safety Items/Processes, and warranty information. Each phase in the life cycle provides a time for determination of what information needs to be recorded in the CSA. And, according to ISO 10007:2003 Configuration Status Accounting formally records and reports requirements for product design, realization, verification, operation and support, the status of proposed changes, and the status of the implementation of approved changes.

Objective of Configuration Verification and Audit according to NASA is ensuring that configuration items have been properly identified, approved, released, and controlled throughout product life cycle, appropriate baselines have been established, and the approved product configuration is reflected in the final product. There are two major types of configuration audits: Functional Configuration Audit (FCA), Physical Configuration Audit (PCA). FCA is a mean to verify that actual functional and performance characteristics of a configuration item meet the specified requirements of that item in its configuration information. In addition, PCA is a mean to verify that physical characteristic of a configuration item matches the specified requirements of the item in its configuration information.

Translating above official definitions into informal terms reveals that configuration management is an approach to ensure properly documented communication between engineering and other parties involved in product realization, production, delivery, and utilization. Configuration management is supposed to ensure just in time availability of needed documents, feedbacks, and change requirements. In other words as Watts (2011) defined Configuration Management is the communication bridge between Design Engineering and the rest of the word. Watts (2011) used Configuration Management (CM) and Engineering Documentation Management (EDM) interchangeably and indicated that EDM is a significant company strategy for TQM, JIT, concurrent engineering, lean manufacturing, and some other approaches, if the company wishes to achieve best-in-class or world-class manufacturing, and proposed a keep simple approach to Configuration Management to make it more suitable for all other industries besides defense and aeronautics industries for which the original CM standard were developed.

PRODUCT LIFECYCLE MANAGEMENT (PLM)

Product launch is the turning point in product lifecycle after which production begins and market feedbacks reveal the actual results of product development efforts. Market feedbacks may lead to modifications in product design to improve design drawbacks. Thus, product design activities are not finished by product launch and more data will be collected and more documents will be generated as the product passes through its lifecycle. Managing these documents and information as well as any other part of product definition information for the entire lifecycle of product is subject of the PLM approach. According to NASA (2008) product definition information is defined as: information that defines the product requirements, documents the product attributes, and is the authoritative source for configuration definition and control. Thus, whatever PLM manages is sets of information that comprises different types of files including CAD files, project files, spreadsheet files, text files, and many other types of files, with different formats, according to related software solutions

that create and retrieve them. Necessity of data exchange between different software solutions calls for integration of software solutions that are applied in the PLM program. In order to manage all aspects of product lifecycle integrating information, people, processes, and business, design and process documents, construction and control of BOM (Bill of Material) and configuration management documents, provision of digital file repository, supporting simultaneous engineering, integration with CAD (Computer Aided Design), CAM (Computer Aided Manufacturing), CAE (Computer Aided Engineering), CRM (Customer Relation Management), SCM (Supply Chain Management), and ERP (Enterprise Resource Planning) software solutions, managing corporate intellectual properties, and project and portfolio management should be handled by the PLM software solution. Based on Technology Evaluation Centers (TEC) (2012) Request for Proposal (RFP) template a typical PLM software solution may cover the following areas as appropriate:

PLM Core:

- Change Management including:
  - Product Data Vaulting and Management
  - Configuration Management/BOM
  - Routing, Approval, and Lifecycle Process
  - Design and Project Collaboration
  - Visualization, Markup and Translation
  - Material Specification Management
  - Bill of Material Management (Packaging)
  - Recipe Management
  - Product Cost Estimation
  - Release to Enterprise
- Product Development and Portfolio Management:
  - Portfolio Management
  - Process and Project Management

- Manufacturing Process Management (MPM):
  - Production Process Planning
  - Production Process Design and Validation
  - Production Modeling and Simulation
  - Ergonomic Evaluation and Simulation
- Ideation and Requirements Management:
  - Ideation
  - Requirements Management
- Service Data Management (Discrete):
  - Product Identification
  - Product As-Built Configuration
  - Product As-Is Configuration
  - Product Service History
  - Product Service Parts
- Regulatory and Compliance:
  - Design for Compliance
  - Management of Hazardous and Controlled Substances
  - Regulatory and Compliance Documentation
  - Managing Recyclables and Controlled Waste
- Product Information Management (PIM):
  - Integration with Back-Office Systems
  - Product Information Repository
  - Data Distribution and Synchronization
  - Employee Productivity
- Application Technology:
  - Platforms
  - Interface
  - Architectural Foundation
  - Application Security
  - Web Forms
  - Functionality
  - Multisite Management
  - Multicurrency Management
  - Software as a Service and Hosting Options

PLM solution lead to shorter time to market, enhanced product quality, reduced costs, savings by using original data repetitively, product opti-mization, waste reduction, savings by streamlining engineering workflow.

PRODUCT DEVELOPMENT TOOLS AND METHODS

Quality Function Deployment (QFD)

Interpreting customer needs and wants into product specification is the major concern of product designers and Quality Function Deployment is the tool for product designers to turn these requirements, which are implicitly expressed in the voice of customer into technical terms to define product specification. QFD can clear the quality that customer wants and the functions that products should serve and defines explicit relationship between product characteristics and customer requirements.

Quality function deployment is a challenge of enhancing effective and efficient productivity strategies for sustaining high quality and marketing competitiveness. It is a planning system that addresses many facets of quality achievements. QFD techniques achieve tangible quality improvements with sustainable productivity by interpreting customer's needs and requirements implied in the voice of customer into product specification for achieving cost effective top quality end products and prompt services. QFD is a mean to translate whats into hows, to interpret colloquial and non-technical language of customer into technical terms of product specifications. QFD is applicable and has been applied in vast variety of design, manufacturing, and service cases. There are certain evidences of QFD application in design of medical devices, construction projects, software engineering, evaluation of technical textbooks, lean manufacturing systems and many other industrial and service cases. Readers are referred to QFD Institute publications to learn more about examples of such evidences.

In QFD applications Whats are translated into Hows through a series of matrices in which defined Hows of the first matrix becomes Whats of the second one, defined Hows of the second matrix becomes Whats of the third one and so forth, until some quantitative and measurable figure is reached.

The first matrix is usually Product Planning Matrix followed by Concept Selection Matrix, multiple Subsystem/Subassembly/Part Deployment Matrices, Process Planning matrix and Production Planning Matrix.

Failure Modes and Effect Analysis (FMEA)

Failure Mode and Effects Analysis (FMEA) is a tool to analyze available failure mode data and based on severity and probability of occurrence of failures and probability of detection of defects that may cause failure of the product during its lifecycle and apply appropriate provisions in the design of products or process to prevent occurrence of the failure in the future. Root Cause Analysis (RCA) and Fault Tree Analysis (FTA) are common tools to find defects or shortcomings that may lead to a failure. Occurrences of failures might be due to poor design or bad processing of the product. Design Failure Mode and Effects Analysis (DFMEA) is the method to prevent design causes of failures, and Process Failure Mode and Effect Analysis (PFMEA) is the method to prevent process causes of failures. For instance, in many occasions overheating might be the cause of failure in the product. Design engineers design cooling systems to take away the heat generated in the product via circulation of air or a liquid coolant. In this case, causes of failure due to poor design might be inappropriate dimensions, specifications, or characteristics of the cooling system. In addition, failure causes due to bad processing might be bad manufacturing, assembly, or installation of an appropriately designed cooling system.

When all of the predictable potential failure modes and their root causes are known and listed, the time for scoring their severity based on their predictable consequences, their occurrence rate, and their probability of detection via design and process controls. Scoring is done by assigning a number from 1 to 10 to severity, occurrence rate, and detection probability so that 1 represent the best case and 10 indicates the worst case and in between numbers indicate degrees of severity, occurrence, and detection of failure modes from low to high. For instance if safety is the most important concern of the designers as it is in automotive industry and consequence of a failure is hazardous without warning, as defined by Chrysler Corporation, Ford Motor Company, and General Motors Corporation (1993), that failure will be ranked 10, the highest severity score, and if a failure has no effect it is ranked 1) the lowest severity score, and scores for in between consequences of failures as defined by Chrysler Corporation, Ford Motor Company, and General Motors Corporation (1993) are: 2) for very minor effects, 3) for minor effects, 4) for very low effects, 5) for low effects, 6) for moderate effects, 7) for high effects, 8) for very high effects, and 9) for hazardous with warning effects. Occurrence rate and detection probability scoring numbers are defined in a similar manner to indicate low to high by 1 to 10. Once these three elements of FMEA analysis are scored, RPN (Risk Priority Number) is calculated by multiplication of these score numbers to indicate a measure for prioritizing predicted potential failure modes. Failure modes with higher RPNs have higher priority for actions to reduce their RPN well below a predefined acceptable limit. Aim of these actions might be decreasing occurrence rate, increasing detection probability, or a combination of both.

FMEA predicts failure modes and reduce risk of product or system failure due to these causes.

Geometric Dimensioning and Tolerancing (GD&T)

Realizing concepts into useful products is concern of design engineers and this cannot be accomplished without visual aids. These visual aids include sketches, drawings, models and any other tool or method that is considered useful to help audience and customers of design engineer to have a comprehensive understanding of the future products during design and development period. Whatever visual aid is used should have same interpretation in the mind of proposed users of design engineer's work. Among these drawings play a major role since drawings contain quantitative and descriptive figures known as dimensions and tolerances to communicate fit, form and function of the product and it's components to manufacturing and inspecting people. Geometric Dimensioning and Tolerancing (GD&T) is the symbolic language applied in this prospect.

Back to "over-the-wall" age, which linear tolerancing was the state of the art, different parties engaged in product development had their own view and interpretation of the drawing according to their major concerns and whatever made their job done easier and faster.

In those days product engineer defined dimensions and tolerances with inadequate communication with manufacturing and inspecting people, and they had to come up with these tolerances and in the case of dispute engineering change orders increased cycle time and product costs. Manufacturing engineer concerned process capability and ease of assembly. Inspecting people concerned capabilities of measurement methods and equipment.

Effect of concurrent engineering approach on engineering design especially in the fields of mechanical engineering and manufacturing was application of Geometric Dimensioning and Tolerancing to replace conventional linear tolerancing. Linear tolerancing was based on limits of size and

this way control of orientation or location relationship between features of a part was not possible.

GD&T by giving a clear definition of geometric concepts classified as orientation, location, form, profile and runout and descriptive concepts classified as modifiers along with appropriate symbol for each concept and feature and a rational approach to apply them, enabled different parties engaged in a dimensional management process to have a common language and consequently common understanding of fit, form and function of the product and it's components:

- Orientation concepts are Angularity, Parallelism, and Perpendicularity.
- Location concepts are Concentricity, Position, and Symmetry.
- Form concepts are Cylindricity, Flatness, and Straightness.
- Profile concepts are Profile and Profile of a Line.
- Runout concepts are Runout and Total Runout.

Modifying concepts are Maximum Material Condition (MMC), Least Material Condition (LMC), Projected Tolerance Zone, Free State, Tangent Plan, Diameter, Spherical Diameter, Radius, Spherical Radius, Controlled Radius, Reference, Arc length, and Statistical Tolerance.

Basic idea behind GD&T is first determining datum features, features (physical portions of a part like surfaces, holes or slot) selected as an origin for locating parts on datum simulators (such as machine tool tables, surface plates, chucks, fixtures and jigs) for manufacturing and measuring purposes. This way designer should keep in mind manufacturing process and measuring procedure during design process.

Second step in GD&T practice is defining nominal distance and orientation (angular relationship) of other features according to determined datum features.

Third step is specifying tolerance zones and/or boundaries and conformance conditions for considered features. Tolerance zones are virtual geometric boundaries corresponded to controlled features shape defined by a) two parallel planes, for width type features like plane or profiled surface features, b) a cylindrical plane or two coaxial cylindrical planes with radius or radial distance equal to specified tolerance, for axes and cylindrical type features, c) a spherical plane with radius equal to specified tolerance, for spherical type features like outer surface of a ball. This way designer should be well aware of manufacturing process capability, i.e. Design For Manufacturability (DFM).

Forth and last step is regarding tolerance interactions between parts that are supposed to be assembled in the assembly process. This will lead designer to take different aspects of assembly like ease of assembly and cost into consideration while defining tolerances, i.e. Design For Assembly (DFA).

Successful application of GD&T calls for concurrent design and engineering teams consisting of representatives from responsible functions like design, manufacturing (process and tooling), quality, and any other function like purchasing and sale as appropriate.

Detailed explanation of GD&T definition and rules is beyond scope of this chapter. Reader can see available standards, literature, and Internet resources for more information on this subject.

Design of Experiments (DOE)

Sometimes designers face the challenge of understanding interactive effects of several factors to find more effective factors and focus efforts on them to optimize the design. The first approach to the problem seems to change one factor at a time while keeping the other factors constant. This is neither a feasible nor even a practical idea because in practice leads to infinite number test runs that may give unknown and unpredictable ambiguous results. Fisher (1935) invented and used design of experiments for the first time to draw valid conclusions on natural factors in agriculture like ambient temperature, soil condition, and rain fall. Afterwards this method was used in other industries successfully. Deming (1982) in the early 1950s thought this method, along with other statistical methods, to Japanese engineers. One of these Japanese engineers, Genichi Taguchi developed a version of design of experiments to focus on product robustness and cope with influence of uncontrollable factors or noise in product.

Telford (2007) defined principles of design for experiments as: randomization, replication, blocking, orthogonality, and factorial experimentation and described these principles as solutions to problems that are due to two types of nuisance. According to Telford (2007) randomization protect against distortion of results by an unknown bias, replication increases precision of the experiment by increasing sample size, blocking also increases precision of the experiment but by removing effects of known nuisances, orthogonality leads to uncorrelated and more easily interpretable factor effects, and factorial design vary all factors simultaneously and orthogonally.

Factorial design considers two levels of low and high for every factor and factor effects are the difference between averages of runs at these low and high levels. For instance if a designer needs to know effect of temperature and pressure on volume of a balloon he or she can run four sets of measurements as:

1. Measuring volume (V1) at *low* pressure and *low* temperature
2. Measuring volume (V2) at *low* pressure and *high* temperature
3. Measuring volume (V3) at *high* pressure and *low* temperature
4. Measuring volume (V4) at *high* pressure and *high* temperature

Effect of pressure is (V4+V3)/2-(V2+V1)/2

Effect of temperature is (V4+V2)/2-(V3+V1)/2

This was a very simple example of an experimental design. As the number of factors increase, the number of test runs will increase exponentially and more sophisticated statistical softwares are needed to calculate and represent effects.

Design of experiments is a powerful tool for improvement of product and process.

TAGUCHI METHOD

Taguchi method, developed by Genichi Taguchi, is a modified version of classical design of experiments that focuses on interactive sensitivity of design parameters to avoid source of nuisance rather than controlling nuisance. That is why this method is named robust design or parameter design. Based on available literature concerning Taguchi method like Taguchi (1986), Taguchi (1993), Bhutta (2003), Olberding, Williams, Schreiner, and Paulsen (2009) this method closely depends on robust design, loss function, signal to noise ratio, and orthogonal array experiments.

Robust design means embedding quality in product at development stage and avoiding inspection and screening after production.

Loss function is Taguchi's definition of cost of quality based on a quadratic function of satisfaction level of customer and customer dissatisfaction is due to product deviating from its target function.

Signal to noise ratio is quantified measure of influence of unwanted factors, i.e. noise, in target functions of product.

Orthogonal array experiments are sets of test runs to find effects of influencing factors on target functions of product.

Taguchi method is an effective approach to build quality in product.

REFERENCES

American National Standard Institute. (2011). *ANSI/EIA-649-B-2011 configuration management standard*. Washington, DC: ANSI.

Bhutta, K. (2003). *Taguchi approach to design of experiments*. Houston, TX: Southwest Decision Sciences Institute.

Billinton, R., & Allan, R. (1992). *Reliability evaluation of engineering systems: Concepts and techniques*. New York, NY: Plenum Press.

Bozovsky, I. (1961). *Reliability theory and practice*. Englewood Cliffs, NJ: Prentice-Hall.

Chrysler Corporation. Ford Motor Company, & General Motors Corporation. (1995). *Advanced product quality planning and control plan (APQP) and control plan: Reference manual*. Detroit, MI: Chrysler Corporation.

Clark, K. B., Chew, W. B., & Fujimoto, T. (1987). Product development in the world auto industry. *Brookings Papers on Economic Activity, 3*, 729–771. doi:10.2307/2534453

Cooper, R. G., Edgett, S. J., & Kleinschmidt, E. J. (2012). *Portfolio management: Fundamental for new product success*. Retrieved from http://www.stage-gate.net/downloads/working_papers/wp_12.pdf

Crawford-Mason, C. (Producer), (1980, June 24). *If Japan can ...Why can't we?*. New York, NY: NBC.

Crow, K. (2001). *Product development strategic orientation*. Retieved from http://www.npd-solutions.com/strategy.htm

Deming, W. (1982). *Out of the crisis*. Cambridge, MA: MIT Press.

Department of Defence. (2002). *MIL-HDBK-61B military handbook configuration management guidance*. Retrieved from http://www.everyspec.com

Earle, J. H. (2008). *Engineering design graphics with AutoCAD 2007*. Upper Saddle River, NJ: Prentice Hall, Inc.

Fisher, R. (1935). *The design of experiments*. Edinburgh, UK: Oliver & Boyd.

Harry, M., & Lawson, R. (1992). *Six sigma producibility analysis and process characterization*. Reading, MA: Addison-Wesley Publishing Company, Inc.

International Organization for Standardization. (2003). *ISO 10007:2003 quality management systems -- Guidelines for configuration management*. Retrieved from http://www.iso.org

Jackson, T. (1995). *Implementing a lean management system*. Portland, OR: Productivity Press.

Khandani, S. (2005). *Engineering design process*. Retrieved from http://www.iisme.org/etp/HS%20 Engineering-%20Engineering.pdf

Levinston, E. (1964). System reliability analysis, introduction. In Rothbart, E. (Ed.), *Mechanical Design and System Handbook*. New York, NY: McGraw-Hill.

Massachusetts Institute of Technology. (1980). *International motor vehicle program (IMVP)*. Retrieved from http://web.mit.edu/ctpid/www/imvp/

Mohan, T. (1992). Deep inside the black box: Early case study findings on technology planning in product development projects. In Feldman, Hustad, & Page (Eds.), *Managing Product Development: Winning in the 90s: Proceedings of the PDMA International Conference*, (pp. 57-65). Chicago, IL: PDMA.

Morgan, J., & Liker, J. (2006). *The Toyota product development system*. Portland, OR: Productivity Press.

NASA. (2008). *NASA-STD-0005 NASA configuration management (CM) standard*. Washington, DC: NASA. Retrieved from http://www.standards. nasa.gov/documents/viewdoc/3315133/3315133

Night, C., Jervis, E., & Herd, G. (1955). Terms of interest in the study of reliability. *I.R.E. Transactions on Reliability and Quality Control, 5*, 34–35. doi:10.1109/IRE-PGRQC.1955.5007222

Ohno, T. (1988). *Toyota production system: Beyond large scale production*. Portland, OR: Productivity Press.

Olberding, J., Williams, B., Schreiner, A., & Paulsen, J. (2009). *Robust engineering*. Iowa City, IA: Quality Control Mini Culture.

Oxford. (2007). *Advanced learner's dictionary* (7th ed). Oxford, UK: Oxford University Press.

Plevyak, H. M., Jr., & Pistolessi, J. F. (1994). *USAF the F-22 integrated product development (IPD) implementation improvement plan*. Retrieved from http://www.mitre.org/work/sepo/toolkits/ippd/ examples/F22_IPD_ImpTemplate.doc

Pugh, S. (1996). *Creating innovative products using total design*. Reading, MA: Addison-Wesley Publishing Company.

Rassouli, H. (2011). *Quality function deployment*. Retrieved from http://www.brighthub.com/engineering/mechanical/articles/94890.aspx

Rassouli, H. (2011). *GD&T - Turning concepts into products*. Retrieved from http://www.brighthub. com/engineering/mechanical/articles/36159.aspx

Taguchi, G. (1986). *Introduction to quality engineering*. White Plains, NY: Asian Productivity Organization.

Taguchi, G. (1993). *Taguchi on robust technology development: Bringing quality engineering upstream*. New York, NY: ASME Press. doi:10.1115/1.800288

Telford, J. (2007). A brief introduction to design of experiments. *Johns Hopkins APL Technical Digest, 27*(3).

United States Department of Defense. (1998). *DoD integrated product and process development handbook*. Washington, DC: DoD.

Watts, F. (2011). *Engineering documentation control handbook: Configuration management and product lifecycle management*. Waltham, MA: Elsevier.

Welker. (1966). System effectiveness. In W. Ireson (Ed.), *Reliability Handbook*. New York, NY: McGraw-Hill.

Welsh, J. (2003). *Jack: Straight from the gut*. New York, NY: Grand Central Publishing.

Wheeler, D. J. (2004). *The six sigma practitioner's guide to data analysis*. New York, NY: SPC Press.

Womack, J., & Jones, D. (1996). *Lean thinking: Banish waste and create wealth in your corporation*. New York, NY: Simon and Schuster.

Womack, J., Jones, D., & Rose, D. (1991). *The machine that changed the world: The story of lean production*. New York, NY: Harper Perennial.

Chapter 5
Supply Chain Analysis

Mohammad Anwar Rahman
University of Southern Mississippi, USA

ABSTRACT

This chapter focuses on supply chain strategies that benefit organizations to meet the global challenges and business success. The chapter supports understanding of functional coordination in supply chain management, importance of supply chain partnership and performance measures, bullwhip cause-effect and consequences, business promotion strategies, and approach to synchronize operation schedule to improve supply chain operations. Bullwhip effect is often seen as a challenge to improve supply chain performances. It also discusses the strategies to reduce bullwhip effect, various well-known contract agreements that are currently practiced by many supply chain organizations, its impact and benefits on supply chain success.

INTRODUCTION

In today's competitive market, business success depends on customers' satisfaction, product quality, timely service and delivery, and competitive price. The supply chain integrates industry partners such as product manufacturers, suppliers, warehouses, transporters, and retailers together to form a network to satisfy the customers' request efficiently. Simchi-Levi et al. (2007) describes supply chain as synchronized business functions where raw materials are procured and items are produced at one or more factories, shipped to warehouses for intermediate storage, and then shipped to retailers or customers. The main focus of the supply chain partners is to develop strategies that benefit their organizations, reduce system wide costs, satisfy service level requirements and meet the global challenges for business success. The strategies defined in a supply chain as such that the production of the products and the distribution emerged in the right quantities, to the right locations, and at the right time. It is essential to monitor the implementation of the strategies and make the necessary adjustment for the strategy to be effective.

DOI: 10.4018/978-1-4666-3658-3.ch005

Supply chain coordination is the sharing of a set of goals and objectives with all partners. Demand variations and global market challenges often seen as the main challenge to the partners to improve the supply chain performance. The functional coordination, joint business promotion strategies, synchronized operations, strategies to reduce bullwhip effect, and various well-known contractual agreements currently practiced by many supply chain organizations, are the central focus for supply chain success.

Following topics are covered in this chapter:

- Importance of functional coordination in supply chain management.
- Inventory strategies and performance measures.
- Bullwhip effect in supply chain.
- Causes of Bullwhip effect in supply chain.
- Bullwhip effect reduction strategies.

Functional Coordination in Supply Chain Management

In the current state of the global economy and volatile market environment, companies in a supply chain require to act together. The coordination among supply chain trading partners is crucial for cost mitigation and revenue maximization. The central focus for companies is to bring products from sources to markets together. Companies share and manage product procurement, marketing and shipping in collaboration. Information technology allows supply chain partners to share demand and inventory data quickly and inexpensively. However, Cachon and Fisher (2000) describes that implementing information technology to accelerate and smooth the physical flow of goods through a supply chain is significantly more valuable than using information technology to expand the flow of information. The performance of the entire supply chain depends on the coordinated effort of each supply chain partners. Lack of coordination can result in one partner in a supply chain shouldering the majority costs, risks the price escalating. Coordinated contracts allow the risk of rising prices to be shared across the supply chain, as well as, the customer.

In supply chain, companies connect the strategic partners to produce and procure low costs but high quality products. The companies find the best logistics transport and support for reliable and faster delivery. The optimal supply chain coordination does not always mean the fastest operation; it may adequately mean to provide time sensitive delivery or pickup, quality product with low cost, and well-managed customer service. Figure 1 illustrates the relationship between supply chain partners and the product flows across a supply chain.

Variability in any service implies additional risks and uncertainty. The larger the uncertainty

Figure 1. Basic supply chain flow

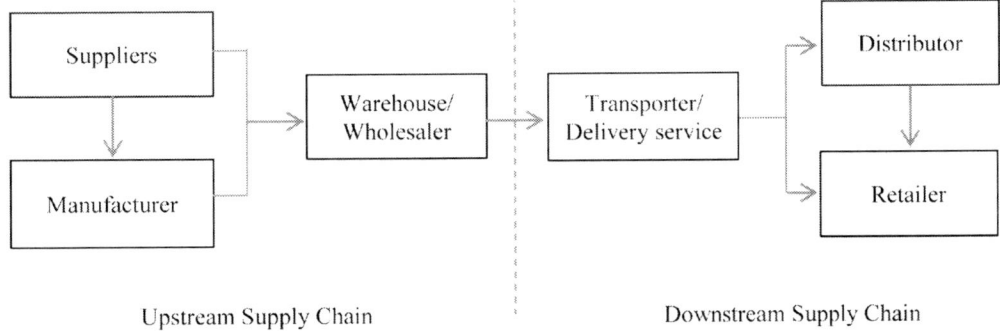

Upstream Supply Chain   Downstream Supply Chain

in a supply chain, the greater the costs for safety inventories, time in transit, or cost of expedited deliveries. In a coordinated supply chain, the following nine issues are essential to harmonize the operation. (*1*) growth in the global market, (*2*) price volatility in markets, (*3*) short product life, (*4*) demand unpredictability in markets, (*5*) increasing capacity utilization, (*6*) managing outsourcing, (*7*) managing inventories, (*8*) increasing e-business, and (*9*) high level of customer services. The supply chain coordination capitalizes the above issues for effective growth and planning strategies to reduce the variability.

Example: Supply Chain Magnitude

National Semiconductors is one of the world's leading manufacturers of semiconductors. National Semiconductors is a global production network, integrates a wide variety of tasks related to capital investment, technology and labor and product. The industry is constantly expanding and modifying to its current form to achieve the competitive advantages. Table 1 highlights the

production and supply chain features of National Semiconductor system (cf. Rodriguez, 2009).

INVENTORY STRATEGIES IN SUPPLY CHAIN

A supply chain focuses with different types of inventories. Inventory is one of the most valuable assets for a company. The skills necessary to manage inventories, affects the overall efficiency of a supply chain. Effective inventory practices not only reduce inventory costs, but also save material handling time, storage spaces, and wastes. The classification of inventories and strategies can be based on functionality and position in the supply chain. A product considered as a raw material inventory when it is required for production or processing. Raw material inventories include parts, raw products, and subassemblies that pass through the manufacturing operations. Work-in-process inventory is the materials either in waiting or within the process. Finished goods inventory is the end item accumulate after the manufacturing

Table 1. Production and supply chain features of national semiconductor system

Features	National Semiconductors Supply Chain
Product range	More than 10,000 different products.
Product type	Wireless handsets, display and imaging technologies, information infrastructure and information access devices.
Production quantity	4.2 billion Chips manufacture each year.
Chips production location	Four in the US, one in Britain and one in Israel.
Global distribution center Chips shipping location	Four in North America, one in Europe, & five in Southeast Asia. 3,800 customers worldwide.
Distribution system	Distribution cost is 1.2% of sales.
	Final product ships to hundreds of facilities all over the world
	20,000 different routes.
	700 logistics employees.
	42 Freight Forwarders.
	15 different airlines involve.
Order cycle lead time ■ Lead time in the Past ■ Current lead time	Delivery cycle was 9 to 25 days.
	2 days for Asia, 2-4 days for North America, 2-5 days for Europe.

or production process. Finished goods then transport to distribution centers and store, consider as the centralized inventory. Finished products wait inside the transportation considered as in-transit inventory. Table 2 presents summary of different types of inventories.

Economic Order Quantity

Effective inventory management is critical and essential for supply chain embellishment and success. A company must determine the appropriate amount of inventories and ordering quantity across all sections. First, it needs to develop a system to keep track of the inventory on hand and on order. Next, it needs a strategy to estimate reliable demand forecast by updating demand information continuously as it occurs to minimize forecast errors. There is various techniques and software available for inventory calculation and management. A company must also consider reasonable estimates of inventory holding costs, ordering cost, and shortage costs to determine the optimal quantity of inventory. Knowledge of lead times and lead time variability information are also important as these relate to the competitiveness in the market and customer demands satisfactorily.

The economic order quantity modeled is the simplest form of inventory cost functions that determines the optimal ordering quantity. The cost associated to EOQ model can be classified into two broad categories: (*1*) The order-processing cost, and (*2*) inventory holding cost. The order processing cost incurs when investors procure the product, and the holding cost is associated with carrying product for a period of time. Solving these two cost functions using partial derivatives with respect to the relevant quantity provides the economic ordering quantity. Following notations considered in this cost model. Figure 2 illustrates an inventory system with time:

- **D:** Annual demand, units
- **Q:** Order quantity per batch, units
- **S:** Order processing cost, \$/order
- **C:** Unit price, \$/unit
- **I:** Proportion of item cost, %
- **H:** Holding cost, \$/unit/year (or, $H = I \times C$)
- **T:** Time period between orders, unit time (or, fraction of a year)
- **SS:** Safety stock, units
- **L:** Lead time, unit time (or, fraction of a year)

Table 2. Different types of inventories and positions

Major Inventory	Position
Raw material (RM) inventory: • Inventory of purchased parts & raw materials • Cost: RM Inventory cost comprises the cost of materials purchased from upstream suppliers	Move from the supplier to manufacturing facility
Work-in-process (WIP) inventory: • Inventory of partially processed goods • Cost: WIP inventory cost consists of raw materials cost and allocated portion of manufacturing and labor cost	Within the manufacturing facility
Finished-goods inventory: • Inventory of finished products • Cost: Finished goods inventory cost is a culmination of costs of raw materials, manufacturing, and handling cost.	Warehouse – manufacturer or distributor
In-transit inventory: • Goods-in-transit bound towards distributors, warehouse, retailers or customers • Cost: In-transit inventory cost is the transportation costs associated with each produced unit	Inside a transport, such as train, truck, cargo

Figure 2. Inventory system with time

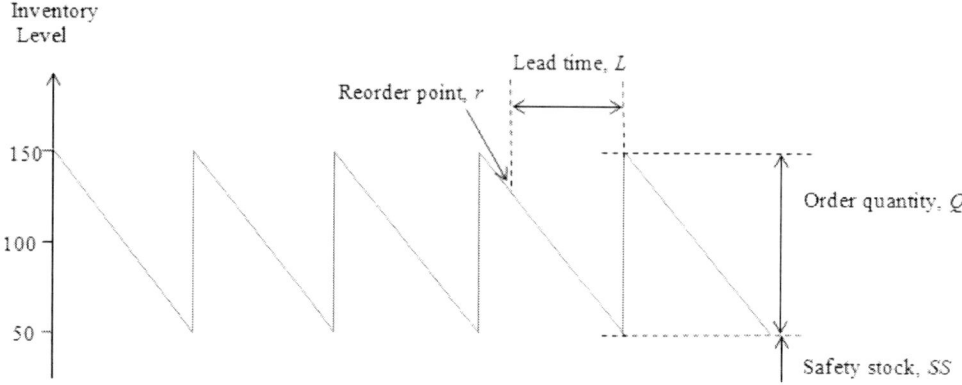

The inventory reaches its high level when purchased order arrives, and then gradually depletes with demand over the selling period. Once the inventory reaches to a preset level, called the reorder point, *ROP*, new order dispatches, which arrive after the lead time, *L* period. The safety stock is the minimum level of inventory. The reorder point is selected as such that the issued order must arrive prior to safety stock level. Once the new order arrives, the maximum level of inventory reaches up to $Q + SS$.

Following is the cost function:

Purchase cost $= C \times D$

Average inventory level = average of maximum and minimum inventory level

$$= \frac{1}{2}[(SS + Q) + SS] = SS + Q/2$$

Annual holding cost $= (SS + Q/2)H$

Annual order processing cost $= (D/Q)S$

Total Annual Cost = Purchase cost + Holding cost + Order processing cost

$$= C \times D + (SS + Q/2)H + (D/Q)S$$

Taking the first derivatives and setting it to zero, the Optimal Order Quantity (*EOQ*) becomes

$$EOQ = \sqrt{2DS/H}$$

Reorder point, *ROP* = demand during the lead time + safety stock $= D \times L + SS$

Problem

In an electronic store, the demand of an item is 300 per year. The order processing cost per order is $150, and unit cost of the product is $500. Inventory holding cost is 20% of unit cost per year. The product delivery lead time is one week period. The store manager wishes to hold 5 weeks of demand as safety stock due to demand uncertainty.

- How many units should it order in one shipment?
- What is the safety stock level, *SS*?
- What should be the reorder point (ROP)?

Answer

Holding cost, $H = I \times C = 20\% \times 500$

The economic order quantity,

$$EOQ = \sqrt{\frac{2DS}{H}} = \sqrt{\frac{2 \times 300 \times 150}{500 \times 20\%}} = 30 \text{units}.$$

Safety stock (*SS*) covers 5 weeks of demand, *SS* = 300 × (5/52) = 29 units.

Reorder point, *ROP* = *D* × *L* + *SS* = 300 × (1/52) + 29 = 35 units

The store manager should reorder whenever the inventory reaches at thirty five units. Inventory practices in the supply chain help the flow of finished goods to the consumer. There are other contemporary inventory practices in supply chain such as (*1*) Pull-Based Replenishment with Suppliers, (*2*) Virtual Inventory Bin, (*3*) Postponement. Finally, this chapter will discuss efficiency measure of supply chain through two common measures: (*1*) inventory turnover and (*2*) weeks of supply.

Centralized Inventory

The centralized inventory intends to consolidate inventories in one location or a small numbers of locations central to all retailers. There are a number of benefits to centralizing inventory in a supply chain. The benefits of centralized inventory locations include the minimization of shipping cost throughout the supply chain.

Pull-Based Replenishment

The pull-based replenishment policy is a demand driven strategy. In this policy, the actual customer demand drives the demand information throughout the supply chain, which is more reliable than forecasted demand. This implies that demand at the upstream processes stimulates the downstream demand. The example of a pull-based replenishment method is the Kanban system that utilizes a system of cards and reacts upon the signal to upstream processes to fulfill the material or com-

ponents requirement. Pull-based replenishment reduces lead times, inventories, and variability across the supply chain. The pull-based method of replenishment reduces the amount of inventory accruing costs in warehouses. Pull-based systems are difficult to implement when the lead-time is long. The advantage of economies of scale in the pull-based system can be achieved if production and shipping schedule are planned in advance.

Virtual Inventory Bin

The virtual inventory bin method is the practice to reallocate and redirect inventory while it is in transit. The virtual inventory bin is treated as available inventory, allows a manager to quick response to higher demand destinations that may have a greater need or profit margin. Virtual inventory bins implies lower safety stock, reduced delivery cost, and maximized revenue. Goods in transit allow a highly responsive supply chain because shipments can be rerouted and reallocated while on the road.

Postponement

Postponement is the strategy common at the manufacturing sectors. Manufacturers produce general-purpose items as the subassemblies. The production of the final products is delayed until the customer orders. In postponement practice, the quantity of generic products is determined based on the aggregated demand of all finished items of the similar product type. Often, the aggregated quantity is more accurate and allows manufacturers to reduce overstocking. Once customer orders a product online, the postponement inventory practices allow the final product to be made after the order. The product is then assembled in a short period and ship immediately to the customer. This practice creates an inventory of subassemblies that can be interchanged between products, and thus, reduce excess production and overstocking.

Efficiency Measures

Inventory turnover is a quantitative measure to determine the inventory management effectiveness. Effective inventory management requires a constant record of on hand inventory and strategies to keep the inventory at the lowest possible level by increasing the annual inventory turnover ratio. The computation of inventory turnover ratio is as follows:

$$\text{Turnover ratio} = \frac{\text{Annual Sales}}{\text{Average Invetory Level}}$$

Managers seek to attain a high annual inventory turnover ratio as it is a measure of growth for a company. High inventory turnover indicates that high flow of materials, thus, little capital tied up in inventory and lesser risk of product obsolescence.

Weeks of Supply

Weeks of supply refers to a method of tracking inventory of the retailers or manufacturers who are in need of product. This method is based upon periodic inventory review system by tracking inventory periodically (say, weekly or biweekly) and allow managers to know the appropriate amount of inventory on hand. This method is used to minimize the inventory on hand, thereby reducing the amount of capital invested in inventory.

BULLWHIP EFFECT IN SUPPLY CHAIN

Product demand is not always constant; it largely varies due to change of customers' behavior and several other factors including the local and global economy, contemporary fashion, duration of peak sale season, weather and so on. Consequently, the demand fluctuates at the retail store, gradually propagate towards the suppliers, and then, at the manufacturer. The effect of demand variability that occurs at the customer level grows sharply over the wholesaler, and dramatically over the distributer and finally peaks at manufacturer facility. Since demand variability initiates from downstream at the customer level, gradually reach to upstream at the manufacturer level, resemble the phenomenon of a bullwhip, called the bullwhip effect. Figure 3 shows how small changes in the customer demand form a bullwhip effect in supply chain. Although bullwhip effect is perceived as an unavoidable effect of demand variation, there are substantial studies and efforts initiated by the researchers and companies to find the root cause of the demand variation, and initiative to minimize the ripple effect.

The bullwhip effect occurs due to overreaction of demand fluctuations, not because of the amplification of demand. When demand fluctuates at the customer level, there create a mismatch between the supply and demand. The ordering quantity either makes excessive inventory or shortages, which propagate to the bullwhip effect

Figure 3. Bullwhip effect in supply chain due to demand variability

in the supply chain. The essence of the bullwhip effect is the orders amplification at the manufacturers' level due to changes made at the buyer. Demand amplification and production fluctuations increase waste of resources and costs. For a large supply chain, it is difficult to control demand amplification and the bullwhip effect.

Identifying ways to control the bullwhip effect requires the experience of how to rapidly react to market and demand changes, agile decision making. Innovative companies have found ways to control the bullwhip effect and improve their supply chain performance by coordinating information and planning along the supply chain. Bayraktara et al. (2008) defines the bullwhip ratio as the ratio of variance of the orders realized by the manufacturer to the variance of the demand observed by the retailer.

$$Bullwhip\ ratio = \frac{Var(Order)}{Var(Demand)}$$

Traditionally, there is bullwhip effect, when bullwhip ratio > 1.

Causes of Bullwhip Effect in Supply Chain

There are various reasons for bullwhip effect in supply chain. One of the main causes of the bullwhip effect is the variability in demand. In a supply chain, forecasts are based on the downstream customer's demand. If there is a variation at the customer's demand, this results large variations as it moves to the upstream in the supply chain. Following are the primary reasons for the bullwhip effect in supply chain (cf. Cachon & Terwiesch, 2009):

- Demand forecast update
- Long lead time
- Decentralized information system
- Order batching

- Shortage gaming
- Trade Promotion in the supply chain

Demand Forecast Update

In a supply chain, product demand forecast is common to every company to prepare for its production planning, capacity planning, inventory control, warehousing and material requirements planning and scheduling. Bullwhip may occur when demand forecast largely depends on previous sales data, without reflecting current demand pattern and the experts' judgment about the future demand. For instance, a manager who is responsible to order products from suppliers and estimate demand using data extrapolating forecasting process, such as exponential smoothing or autoregressive method, will not be able to incorporate the most recent events in the forecast. Therefore, there will remain differences between actual demand and supply, which will cause order fluctuation, lead to bullwhip effect in the supply chain.

Long Lead-Time

Most of the products procurement associates with lead-time. In supply chain, the delay time between the order dispatch and the arrival is called lead-time. The dispatched order quantity reflects the lead-time demand, future demands, as well as the necessary safety stocks to maintain a customer service level. The demand quantity should update according to the current sales data and the market trend. For long lead times, safety stocks and lead-time demand are large quantity. If forecast period is long, the demand may considerably fluctuate and forecast for the lead-time period becomes error prone. This may cause larger differences at the order quantities and the lead-time demand. Therefore, variability increases dramatically in the supply chain.

Base stock quantity and safety stock both depend on average demand, lead time, and reorder point. Equations are the following. Base stock level= average customer demand, *D*, multiply with lead time, *L* and reorder point, *ROP*.

Base stock level = $D(L+ROP)$

Safety stock = safety factor, Z (Normal distribution), multiply with standard deviation of customer demand, STD, and multiply with square root of both lead time, *L* and reorder point, *ROP*.

Safety stock level = $Z \times STD \times \sqrt{(L+ROP)}$

Decentralized Information System

In a decentralized information system, companies do not coordinate on their sale information and demand forecasts. Technology benefit is a prevalent throughout all bullwhip reduction strategies. In a supply chain, a decentralized system strategy is unable to share information between upstream and downstream companies. It is difficult to establish business plans, demand forecast, and replenishment operations between the participants within the total supply chain enterprise. The benefit of collaboration between supply chain partners of having fewer inventories, reduced transportation costs, and product availability cannot be achieved. Therefore, demand variation increases multiplicatively towards the upstream trading partners and cause bullwhip effect in supply.

Order Batching

In order batching, orders are aggregated into batches for cost benefit (aggregate ordering, transportation, loading, unloading and warehousing cost). This encourages the buyers to procure products in large quantities more than the actual demand. Buyer then suspends buying for next few periods. Consequently, supply chain suddenly receives a large order at one period, and no orders at the next several periods, which cause variability increase in supply chain, leads to bullwhip effect.

Shortage Gaming

Shortage gaming occurs if there is any sudden change of demand or sudden influx of sudden event changes such as unexpected weather, calamities. Buyers intend to order more than the required quantity with the anticipation of shortages in the near future. The buyers respond by dispatching large orders more than the actual demand to secure that they will receive adequate inventory to manage the changes. Shortage gaming leads to a bullwhip effect in supply when suppliers realize a drastic overinflated demand.

Trade Promotion

Trade promotion is the discount wholesale price offer to the retail chains for a short period. Trade promotion often stimulates retailers to buy on discount deal and make them inspired before making a purchase. Consequently, retailers purchase more items than the actual demand that may perceive in that short period. For illustration, consider a single retailer and a single supplier case. Supplier may be a manufacturer, or a distributor, or a wholesaler and assume retailer procures items from its supplier on regularly basis. If the supplier stimulates sales of a particular brand by Trade Promotion, the retailer will order promotional item more than their favorite brand. Retailer will buy in bulk and expect the demand of the promotional brand to grow. However, if demand remains same, the retailer expenditure will increase not only by purchasing more quantity, but also to keep the excess inventory.

Problem

Suppose a distributor agent of a manufacturing company sells a branded product 'X' at a regular price of $50 per item. The distributor offers 10% discount twice a year during early spring, and early fall for one month period to stimulate the sales. Suppose, the early spring, and the early fall promotional sales begin at February and August, respectively. Consider retailer sales on average 100 units per month and carry half of a month worth of inventory. For simplicity, assume the transportation time is negligible, therefore, the products arrive instantaneously (within a few hours) once the retailer dispatches the order. The retailers annual holding cost rate is 20 percent of the inventory value.

Let's compare the retailer's profit with two different ordering strategies, without considering the transportation cost. With the first strategy, the retailer orders every month according to the regular customer demand throughout the year, considered as *demand-pull*. In the second strategy, with the trade promotion, retailers order twice per year at a substantial quantity. This strategy is based on the fact of *forward buying*, as these orders contain future demand. Figure 4 shows the on-hand inventory over the period of one year with both ordering strategies.

Retailers Strategy with Demand-pull:

- Order dispatches per month = 100 units
- Safety stock (half of a month inventory) = 50 units
- Maximum inventory level = Order Quantity + Safety Stock = 100 + 50 = 150 units

- Average inventory = Average of Maximum and Minimum Inventory = 100 units
- Promotional price = $50 – (10%) × $50 = $45.00
- Total cost during promotional period (2 months) = 100 × $45 × 2 = $9,000
- Total cost without promotional period (10 months) = 100 × $50 × 10 = $50,000
- The weighted average inventory = ($9,000 + $50,000)/12 = $4916.67
- Annual holding cost (20%) = $7,375 × 20% = $986.33
- Total annual cost = $5900

Retailers Strategy with Forward Buying:

- Order dispatches for six month = 600 units
- Safety stock (50% of the ordered quantity) = 300 units
- Maximum inventory level = Half year Order Quantity + Safety Stock = 600+300=900 units
- Average inventory = Average of Maximum and Minimum Inventory = 600 units
- Total cost during promotional period (2 months) = 600 × $45 × 2 = $54,000
- The weighted average inventory = ($54,000)/12 = $4500

Figure 4. 'On-demand' and 'forward buying' in a supply chain

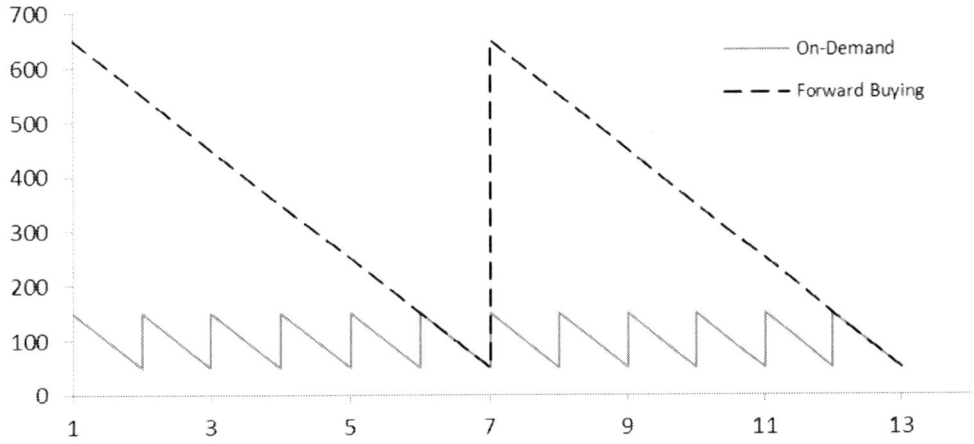

- Annual holding cost (20%) = $4500× 20% = $900
- Total annual cost = $5400

The above analysis indicates that forward buying strategy is more profitable than demand-pull strategy. However, forward buying option causes a large fluctuation of demand in the supply chain. If the supplier would offer the discount price for year round, the supplier would lose its revenue. A similar analysis as above shows that retailer receives the same benefit as it is in the trade promotion. Therefore, a trade promotion for a short period does not benefit suppliers, or the retailers. It instead causes a Bullwhip effect in the supply chain.

Demand-pull Strategy with Discount Price:

- The weighted average inventory = $4500.00
- Annual holding cost (20%) = $900
- Total annual cost = $5400

BULLWHIP REDUCTION STRATEGIES

There are several ways to mitigate the bullwhip effects in supply chains. It requires to overview the perspective of systems thinking and coordinated efforts within and between the supply chain partners. Production and/or product procurement lead times always plays a crucial role in bullwhip effects and supply chain competitiveness. Following are the practices emphasized to reduce the bullwhip effect:

- Data sharing
  - Electronic data interchange
  - Collaborative planning, forecasting, and replenishment
  - Smaller orders (frequent deliveries of smaller order)
- Reduce lead times
- Minimizing forecast error
- Synchronization

Data Sharing

Since decentralized information, order batching, demand surge contribute to the bullwhip effect, companies need to devise strategies that lead to share information, data exchange and smaller and more frequent order dispatching. Data can be shared mainly in two forms: Electronic Data Interchange and Collaborative Planning, Forecasting, and Replenishment.

Electronic Data Interchange (EDI)

EDI is an excellent way to share data. EDI helps to create automatic replenishment procedure by dispatching the order when the stock reaches the reorder point.

Collaborative Planning, Forecasting, and Replenishment (CPFR)

CPFR is the sharing of current information between the supplier and buyer. By sharing information between the supply chain partners, both parties know exactly what is selling, what is in stock, how much is in the inventory and how to adjust their forecasts utilizing the most recent data. The goal for sharing information is to know which products to be procured, at what time. The information sharing helps to smooth flow of products and reduce shortages, both reduce the bullwhip effect in supply chains.

Smaller Orders

EDI helps to smooth flow of products and create smaller more frequently orders and deliveries by working with suppliers. Small orders and reduced batch orders increase information quality and alleviate the bullwhip effect in supply chain.

Minimizing Forecast Error

Minimizing the forecast error reduces the bullwhip effects. At the planning level, forecast model should be updated as new data becomes available. The centralized data information, such as EDI, and advanced forecast techniques, such as Bayesian techniques enable the most recent demand observant to incorporate in the forecasting process. The effective integration of data update reduces forecast errors and demand uncertainties, and avoids bullwhip in the supply chain.

Synchronization

Supply chain synchronization requires that people, schedules, internal and external entities and processes be designed and developed with synchronization in mind. Many supply chain experts suggest use electronic catalogues, standard transactions, purchase orders report, shipment notices, and supply chain co-ordination for dynamic process modification. In the synchronization process, the above dimensions should be linked together to the physical movement, order timing, delivery timing such as, quantities of raw materials to be purchased, and process monitoring and reporting such as, production flow rates, and buffer inventories.

SUMMARY

There is an increase in competition and globalization. The market competition and product demands change at a faster rate. Companies need to make a decision with shorter notice and less information. Coordinated efforts among the partners improve supply chain performances. Decentralized information system, self-decision making does not increase supply chain efficiency. In addition, the propagation of demand variability causes bullwhip effect in supply chain. At the strategic level, this remains a challenge to supply chain partners to determine the root causes of the demand variability. All trading partner should realize that the competition is between the supply chains, not individual firms. Trust and coordination are the keys to success. Distorted information from one end of a supply chain can lead to enormous bullwhip effects at the other end. There are many new trends in technology, which can alleviate some of the challenges that supply chains currently experience. Electronic commerce helps to reduce costs and time associated with supply chain relationships. The satellite network enables suppliers, manufacturers, and distributors to link together regardless of the company location.

Some of the many career positions in the supply chain field are the followings: Supply Chain Management Analyst, Commodity Manager, Customer Service Manager, International Logistics Manager, Logistics Services Salesperson, Production Manager, Sourcing Analyst, Logistics and Material Planner, Systems Support Manager (MIS), Transportation Manager, Process Analyst, Scheduler, Purchasing Agent, etc.

REFERENCES

Bayraktara, E., Lenny Koh, S. C., Gunasekaranc, A., Sarid, K., & Tatoglu, E. (2008). The role of forecasting on bullwhip effect for E-SCM applications. *International Journal of Production Economics*, *113*, 193–204. doi:10.1016/j.ijpe.2007.03.024

Cachon, G., & Fisher, M. (2000). Supply chain inventory management and the value of shared information. *Management Science*, *46*, 1032–1048. doi:10.1287/mnsc.46.8.1032.12029

Cachon, G., & Terwiesch, C. (2009). *Matching supply with demand: an introduction to operations management* (2nd ed.). New York, NY: McGraw-Hill/Irwin.

Rodrigue, J. P. (2009). *The geography of transport systems* (2nd ed.). Abingdon, UK: Routledge.

Simchi-Levi, D., Kaminsky, P., & Simchi-Levi, E. (2007). *Designing and managing the supply chain* (3rd ed.). New York, NY: McGraw-Hill/Irwin.

Chapter 6
Intellectual Property Rights in Semi–Conductor Industries:
An Indian Perspective

Satish Chandra Tiwari
NSIT, India

Maneesha Gupta
NSIT, India

Mohammad Ayoub Khan
Government of India, India

A. Q. Ansari
Jamia Millia Islamia, India

ABSTRACT

Fundamentals of intellectual property rights are provided. In addition, the trends of patenting and patented technologies in India in different areas of semi-conductor technologies are analyzed. The authors discuss many aspects of the patents and patentability. They present patent practices in India covering required forms needed to be filled in order to file a patent. Finally, the importance of patenting and its growth is shown with few year wise statistics.

INTRODUCTION

Everyone has right to the protection of moral and material interests resulting from any scientific, literary, or artistic production of which he is author. Article 27 of universal human rights (Morsink Johannes, 1999)

DOI: 10.4018/978-1-4666-3658-3.ch006

Origin of modern Intellectual Property Rights (IPR) can be traced back to Statue of Monopoly (1623) and Statue of Anne (1710) for patent and copyright law respectively (Brad, 1999). Thereafter, IPR remained present in one form or other until 19th century where its reach spread to the whole world. As said by article 27 of universal human rights, the sole intention was to provide protection to inventor, from others to copy the invention. To the general public, the perception towards IPR had varied from time to time. Initially the justification for the granting of IPR law was to give as little protection as possible in order to encourage innovation. Hence, they were granted only when there was a necessity to encourage innovation, limited in time and scope. Historically, in its initial form, IPR was granted for one to three years. Today, particularly in U.S. the intention behind the grant of IPR is "absolute protection" (Herman, 1911; Library of Congress, 2009). Now justification behind absolute protection is the thought that, creators will not get sufficient incentive of their inventions, unless they are legally owner of their invention. Today every country gives IPR roughly for 20 years. Intellectual property in today's context can be viewed as another type of real property, following all the laws of it. The term IPR has lot of domains concatenated under it; all of them provide protection form copying or reproduction (Herman, 1911). Contrary to today's understanding patents does not specifically begin with invention rather it originated with royal grants by Queen Elizabeth I (1558-1603) for providing monopoly privileges (Mossoff, 2001). Slowly and steadily IPR took legal shape with associated laws and practices defined. Technically usage of the term IP (intellectual property) can be traced back to 1867 where north German confederation's constitution granted legislative power over the protection of intellectual property (Schutz des geistigen Elgentums) to confederation (Hastings Law Journal, 2001). After the merger of Pars convention (1883) and Bens Convention (1893), the term intellectual property was adopted in their new combined title *The united International Burex for protection of Intellectual property.* Further, the organization was relocated to Geneva in 1960 and was succeeded by the establishment of the World Intellectual Property Organization (WIPO) in 1967 by treaty as an agency of United Nations (WIPO, 1967). Another form of intellectual property "copyright" can also be traced back to 16th century.

Now as the time progressed the whole context of IPR has evolved; it is not only relate to protection rather it has substantial amount of share in economic growth. In a joint research by WIPO and United Nations University for investigation of IP Systems impact on six Asian countries found "a positive correlation between economic growth and strengthening the IP system." In the similar path Anti-Counterfeiting Trade Agreement (ACTA) states that "effective enforcement of intellectual property rights is critical to sustaining economic growth across all the industries and globally (Walsh, 2012). Hence, IP and economics go hand in hand. Going by IP products becomes excludable, non-rival intellectual products which was not possible when not going by it. It was estimated that two thirds of large business in U.S. are directly related to IP assets. Similarly, if we analyze IP-intensive industries, they roughly generate 72% more value added (price minus material cost) per employee as compared to non-IP intensive industries (Robert, et al., 2007).

Few basic fundamental IP laws are common among different countries but in general different countries have different opinions and laws that IP in their territory. Hence grant and enforcement of IP's are governed by national laws and also by international treaties applicable where those treaties have their effect in national laws.

As any new thought is appreciated as well as criticized, criticism of intellectual property is as much older as its inception. One of the major arguments related to criticism of IPR is "if one makes a copy of work, the enjoyment of the copy does not prevent enjoyment of the original." Those in free culture movement point that IP monopoly prevent progress and argue that general public interests are harmed by IP monopolies in form of copyright extensions, software patents and business method patent. Similarly, there is criticism on the tendency of IP to expand both in duration and scope (Daniel, 2008; Joseph, 2006).

Intellectual property as of now has different domains under it, which are classified on the basis of their subject coverage as shown in Figure 1. Few of the important ones are discussed in this section.

Copyright

It is a legal concept giving exclusive rights to creator of an original work usually for limited time. In general, it means "right to copy," giving the holder of copyrighter rights to determine who may adopt the work in same or different forms, who may perform it, who may get financially benefit from it and other related rights. It was initially meant to promote creation of new works and protect against illegal copying and printing. Copyright are territorial in nature, which means that they do not extend beyond the territory. Today however, many aspects of national copyright laws have been standardized through international copyright agreement. Even then copyright laws of most countries have some unique features. Duration of copyright is typically whole life of creator plus fifty to hundred years from creator's death or a finite period for corporate creations. As far as jurisdictions are concerned, copyright is enforced as a civil matter and some cases criminal sanctions may apply.

Trademark

A trademark is a symbol, sign or indicator distinctive in nature can be used by individual or business organization to represent their products and services, distinguishing form others. A trademark may be represented by the following symbols:

- **TM (Trademark):** For unregistered trademark (for promotion or branding of goods).
- **SM (Service mark):** Used for promotion or branding of services.
- **R:** Registered trademark.

A trademark can be logo, word, name, phrase, symbol, design, image, or a combination of all of them. There are also non-conventional ones other than aforementioned ones i.e. based on colour, smell, or sound. In case of infringement owner of registered trademark may commence legal proceedings. If the trademark is used in relation to services rather than products, it is protected under Service Mark (SM).

Geographical Indication

If a product possesses certain qualities depending upon geographical location, or is made in accordance with traditional methods or enjoys a certain reputation falls under Geographical Indication (GI). A geographical indication is used to represent the products, which have aforesaid specialization.

Patents

The word patent has its origin from Latin word "patere" which means "to lay upon," i.e. to make

Figure 1. Intellectual property domain

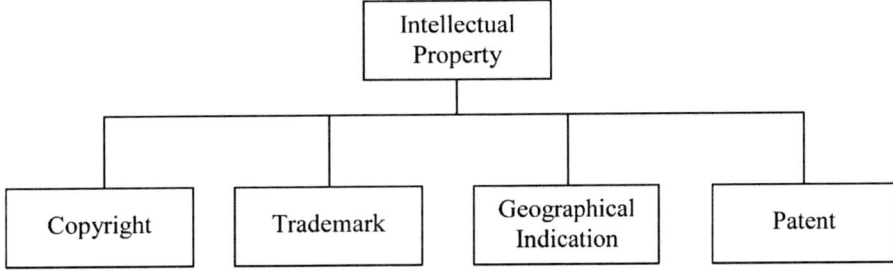

99

available for public inspection. Historically patent can be traced back to 500 B.C. in Greek city where it was there as source of encouragement to all who discover any new refinement in Luxury. The profit arising from patent was secured by inventor for one year span. Filippo Brunelleschi (1421) received a three year patent for barge with hosting gear that carried marble along the Arno River (Christine, 2002). In 1474 the republic of Venice enacted a decree that new inventions and products, once put into practice, had to be informed to republic for obtaining the right to prevent others from using or making them. First English patent was granted by king Hennery VI with a license of 20 years to John Utyam for manufacturing of coloured glass. England then followed Statue of monopoly in 1624 which declared patents can only be granted to newly invented products. During Queen Anne (1702-14) reign, lawyers of the English court developed the requirement that for a patent to be granted, a written description of invention must be submitted (Brad, et al., 1999). Patent systems in many countries until now are based on British Law which can be traced back to Statue of Monopoly. Different countries have adopted patenting procedure at different instance of time. In North America, Samuel Winslow in 1641 was granted the first patent for new process of making salt. Patents in France were granted by monarchy and some institutions like "Maison du Roi." Novelty was examined by academy. There was not any requirement to publish a description of invention and examinations were done in secret.

Disclosure of actual use of the invention was considered to be sufficient. Modern patenting system was developed in 1971 during revolution. Without examination the patents were granted, since it was considered to be inventors natural right.

In U.S. first congress adopted a Patent act in 1970 and under this act first patent was granted to Samuel Heokins for Potash production technique on July 31, 1790 as shown in Figure 2 (Paynter, 1990).

As discussed earlier patents are now regarded as exclusive property. Contrary to the thoughts a Patent is not a right to practice or use the invention, rather it gives right to exclude others from using, making, selling or implementing the patented invention for whole term of patent (Usually 20 years from the date of filing) subject to payment of maintenance fees. It can also be understood as a limited property right offered by government to inventors in exchange of their agreement to share the credentials of patent with public. Since for a limited period of time, invention remain property of inventor, hence, it may be sold, licensed, mortgaged, assigned or transferred or simply abandoned.

What is Patentability?

Still there is not any universal law applicable to all countries that can tell what Patentable is and what's not. Hence, different countries have different perspective on this. Those countries having same set of rules can accept patents as it is whereas those having different perspective require a patent to get filed in their country. There is still variation in rights covered by patent in different countries. For example, research is covered by patents in U.S. except purely philosophical one.

What is Infringement?

Infringement is the violation of patent by any other individual, organization, or group. Even the criteria for patent infringement or violation have variation dependent on country to country. U.S. considers infringement by any making of invention even if that making generates/develops a new invention, which may itself be able to be Patentable. Even if the Patent gives exclusionary rights, however, does not give the owner of patent the right to exploit the patent. This can be more clearly understood by the fact that many inventions are an improvement of prior inventions that may still be covered by someone else's patent (Herman,

Figure 2. First granted patent (Paynter, 1990)

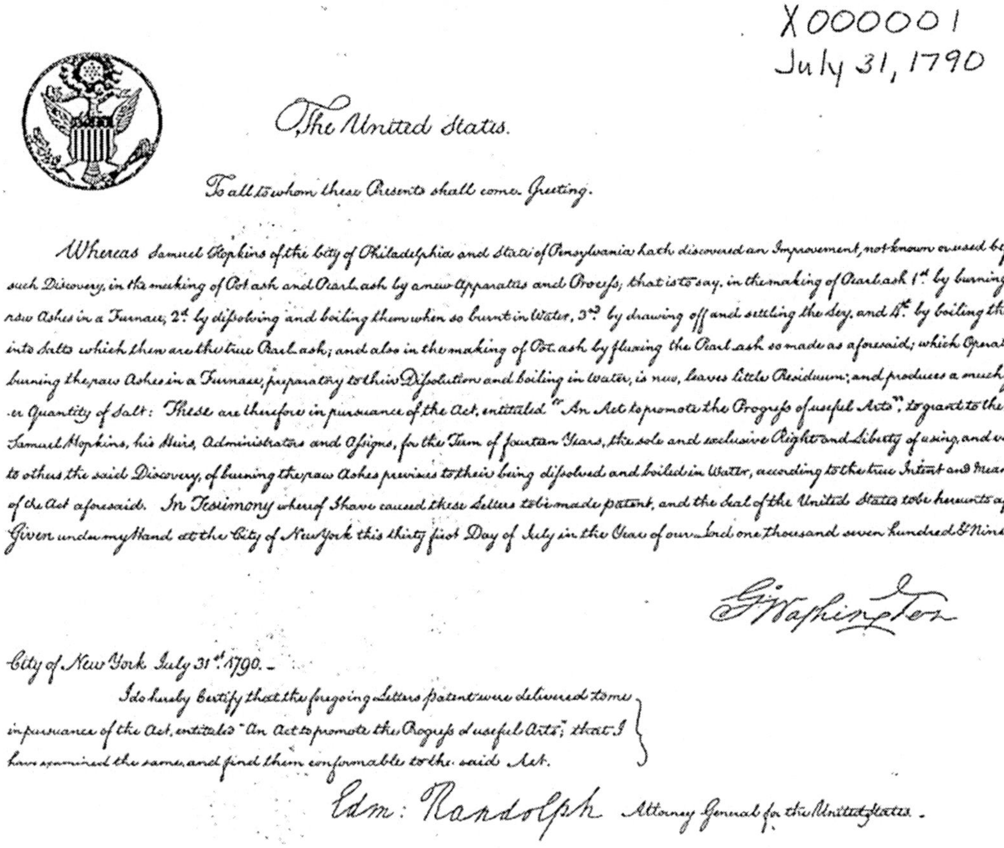

X000001
July 31, 1790

The United States.

To all to whom these Presents shall come Greeting.

Whereas Samuel Hopkins of the City of Philadelphia and State of Pensylvania hath discovered an Improvement, not known or used before such Discovery, in the making of Pot ash and Pearl ash by a new Apparatus and Process; that is to say, in the making of Pearl ash 1st by burning the raw Ashes in a Furnace, 2d by dissolving and boiling them when so burnt in Water, 3d by drawing off and settling the Ley, and 4th by boiling the Ley into Salts which then are the true Pearl ash; and also in the making of Pot ash by fluxing the Pearl ash so made as aforesaid; which Operation of burning the raw Ashes in a Furnace, preparatory to their Dissolution and boiling in Water, is new, leaves little Residuum, and produces a much greater Quantity of Salt: These are therefore in pursuance of the Act, entituled "An Act to promote the Progress of useful Arts", to grant to the said Samuel Hopkins, his Heirs, Administrators and Assigns, for the Term of fourteen Years, the sole and exclusive Right and Liberty of using, and vending to others the said Discovery, of burning the raw Ashes previous to their being dissolved and boiled in Water, according to the true Intent and Meaning of the Act aforesaid. In Testimony whereof I have caused these Letters to be made patent, and the Seal of the United States to be hereunto affixed. Given under my Hand at the City of New York this thirty first Day of July in the Year of our Lord one thousand seven hundred & Ninety.

G Washington

City of New York July 31st 1790.—

I do hereby Certify that the foregoing Letters patent were delivered to me in pursuance of the Act, entituled "An Act to promote the Progress of useful Arts"; that I have examined the same, and find them conformable to the said Act.

Edm: Randolph Attorney General for the United States.

1911). *If an inventor takes a prior patented design and get its improved version patented, he or she can legally build his or her improved design with permission from the original patent holder assuming the original one is still in force.* On the other hand, inventor of improved one can exclude the original patent owner from using improvement.

Convention Countries

Now as we know from the above section that different countries have different observation on what is Patentable and what's not. Conventions are signed between different countries, which agree on specific rules or have same thought. Once a country signs a convention then member countries can avail same benefit in that country as in their own country. At different instance of time, different conventions are made and a county can be member of any number of conventions. Some important conventions are as follows.

Paris Convention (India is Member since 1998) (Paris Convention, 2011; Bodenhausen, 1968)

Convention came into picture in 1883 and amendments were made to it from time to time. This is more related to industrial protection rights but also covers:

- Patents for one year.
- Utility models (not available in India).
- Industrial designs 6 months.

- Trademark for 6 months.
- National treatment to member countries.

The very essence of Paris convention is the fact that it gives equal treatment to applicants from member countries. For priority rights the applicant should file application in all the member countries in same date. However, it can be understood that it is not easy to file application in all the member countries at same time. Hence, to facilitate simultaneous protection in member countries, the Convention has a provision that within 12 months of national filing, if patent applications are filed in member countries, if granted in member countries, the effective date will be the day of national filing.

Patent Cooperation Treaty (PCT, India is a Member) (Article 32, 2011; WIPO, 1995)

Patent convention treaty is international body in which a patent can be filed internationally covering more than one county. Here it can be noted down that; a Patent can be filed to different countries individually which is even fast. Perhaps going by PCT rout have delay about 18 months while it saves money. PCT has universal coverage, covering 130 countries. However there are few important non-PCT countries like Taiwan and Argentina. To have coverage in non-PCT countries a Patent must be filed in those countries individually.

WTO (World Trade Organization, India is Member since 1995) (Article 27.1, 2012; Article 33, 2012)

It is a kind of agreement on trade related aspects related to intellectual properties between member countries. According to this Patent of any field should be available in WTO member countries for any invention and duration of Patent is twenty years.

PATENT PROCEDURE AND PRACTICES

Criteria for Patentability

What a patent must possess or the criteria to be looked in an invention for getting patent on it? Following are some basic criterion that an invention must possess in order to get it patented (IPO, 2011):

- Novelty
- Non-obviousness
- Usefulness

Novelty

An invention will be considered novel if it fulfills following two criteria:

1. It should not be described earlier in written or orally or it should not form the stats of art.
2. It must not be published or used before the date of patent file.

Inventions (Non-Obviousness)

A patent application considered to be non-obviousness, if a person skilled in the art in skilled in the subject matter finds it non-obvious; hence, patent application is license to have invention steps. Guaranty of patent does not have any depending on the complexity or simplicity of an invent step.

Usefulness

An invention must be backed by application, no valid patent can be granted to an invention which takes diminishes usefulness.

General Procedure for Patenting

A patent application satisfying above three major requirements may subject to issuing of Patent. Generally, a patent has tenure of 20 years in almost all the countries. Once an inventor is sure that the patent has satisfies all the above three aspects, a detailed and fine search is required for surety that whether this has been earlier published as patented or not. The search has to be done with the prior at databases. Here patenting companies may help in searching the keywords and finding patents and other publications that are clearly released to the inventor's invention.

Once search is complete inventor may proceed for drafting of the patent. Since drafting of a patent require both legal and technical skills hence skilled experts from patenting agencies may help in defining the boundaries that is covered by this patent and mainly the claims. As discussed earlier patent filling and granting rules vary from country to country, hence a patent must be drafted in accordance with the Patent procedure of individual country. Here we will talk about the procedure specifically followed in India:

1. Filing Patent application
   a. Place of filing
   b. Priority
   c. Specification
   d. Inventor Names
   e. Information of corresponding application in other countries
2. Patent Examination
   a. Examination application
   b. Procedure
3. Patent Publication
   a. Amendment application
   b. Pre-grant Application patent application
   c. Publication of subject matter
   d. Revocation of a patent
   e. Annuities
   f. Restoration

Not Patentable Inventions in India (IPO, 2011)

1. An invention which is frivolous or which claims anything obviously contrary to well established natural laws.
2. An invention subs by Act 38 of 2002 for claim 6 (w. e. f. 2003) the primary or intended use or commercial exploitation of which could be contrary to public order, morality or which causes serious prejudice to human, animal or plant life or health or to the environment.
3. The mare discovery of a scientific principle or the (by act 38 of 2002 section-4 w. e. f. 205-2003) formulation of an abstract theory (or discovery of any living thing or non-living substances occurring in nature).
4. The mare discovery of any new property or [mare new use] for a known substances or of the mare use of known process, machine or apparatus unless such known process results a new product or employee at least one new reactant. [see by patent (Amendment) ordinance 2004 (ord. 7 of 2004) see 3 (a) for "new use" (w. e. f. 1-1-2005).
5. A substance obtained by a mare a mixture resulting only in the aggregation of the properties of the components there of or a process of providing such substance.
6. The more arrangement or rearrangement or duplication of known devices each functioning independently of one another in a known way. (Clause (g) omitted by Act 38 of 2002 see 4 (w. e. f.20-5-2003).
7. Invention reality to atomic energy not patentable.
8. No patent shall be generated in request of an invention reality to atomic energy falling within sub section (1) of section 20 of the Atomic Energy Act 1962 (33 to 1962).
9. A method of agriculture or horticulture.
10. Any process for medicinal, surgical, curative, prophylactic (diagnostic, theraptic) or other treatment of human beings or any process for

similar treatment of animals to render them free of disease or to increase economic value or that of their products.

11. Plants and Animals in whole or any part thereof other than microorganisms but including seeds, varieties and species and eventually biological processes for production or propagation of plants and animals.

12. A computer program per se other than its technical application to industry or a combination with hardware.

13. A mathematical method or a business method or algorithm.

14. A literary, dramatic, musical, or artistic work, or any other artistic creation whatsoever including cinematographic works and television production.

15. A mere scheme or rule or method of performing mental act or method of playing game.

16. A presentation of information.

17. An invention which in effect is traditional knowledge or which is an aggregation or duplication of known properties of traditionally known components or compounds (see Table 1).

STANDARD FOR PATENT DRAFTING

A patent specification draft must have a complete written description of invention along with claims. The description of invention should be like this that a person skilled in that art or from the same field must be able to replicate it after reading the draft. It must completely describe specific embodiments, of the process, machine, manufacture, composition of matter and must also explain the mode of operation or principle wherever applicable. The specification must point out clearly the parts of the process, machine, manufacture, or composition of matter to which the invention or improvement relates. Moreover the description should not be very large, instead it should be confined to spe-

cific improvements and to such parts which are necessarily required to understand the invention.

A typical Patent draft should have the following sections:

1. Title of the Invention.
2. Cross Reference to related applications. (Related applications may be listed on an application data sheet).
3. Background of the Invention.
4. Brief Summary of the Invention.
5. Brief description of the several views of the drawing (if any).
6. Detailed Description of the Invention.
7. A claim or claims.
8. Abstract of the disclosure.
9. Sequence listing (if any).

All the above mentioned sections can be clearly understood by their names themselves. Here the most important section is claims and its drafting. Since it is most important section of a Patent draft we need to explain it further.

How to Write Claims?

The importance of claims can be more understood by the fact that novelty and patentability of invention are judged by claims and patents are granted. Claims form a boundary line of a patent that lets others know what your boundaries are and when they are infringing your rights. Claims are brief descriptions of the subject matter of the invention, eliminating unnecessary details and distinguishing the invention from past inventions related to the same area. Major function of claims is to clearly define the scope and spirit of protection required/granted. The claims are written on a separate page and each claim is written as a single sentence. Only single claim is allowed in utility innovation whereas multiple claims are allowed in patent specifications. However, the multiple claims of a patent must relate to same invention. There are no restrictions on the number of claims in a pat-

Table 1. Various forms at Indian patent office

Form 1	Application for grant of patent.
Form 2	Provisional/complete specification.
Form 3	Statement and undertaking under section 8.
Form 4	Request of extension of Time.
Form 5	Declaration of inventor ship.
Form 6	Claim or request regarding any change in applicant for patent.
Form 7	Notice of opposition.
Form 8	Request on claim regarding of inventor as such in a patent.
Form 9	Request for Publication.
Form 10	Application for amendment of patent.
Form 11	Application for Direction of controller.
Form 12	Request for grant of patent under section 26(I) and 52(2).
Form 13	Application for amendment of the application for patent/complete specification.
Form 14	Notice of opposition to Amendment/Restoration/Surrender of patent/grant of compulsory license of revision of terms there of or to correction of critical errors.
Form 15	Application for the restoration of patent.
Form 16	Application for registration of Title/Interest in a Patent or registration of any document par porting to affect proprietorship of patents.
Form 17	Application of compulsory license.
Form 18	Request/Express request for examination of application for Patent.
Form 19	Application for revocation of Patent for non-working.
Form 20	Application for revision of terms and conditions of license.
Form 21	Request for termination of compulsory license.
Form 22	Application for registration of Patent.
Form 23	Application for restoration of the name in the register of Patent agents.
Form 24	Application for review/setting aside controller's decision/order.
Form 25	Request for permission for making Patent application outside India.
Form 26	Form for authorization of a Patent agent/or any person in a matter of proceeding under the act.
Form 27	Statement regarding the working of patented invention on commercial scale in India.

ent draft; however there are extra fees applicable above some threshold. For example IPO (Indian Patent Office) allows ten claims as the threshold and applicant must pay additional fees if claims are more than ten.

Content of Patent Application

Following are the content of a Patent application as given in "MANUAL OF PATENT OFFICE PRACTICE AND PROCEDURE (Version 01.11) as modified on March 22, 2011" (IPO, 2011).

A patent application shall contain the following:

1. Application for grant of patent in Form-1.
2. Applicant has to obtain a proof of right to file the application from the inventor. The Proof of Right is either an endorsement at the end of the Application Form-1 or a separate assignment.

3. Provisional/complete specification in Form-2.

4. Statement and undertaking under Section 8 in Form-3, if applicable. An applicant must file Form 3 either along with the application or within 6 months from the date of application.

5. Declaration as to inventor ship shall be filed in Form 5 for Applications accompanying a Complete Specification or a Convention Application or a PCT Application designating India. However, the Controller may allow Form-5 to be filed within one month from the date of filing of application, if a request is made to the Controller in Form-4.

6. Power of authority in Form-26, if filed through a Patent Agent. In case a general power of authority has already been filed in another application, a self-attested copy of the same may be filed by the Agent. In case the original general power of authority has been filed in another jurisdiction, that fact may also be mentioned in the self-attested copy.

7. Priority document is required in the following cases:
   a. Convention Application (under Paris Convention).
   b. PCT National Phase Application wherein requirements of Rule 17.1(a or b) of regulations made under the PCT have not been fulfilled. The priority document may be filed along with the application or before the expiry of eighteen months from the date of priority, so as to enable publication of the application. In case of a request for early publication, the priority document shall be filed before/along with such request.

8. Every application shall bear the Signature of the applicant or authorized person/Patent Agent along with name and date in the appropriate space provided in the forms.

9. The Specification shall be signed by the agent/applicant with date on the last page of the Specification. The drawing sheets should bear the signature of an applicant or his agent in the right hand bottom corner.

10. If the Application pertains to a biological material obtained from India, the applicant is required to submit the permission from the National Biodiversity Authority any time before the grant of the patent. However, it would be sufficient if the permission from the National Biodiversity Authority is submitted before the grant of the patent.

11. The Application form shall also indicate clearly the source of geographical origin of any biological material used in the Specification, wherever applicable.

Patents of VLSI Industry Filed in India

In VLSI industry, the patents have a huge horizon. Right from the algorithm level to layout level patents are filed. It's not possible to list out all the patents by this branch i.e. VLSI, still this section provides the basic understanding of type of patents is filed by VLSI industry in India.

Table 2 is a list of few granted patents related to Integrated circuits.

Table 3 is the list of few recently published patents in Integrated Circuit domain.

We have shown few granted patents related to semiconductor memories in Table 4.

We have shown few granted patents related to Integrated Circuits in Table 5.

In the Table 6, we present few Interdisciplinary VLSI technology related patents.

Trends of Patent Filing and Grants in India

Published patents (year wise) having semiconductor or digital or integrated circuit or analog or transistor or system on chip in their abstract (see Table 7).

Table 2. A few granted patents related to digital electronics

SL. No.	Application Number	Patent Number	Title of Invention	Date of Filing(National)
1	94450	94450	A digital circuit arrangement system.	27/06/1964
2	92110	92110	A device for the digital display of data stored in electronic circuits or magnetic matrix memories.	05/02/1964
3	70983	70983	Improvements in or relating to digital summing circuits.	05/03/1960
4	54041	54041	Improvements in circuits for transferring digital information from a first storage circuits.	16/03/1955
5	906/DELNP/2006	249542	"a hardware generator for uniform and gaussian deviates employing analog and digital correction circuits"	21/02/2006
6	1958/KOLNP/2005	238301	Method and digital processing system for automated synthesis of multi-channel circuits	04/10/2005
7	792/CHENP/2004	237717	A digital implementation of multi-channel demodulator circuit for processing a multi-channel analog rf signal	16/04/2004
8	2973/DEL/1997	232347	"a digital circuit architecture for area efficient realization of coefficients"	15/10/1997
9	1367/CHENP/2003	224877	An integrated circuit with analog and digital circuits	29/08/2003
10	2123/DEL/1998	216795	"a digital falltime measurement circuit"	22/07/1998
11	IN/PCT/2001/133/CHE	211942	A digital circuit and a method of communication for a transceiver	29/01/2001
12	933/MAS/1999	200760	A low power digital circuit with enhanced noise immunity	21/09/1999
13	377/CAL/1994	181892	"digital image processing circuitry"	20/05/1994
14	162/MAS/1990	175693	A digital controlled switching circuit	02/03/1990
15	531/MAS/1987	169848	A digital solid-state trip unit for an electrical circuit breker with separable contacts	24/07/1987

Table 3. A few recently published patents related to digital electronics

SL. No.	Application Number	Publication Number	Title of Invention	Date of Filing	Priority Country
1	2280/CHENP/2006	23/2007	Optical flip-flop based read-out arrangement	22/06/2006	EUROPEAN UNION
2	2507/DEL/2008	01/2010	Ultra low power, single edge triggered d-flip/flop (static and dynamic) with reduced transistors	05/11/2008	N/A
3	2519/DEL/2008	32/2011	Ultra low power, double edge-triggered d-flip flop (static & dynamic) with reduced transistors	06/11/2008	N/A
4	2819/DEL/2006	31/2008	'a low power flip flop circuit'	28/12/2006	N/A
5	4114/DELNP/2010	46/2010	"analog scan circuit, analog flip-flop, and data processing device"	09/06/2010	Japan
6	6120/DELNP/2006	35/2007	Phase frequency detector with a novel d flip flop.	19/10/2006	Canada
7	79/DEL/2009	34/2010	Flip-flop circuit with internal level shifter	16/01/2009	N/A
8	8181/DELNP/2009	26/2010	"scan flip-flop with internal latency for scan input"	15/12/2009	U.S.A.
9	8331/CHENP/2010	34/2011	Power saving circuit using a clock buffer and multiple flip-flops	21/12/2010	U.S.A.
10	896/MUM/2010	24/2010	E-t flip-flop	29/03/2010	N/A

Table 4. A few granted patents related to backend VLSI

SL. No.	Application Number	Patent Number	Title of Invention	Date of Filing(National)
1	IN/PCT/2001/00606/DEL	233608	"a semiconductor memory card access apparatus"	06/07/2001
2	IN/PCT/2000/00437/DEL	227539	Semiconductor memory card and data reading apparatus	18/12/2000
3	942/MAS/2000	200712	A semiconductor memory device	03/11/2000
4	35/DEL/1985	162453	Nonvolatile semiconductor memory unit	21/01/1985

Table 5. A few granted patents related to integrated circuits

SL. No.	Application Number	Patent Number	Title of Invention	Date of Filing(National)
1	164/MUMNP/2008	251491	An integrated circuit package and method of packaging integrated circuits to reduce packaging parasitics	30/01/2008
2	1308/MUMNP/2007	245686	An integrated circuit with distributed supply current switch circuits for enabling individual power domains	30/08/2007
3	741/CHENP/2005	241154	System for reducing leakage in integrated circuits	26/04/2005
4	2449/DEL/1997	228263	A process for the fabrication of improved conductor for power thick film hybrid microcircuits/microwave integrated circuits(phmc)	28/08/1997
5	1367/CHENP/2003	224877	An integrated circuit with analog and digital circuits	29/08/2003
6	1094/CHENP/2005	220935	Integrated solid-phase hydrophilic matrix circuits and micro-arrays	01/06/2005
7	1177/DELNP/2003	217703	"viscous protecteive overlayers for planarization of integrated circuits"	29/07/2003
8	IN/PCT/2002/1118/CHE	212834	Method and circuit for providing interface signals between integrated circuits	19/07/2002
9	979/MAS/2001	211081	Method for initializing an integrated circuit and an integrated circuits thereof	05/12/2001
10	216/MAS/2001	197938	A system for high yield and speed enhancement of semicoductor integrated circuits	12/03/2001
11	588/CAL/1995	184391	"method of producing integrated circuits with encased lead frames and apparatus for producing the same"	25/05/1995
12	4/CAL/1979	150616	An integrated circuit stracture particularly for cmos/sos integrated circuits	02/01/1979
13	196/CAL/1975	144488	Improved zener diode for integrated circuits	01/02/1975
14	1769/CAL/1974	142592	A novel device for evaporating sillicon honoxide thin films on glass/alumina substrates used for the fabrication of hybrid integrated circuits	07/08/1974
15	796/CAL/1974	141912	Improved feeder for flash evaporation of nickelchromium powder for the fabrication of thin film hybrid integrated circuits	09/04/1974

In Table 8, we present granted patents (year wise) having semiconductor or digital or integrated circuit or analog or transistor or system on chip in their abstract.

The growth of VLSI technology has impacted every branch of science. Whether it is Biotechnology, Mechanical, Computational etc. each of the branch relies on VLSI to process the jobs easily. Things like pacemaker are proving as miracle by correction of Hearts rhythm using an IC. There are many more interdisciplinary domains where VLSI technology can be applied; hence, there is huge potential of Patents in VLSI industry.

Table 6. A few interdisciplinary VLSI technology patents

SL. No.	Application Number	Patent Number	Title of Invention	Date of Filing(National)
1	82/MUMNP/2005	236949	Micro integrated cardiac pacemaker and cardiac pacing system	28/01/2005
2	2184/CAL/1997	193822	Cardiac output enhanced pacemaker	19/11/1997
3	1152/DEL/1991	185002	"a pacemaker stimulating device for use in the ailment of obesity"	25/11/1991
4	124964	124964	Bifocal demand pacemaker.	22/01/1970
5	121455	121455	Pacemaker to be used internally or externally Of a human to deliver a pressure to stimulate the heart.	21/05/1969
6	925/DEL/1995	191193	"a system for controlling the temperature climate in a variable temperance occupant seat"	23/03/1995
7	884/MUM/2004	210442	Safer car	16/08/2004
8	1466/MAS/1996	193838	Hydraulic vehicle brake system with anti-lock device	20/08/1996

Table 7. Statistics related to VLSI patents published in India

Year	No. of Patents Published	Time duration
2005	2131	Between 1st Jan to 31st December
2006	2135	Between 1st Jan to 31st December
2007	2167	Between 1st Jan to 31st December
2008	2176	Between 1st Jan to 31st December
2009	1294	Between 1st Jan to 31st December
2010	2168	Between 1st Jan to 31st December
2011	2161	Between 1st Jan to 31st December
2012	2145	Between 1st Jan to 15th July

Table 8. Statistics related to VLSI patents granted in India

Year	No. of Patents granted	Time duration
2005	2140	Between 1st Jan to 31st December
2006	2146	Between 1st Jan to 31st December
2007	2166	Between 1st Jan to 31st December
2008	2174	Between 1st Jan to 31st December
2009	2171	Between 1st Jan to 31st December
2010	2168	Between 1st Jan to 31st December
2011	2161	Between 1st Jan to 31st December
2012	2347	Between 1st Jan to 15th July

CONCLUSION

The chapter has presented basic information related to Intellectual Property Rights (IPR). Basic terms, definitions, and scope have been discussed along with their historical importance. In addition, an effort has been made to cover historic aspects as well as different amendments until date. All the different forms of IPR have been covered.

The chapter also provides an insight into IPR for a budding inventor or researcher to know what can be patentable and what cannot. A section has been devoted to patent practices in India covering required forms needed to be filled in order to file a patent. Finally importance of patenting and its growth has been shown with few year wise Statistics.

REFERENCES

Bodenhausen, G. H. C. (1968). *Guide to the application of the Paris convention for the protection of industrial property as revised at Stockholm in 1967*. New York, NY: World Intellectual Property.

Herman v. Youngstown Car Mfg. (1911) Co., 191 F. 579, 584-85, 112 CCA 185 (6th Cir. 1911)

International Intellectual Property Alliance. (2012). *Article 27.1*. Retrieved from http://www.iipa.com

IPO. (2011). *Manual of patent office practice and procedure, version 01.11*. Retrieved from http://www.ipindia.nic.in

Library of Congress. (2009). *A century of lawmaking for a new nation: U.S. congressional documents and debates*. Washington, DC: Law Library of Congress.

MacLeod, C. (2002). *Inventing the industrial revolution: The english patent system, 1660-1800*. Cambridge, UK: Cambridge University Press.

Morsink, J. (1999). *The universal declaration of human rights: Origins, drafting, and intent*. Philadelphia, PA: University of Pennsylvania Press.

Mossoff, A. (2001). Rethinking the development of patents: An intellectual history, 1550-1800. *The Hastings Law Journal, 52*, 1255.

Paris Convention. (2011). *World intellectual property organization (WIPO)*. Retrieved from http://www.wipo.int/

Paynter, H. M. (1990). The first patent. *Invention & Technology*. Retrieved from http://www.me.utexas.edu/~longoria/paynter/hmp/The_First_Patent.html

Ravicher, D. B. (2008). *Protecting freedom in the patent system: The public patent foundation's mission and activities*. Retrieved from http://events.stanford.edu/events/50/5004/

Shapiro, R., & Pham, N. (2007). *Economic effects of intellectual property-intensive manufacturing in the United States*. Retrieved from http://www.sonecon.com/docs/studies/0807_thevalueofip.pdf

Sherman, B., & Bently, L. (1999). *The making of modern intellectual property law: The British experience, 1760-1911*. Cambridge, UK: Cambridge University Press.

Stiglitz, J. (2006). *Making globalization work*. New York, NY: W.W. Norton and Company.

Walsh, J. (2012). *Europe's internet revolt: Protesters see threats in antipiracy treaty*. The Christian Science Monitor.

WIPO. (2011). *Article 32*. Retrieved from http://www.wipo.int/pct/en/texts/articles/a32.htm

World Intellectual Property Organization. (1995). *The first twenty-five years of the patent cooperation treaty (PCT) 1970-1995*. Washington, DC: WIPO Publications.

WTO. (2012). *Article 33*. Retrieved from http://www.wto.org

Chapter 7
Defect Trend Analysis of MI–172 Helicopters Through Maintenance History

Mudassir Hussain
Centre for Advanced Studies in Engineering, Pakistan

Irfan Anjum Manarvi
HITEC University, Pakistan

Assad Iqbal
Bahria University Islamabad, Pakistan

ABSTRACT

MI-172 helicopters are a variant of MI-17/171 helicopters. These helicopters were inducted in the organization, which was originally designed and tuned to the MI-17/171 series helicopters. Since the induction was not premeditated, it resulted in diverse problems that have accumulated over a period of 3 years. This chapter focuses on the defect trend analysis and identification of root causes of the accumulated problems. Technical and flying data is collected from the user Squadron and analyzed by statistical tools. The chapter is helpful in eradicating the existing problems and suggesting pragmatic solutions for an overall improvement of the maintenance setup.

INTRODUCTION

MI-172 is a medium category helicopter of Russian origin. It is a modified and upgraded version of military MI-17 multipurpose helicopter. It is a custom made machine having different engines, avionics suite, safety equipment and interior furnishings etc. Its divergent equipment is a major cause of its diverse defects; which are not being faced by MI-17/171 fleet. The more pronounced causes of unusual faults are varied manufacturing, non-availability of logistics backup, non-existent technical training and poor technical support. Its maintenance and repair operations are quite unusual and time consuming which add towards increased down time.

DOI: 10.4018/978-1-4666-3658-3.ch007

Diagnosis of an unusual fault is mostly ineffective and expensive and at times results in "No Fault Found (NFF)" situations which heavily contribute towards enhanced maintenance costs (Wu, Liu, Ding, & Liu, 2004). A continuous system monitoring is extremely helpful in anticipating the unexpected flaws and their subsequent communication/rectification (Owotoki & Mayer-Lindenberg, 2005). Integrated Diagnostic System (IDS) can be extremely helpful in improving the overall maintenance efficacy at all levels and reduction of uncertainty in the defect isolation process (Wylie, Orchard, Halasz, & Dube, 1997). Aircraft maintenance is an intricate and collective effort, where an individual's blunder may spoil the whole show and reverse the combined achievement (Yanjie, Zhigang, & Bifeng, 2005). The quantitative analysis of Direct Maintenance Man Hour/Flying Hour (DMMH/FH) is represented through a "Spoon-shaped Curve" which divides the aircraft life in four stages i.e. prelim, developing, mature and aging malfunctions (Yanjie, Zhigang, & Bifeng, 2002). Similarly, Airplane Health Management (AHM) is a proactive maintenance approach, which utilizes historical data for reduction of scheduled interruptions and thus enhances overall maintenance performance (Osmanbhoy, Runo, & Mallasch, 2010). The historic maintenance record/information of previous years' defects can assist the maintenance teams to better preplan the impending maintenance actions for their fleet (Manarvi & Umer, 2009). The systems that are critical for serviceability of the aircraft include Power-plant, Airframe, Electrical and Instrument trades and defects of these systems will increase the aircraft down time, burden the maintenance echelons and would require better planning by the logistics departments (Tariq & Manarvi, 2010).

Therefore, the current research is aimed to evaluate the inherent flaws of organizational maintenance system by reviewing the historic record of defects of three years. Data of flying hours and defects is gathered from the user Squadron and converted into appropriate numerical format for statistical analysis and bring out certain consequential inferences.

METHODOLOGY

Flying and defects related data of MI-172 helicopters is composed for a period of three (3) years for the entire fleet of three (3) helicopters. Defects are categorized under the five (5) trades of helicopter i.e. Engine, Airframe, Electrical, Radio and Instrument. The data is used for defect trend analysis, relationship between flying hours and defects and detailed evaluation of root causes of various snags in the maintenance setup. The sequence which is followed for this study is shown as Figure 1.

The data is arranged in Microsoft Office Excel Data Sheet in a tabulated form having the following variables:

1. Date of defect (Month/Year)
2. Details of defect
3. Major trade of defect (Helicopter System)
4. Assessment that whether the encountered defect can arise on MI-17/171 series or not?
5. Is any component replacement involved in the rectification process or not?
6. Irrespective of the fact that whether any component is used for rectification or not, is it available in the stock?
7. Whether the requisite Technical Manual is available in the maintenance setup?
8. Whether the requisite Technical Expertise is available in the maintenance setup?

The collected data is converted into numerical form for an easy identification by statistical tools and subsequent analysis. The statistical analysis is performed through MS Excel and StatPro and inferences are drawn for an overall betterment of the maintenance setup. The analysis and findings are expected to positively contribute towards helicopter maintenance efforts at an international forum.

Figure 1. Sequence diagram for the research

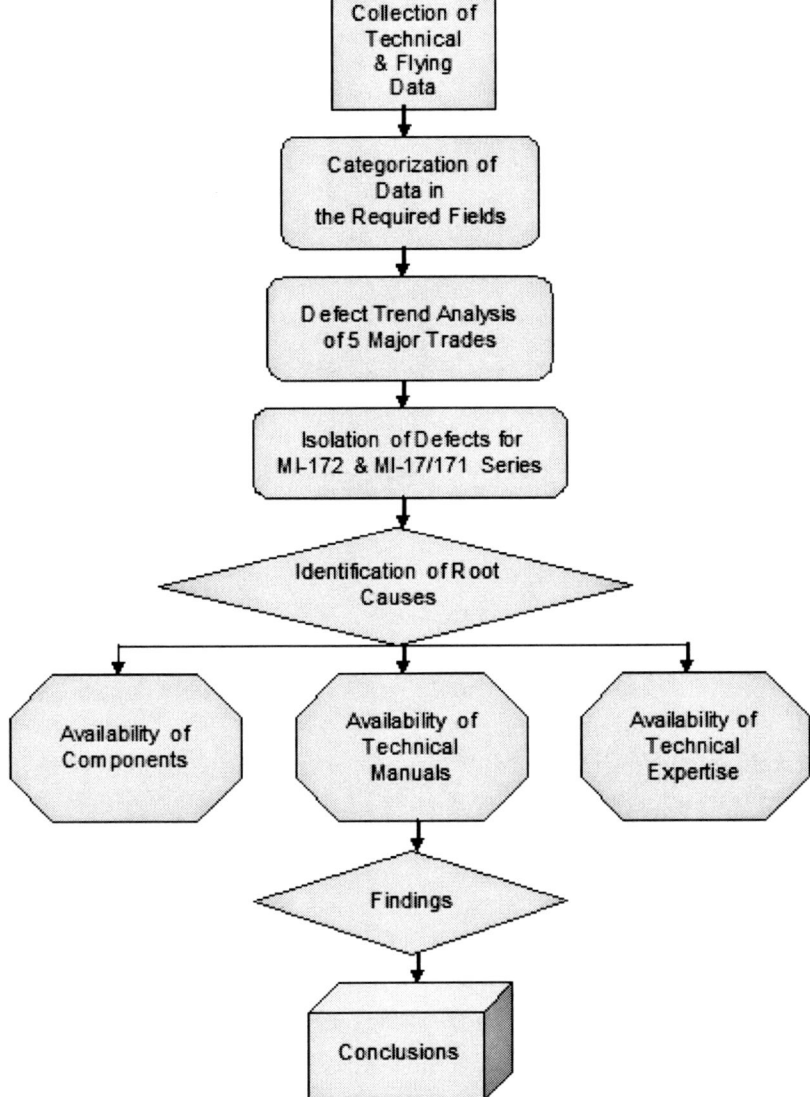

DEFECT TREND ANALYSIS

Categorization of Defects

For defect categorizing, aircraft historic record i.e. PAF-EME-E (53) and PAF-EME-E (54) are thoroughly scrutinized. All substantial aircraft defects are included in the analytical data for three (3) years, i.e. from 1 August 2008 to 31 July 2011. The defects are tabulated according to the date (month/year) and major helicopter systems (trades) i.e. power-plant, airframe, electrical, radio, and instruments. A total of 161 records have been collected and analyzed in this chapter. Initially, total defects of all trades for a year have been summed up and plotted as shown in Figure 2.

The following is inferred from the Figure 2:

- In year 1 (1 August 2008 to 31 July 2009), a total number of 41 defects of all trades

Figure 2. Yearly summary of aircraft defects

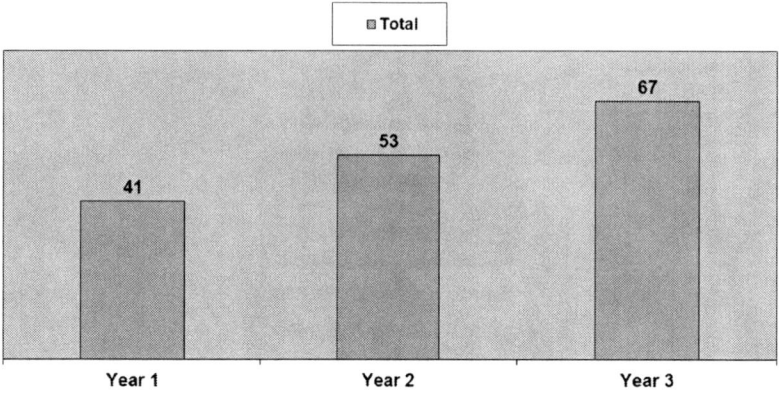

have been experienced. In year 2 (1 August 2009 to 31 July 2010), there are 53 defects of all trades in total. In year 3 (1 August 2010 to 31 July 2011), 67 defects of all trades are faced.

With passage of time, there is a fault growth tendency for MI-172 helicopters and it can be initially inferred that aircraft ageing is a contributory factor towards this ascending trend (positive upwards inclination).

After this elementary study, the defects have been filtered according to the five (5) basic helicopter trades, i.e. power-plant, airframe, electrical, radio and instrument for further analysis. Then, defects for each trade are summed up for an overall period of 36 months and plotted as a pie chart as shown in Figure 3.

The following is inferred from the Figure 3:

1. In the overall period of 36 months, the highest number of defects occurred in the airframe system which is 32% of 161 defects. It is closely followed by instrument system which is 29%. Then the radio, engine and electrical systems are in the descending order.

2. In the overall period of 36 months, electrical system has been recognized as the most unfailing in comparison to the other trades i.e. 10% only.

Afterwards, the already filtered defects of each trade are year-wise compared to find out some useful deductions. Total defects of each trade for a particular year are summed up and side by side plotted as shown in Figure 4.

Figure 3. Three years' summary of each trade defect

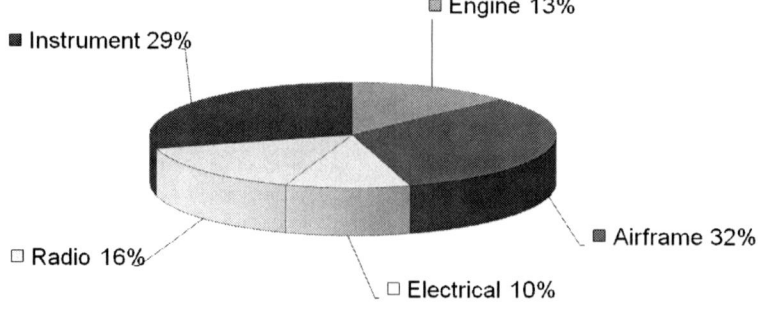

Figure 4. Yearly analyses of each trade defects

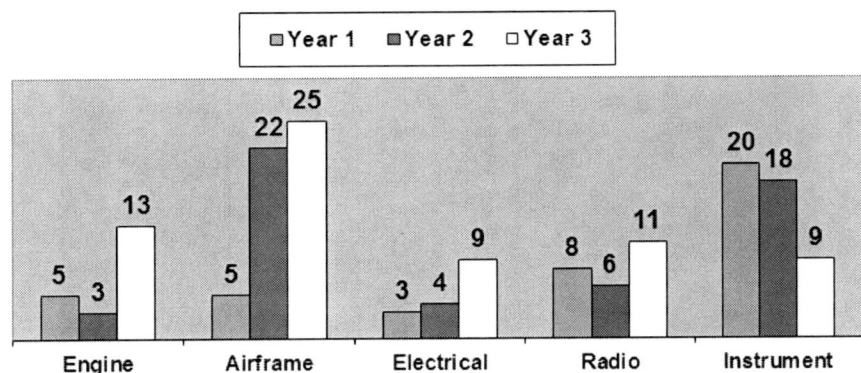

The following is deduced from Figure 4:

1. In year 1, the highest number of defects is observed in the instrument system, which is correspondingly followed by radio, airframe, engine and electrical systems.
2. In year 2, the highest number of defects is encountered in the airframe system, which is followed by instrument, radio, electrical and engine trades respectively.
3. Again in year 3, number of airframe defects remained high; which are followed by engine at number 2 and radio at number 3 positions respectively. Instrument and Electrical systems defects remained the lowest.
4. In the overall period of 3 years, electrical system has less failing tendency and is found to be the most dependable.
5. Over a period of three (3) years, airframe system defects remained at number 1 position (52) and are closely followed by instrument system defects (47).
6. The numerical/graphical representation clearly demonstrates that somehow or the others, the operator/maintenance crew remained helpless to arrest the rising defects of airframe and instrument systems. It is perceived that the reasons to this shortfall could be non-availability of requisite logistics support or technical proficiency of the maintenance personnel.

These defect trend deductions are further analyzed in comparison with flying hours so as to evaluate certain correlations with each other.

DEFECTS VIS-À-VIS FLYING HOURS

Month-wise data of flying hours is noted from aircraft historic record i.e. PAF-EME-E (52). The data is gathered for 36 months corresponding to the defects data, so as to find out certain relevant connection between the two (2) variables. Initially, total flying hours for a year are summed up and plotted as shown in Figure 5.

The following is deduced from Figure 5:

1. In 1ˢᵗ year (1 August 2008 to 31 July 2009), a total of 266.7 hours have been flown. In 2ⁿᵈ year (1 August 2009 to 31 July 2010), a total of 576.8 hours have been flown. In 3ʳᵈ year (1 August 2010 to 31 July 2011), a total of 716.5 flying hours have been generated.
2. The plot crystallizes that every passing year has brought a remarkable increase in the total number of flying hours for MI-172 helicopters which is a positive upwards inclination.

Since the data is collected on monthly basis; hence it is considered appropriate to further sift this data, so as to discover certain peculiar flying trends at a particular point of the year. For this

Figure 5. Yearly summary of flying hours

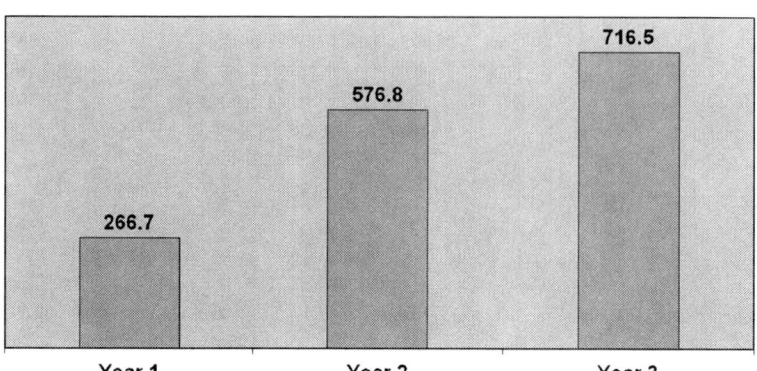

purpose, data of all three (3) years is analyzed for corresponding months and the resulting graph is shown below:

The following results have been deduced from Figure 6:

1. In 1st year, the lowest flying hours have been generated in the months of February and March. The highest figure of flying hours is noted for the month of December and for remaining months of the year, the trend remained variable.

2. In 2nd year, the lowest flying hours have been generated in the months of September and March. The highest figure of flying hours is recorded for the month of April and for remaining months of the year, the trend remained variable.

3. In 3rd year, the lowest numbers of flying hours have been produced in the months of November, February and July. During this year, the highest figure of flying hours is endorsed for the month of August and for remaining months of the year, again the trend remained variable.

4. The preceding analysis depicts that there has been no set outline of flying hours for any particular part of a year. It is therefore deduced that the flown hours may be dependent on the released flying task, impending mission requirements and availability (serviceability) of the helicopters.

Figure 6. Data of monthly flying hours

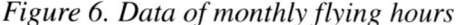

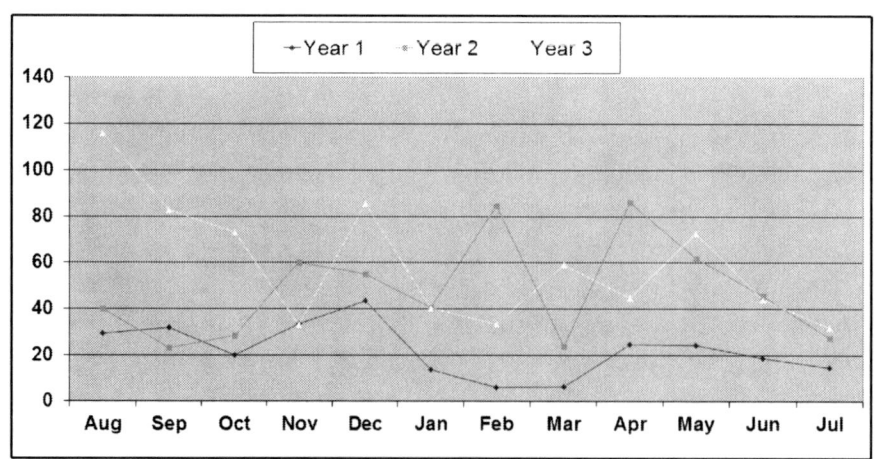

After these fundamental analyses of flying hours, a comparative study of defects vis-à-vis flying hours is accomplished and a line chart is drawn to identify the relationship between these two major variables. This line chart is shown as Figure 7.

Figure 7 provides the following results:

1. For the analyzed data of three (3) years, there is a corresponding increase in the number of defects and generated flying hours.
2. With passage of time, the trend lines for defects and flying hours present a matching upward shift. It provides a hint that both these variables are closely linked with each other.
3. Initially, on the basis of Figure 2, it has been deduced that ageing of aircraft is a contributory factor towards the ascending tendency of faults. However, Figure 7 helps to refine that uphill trek of faults is not dependent on ageing aspects only rather increase in aircraft operation is also equally responsible to this shift.
4. Value of correlation between the number of defects and flying hours is calculated by using StatPro scatter plot. The value comes out to be 0.966; approaching 1; which shows an almost perfect correlation between the selected variables over the year-wise sifted data.

Now, it is an appropriate time to find certain close association between the number of flying hours and defects for cumulative values of months for three (3) years data. For example, flying hours and defects of January occurring in all three (3) years are summed up and analyzed through stacked line excel chart as shown in Figure 8.

The following deductions are drawn from Figure 8:

1. For cumulative month-wise analysis of flying hours and defects; the stacked trend lines follow a close and uniform path.
2. The month-wise trends of defects versus flying hours help to confirm our previous deductions which are based on year-wise comparison of the same two (2) variables.
3. It is now established that a direct and positive correlation exists between these two (2) variables. This correlation was 0.978. The value is pretty close to 1 which helps to identify that an almost perfect relationship exists between the number of generated flying hours and defects.

Figure 7. Defects viz-a-viz flying hours

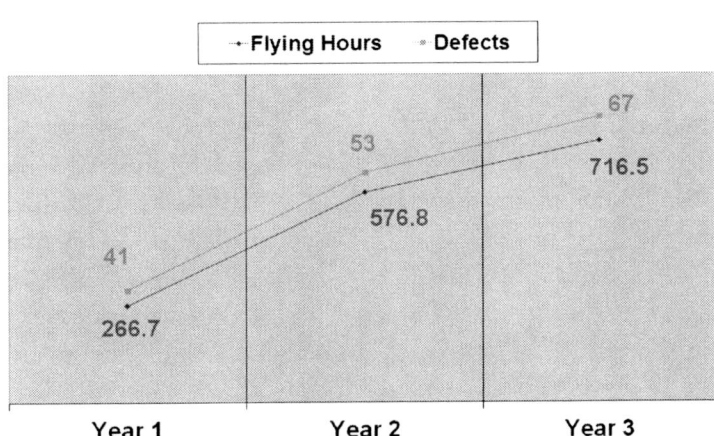

Figure 8. Cumulative monthly defects viz-a-viz flying hours

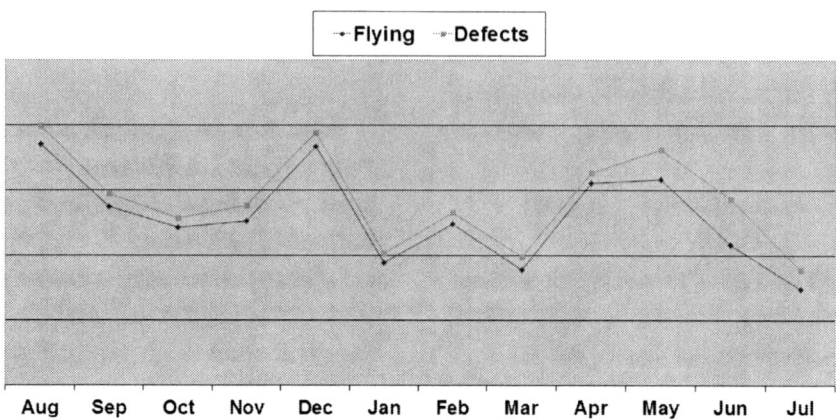

At this point of time, it is considered feasible to evaluate the effect of each major system on overall defects so as to further refine the defect trend analysis.

MAJOR SYSTEMS' DEFECT ANALYSIS

The five trades which are also termed as major helicopter systems include Power-plant, Airframe, Electrical, Radio and Instrument. These are imperative for serviceable status of the helicopter. Their defects can endow longer down time and enhance maintenance burden. The further analysis includes segregation of all trade defects and years through separate charts as illustrated in Figures 9-11 and lead to the subsequent deductions:

1. During 1ˢᵗ year (1 August 2008 to 31 Jul 2009), minimum defects occurred in the Electrical system and maximum in the Instrument system.

2. During 2ⁿᵈ year (1 August 2009 to 31 Jul 2010), minimum defects occurred in the Power-plant system whereas maximum were recorded for the Airframe system.

3. During 3ʳᵈ year (1 August 2010 to 31 Jul 2011), minimum defects occurred in the Electrical and Instrument systems whereas maximum are noted for the Airframe system again.

4. In the overall period of three (3) years, minimum defects are encountered by the Electrical system and it proved to be the most dependable system. For an overall period of 36 months, there are 24 months (2/3ʳᵈ spectrum) in which no defects are faced by this system. As a matter of fact, certain cogent reasons exist for this system to be so dependable over the entire spectrum. An insight in the system illustrates that it is much similar to the electrical system of MI-17/171 series; hence, it is better maintained during scheduled inspections.

5. The highest number of defects is observed in the Airframe system and it remained at rank one (1) for two (2) consecutive years. A comparatively large value occurred in February 2010 and the highest crest appeared in June 2011. It is imperative to argue that Airframe system is supposed to be the most dependable fragment, but it illustrated the most failing trend which in-fact is owing to a few occurrences which resulted in damage to the structure.

Figure 9. Monthly defects of each trade for year 1

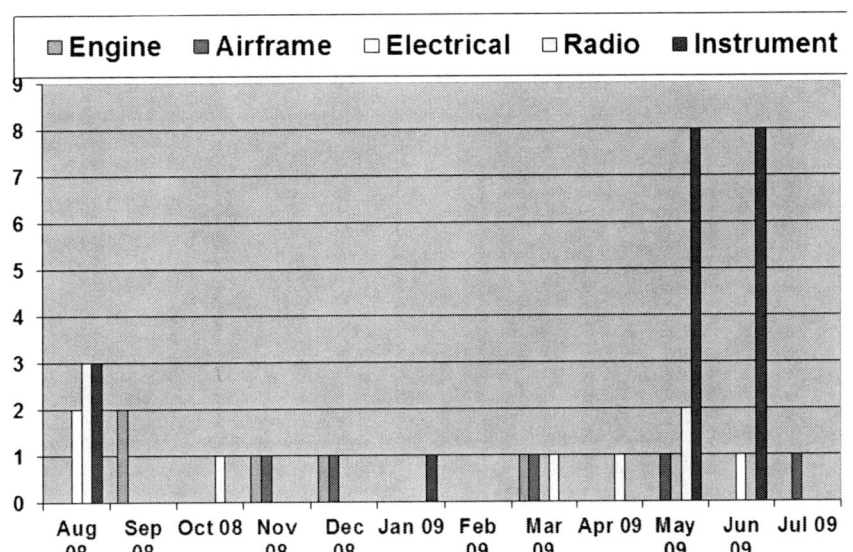

Figure 10. Monthly defects of each trade for year 2

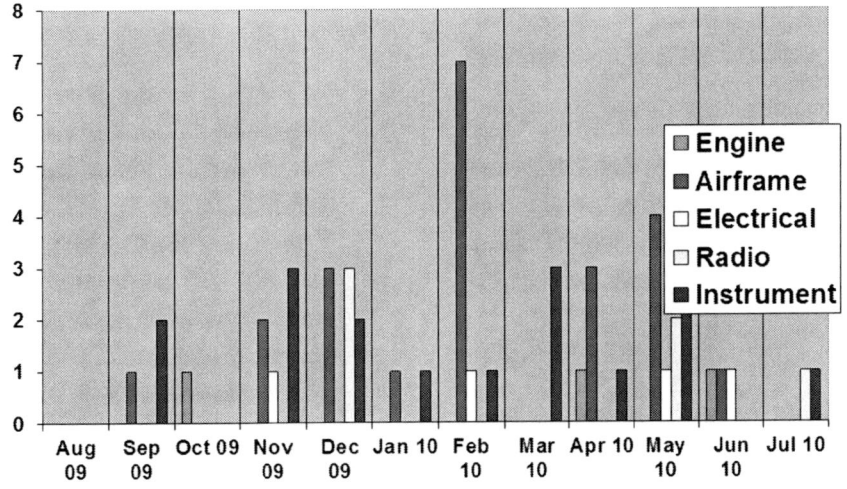

6. Instrument system remained at rank two (2) in terms of highest number of defects and the highest crests are observed in May and June 2009. Instruments are supposed to be a highly dependable aspect of flying machines especially when IFR flying is carried out where no direct visual connection exists with the ground. In MI-172 helicopters, such a high failure rate is alarming and necessitates a perceptive analysis to assess the actual root causes.

7. Radio and Engine systems showed a mediocre trend over the entire period. A small peak in Engine system defects is noted for June 2011, which is owing to an occurrence.

8. The defects' trend lines of five (5) helicopter systems over a period of three (3) years are heterogeneous. These prohibit any infer-

Figure 11. Monthly defects of each trade for year 3

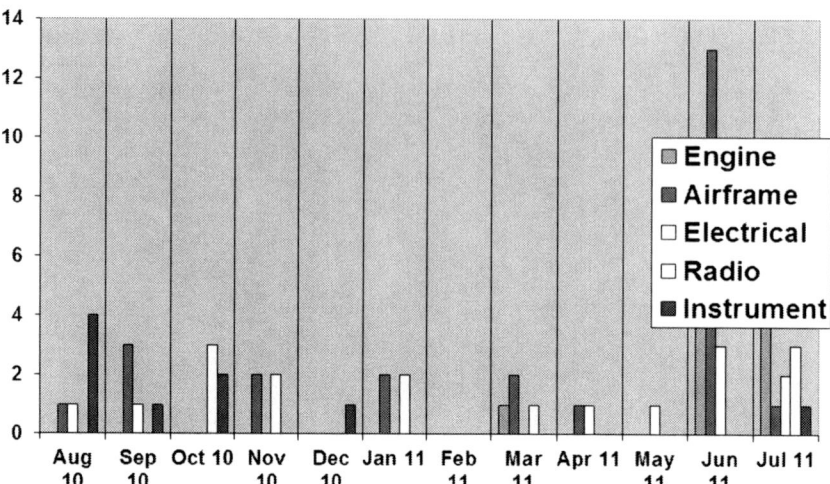

ences, which can be applied to arrest the defect tendencies. These can neither be associated with the extreme climatic conditions nor with the assorted terrain circumstances; hence offer absolutely asymmetrical tendencies.

Reaching this phase, it is recognized that further refined analysis is required to pinpoint the root causes of various trades' defects. Identification of these snags can positively contribute towards the organization.

IDENTIFICATION OF ROOT CAUSES

Isolation of Major System Defects for MI-172 and MI-17/171 Helicopters

It is initially discussed that MI-172 helicopters are a custom product and possess quite diverse equipment than their descendent MI-17/171 helicopters. Here, it is essential to establish that how this diverse equipment contributes to a high rate of defects and increased downtime. In the master data sheet, the faults which can arise on MI-17/171 series helicopters are differently coded for isola-

tion. These defects are then segregated by using Histogram characteristics of StatPro as below:

The following is inferred from Figures 12 and 13:

1. According to the given coding; it can be established from the Histogram that there are 95 defects which can exist both on the MI-172 & MI-17/171 series helicopters. There are 66 faults which are associated with the MI-172 helicopters only. This huge disparity is owing to the deviating equipment installed on MI-172 Helicopters.
2. A pie chart is drawn to represent the assessed frequency and it is shown that there are 41% defects which can occur on MI-172 helicopters only.
3. These figures provide guidance that a separate maintenance setup is required to handle the diverse defects and enhance serviceability status of these contrasting machines.

Identification of Defects Involving Component Replacement

Again, StatPro's Histogram characteristic is used to isolate the defects which involve component

Figure 12. Isolation of defects on MI-172 and MI-17/171 series helicopters

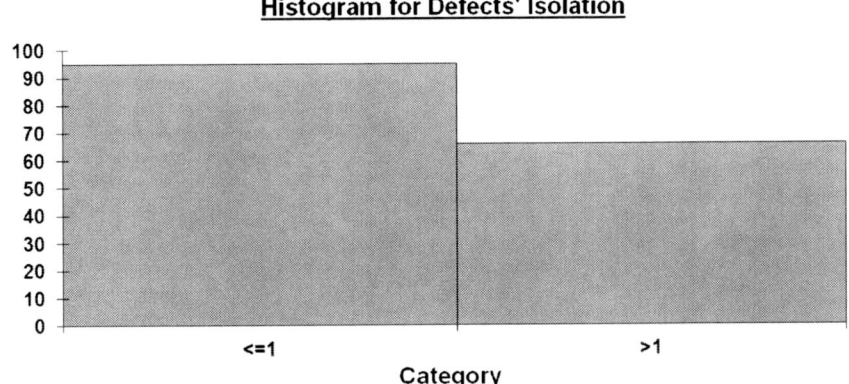

replacement. The results are shown in Figures 14 and 15.

The following is deduced from Figures 14 and 15:

1. The assigned coding for Histogram depicted that there are 38 defects which involve component replacement. The remaining 123 defects involve no component replacement.
2. In terms of percentage, 24% defects involve replacement of components.

Requirement of Components

It has already been established that MI-172 helicopters; by virtue of varying defects; require a different maintenance setup. However, nothing regarding requirement of logistics support (spare stock) has yet been highlighted. In order to comment on this aspect; a further analysis of spare stock ex MI-172 helicopters is carried out. For this analysis, it is assumed that every defect occurring on MI-172 helicopters necessitates a component replacement and hence availability has been assessed for each defect as depicted below:

The following is inferred from Figures 16 and 17:

1. The assigned coding helps to calculate that out of 161 required components (161 defects); 67 components are available from MI-17/171 stock and 94 are not available.
2. In terms of pie chart, for 161 encountered defects, there are only 42% components available and 58% do not exist in the logistics support.
3. As a matter of fact, since no planning exists for the maintenance support, the same holds good for logistics backup as well.
4. Due to non-availability of components; for rectification of certain identified defects; the components are interchanged within the fleet.

Figure 13. Percentage of defects on MI-172 and MI-17/171 series helicopters

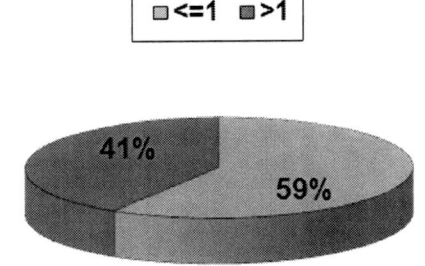

Figure 14. Isolation of defects involving component replacement

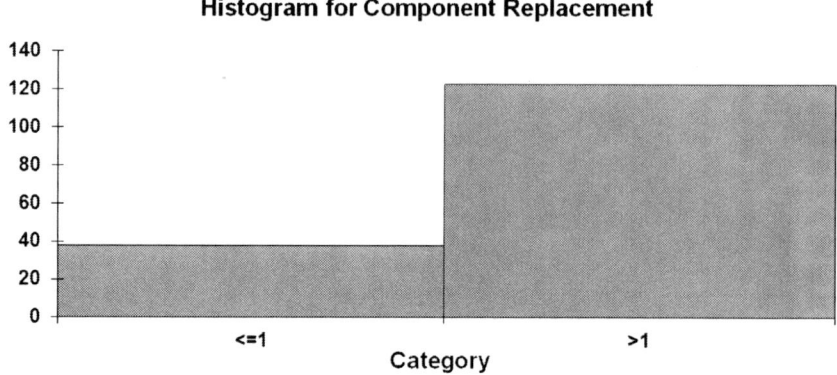

Histogram for Component Replacement

Requirement of Technical Manuals

Aircraft maintenance actions are largely dependent on Technical Manuals (TMs). TMs provide troubleshooting guidelines and guarantee the resultant corrective actions. Their non-availability is catastrophic for a specific maintenance setup; hence, a comprehensive analysis of their availability is required for MI-172 helicopters. Data is gathered and converted in graphical form as below:

The following is inferred from Figures 18 and 19:

1. The assigned coding helps to calculate that against 161 encountered defects; TMs are available for 93 only. For 68 defects, there are no authentic guidelines available in the organization.
2. As per MS Excel pie chart representation, there are no technical publications available against 42% defects, which is a huge impediment to the entire maintenance setup.
3. Thus non-availability of TMs adds to the already identified grey areas i.e. non-availability of a comprehensive maintenance setup and logistics backup.
4. Even with non-availability of TMs, certain defects are being rectified through hit & trial or previous expertise.

Requirement of Technical Training

Training of technical personnel is an essential fragment of an efficient maintenance setup. In order to assess the acquired expertise level of technicians against the existing defects, a statistical analysis was performed. The results are depicted below:

The following is deduced from Figures 20 and 21:

1. The histogram depicts that out of 161 defects; 84 can be resolved through inherent technical training whereas 77 are diverse to the existing MI-17/171 series and require dedicated training.

Figure 15. Defects involving component replacement

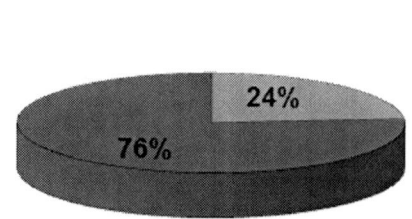

Figure 16. Availabilty of components

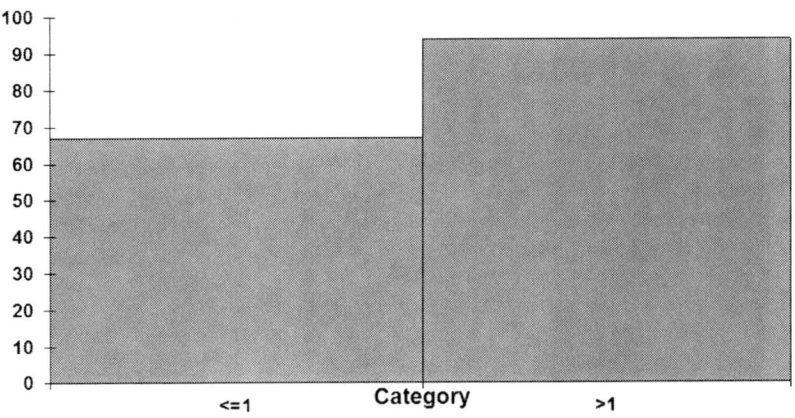

2. In terms of percentage; 48% defects; owing to peculiar nature, demand a separate training or expertise for an efficient resolve.

3. This inference adds to the previously identified list of concerns and brings another prime root cause in lime light.

4. Even with non-availability of technical expertise, certain defects are being rectified through previous expertise or trial & error techniques.

MI-172 defect trend analysis and identification of principal issues is based on a comprehensive data of 36 months and it helped to arrive at the following findings:-

FINDINGS

The aforesaid research is carried out to discover the defect trend analysis and root causes of major snags in the maintenance setup of MI-172 helicopters. It has assisted to reach at the following findings:

1. The research gives a comprehensive defect trend analysis of MI-172 helicopters for the five major trades i.e. Engine, Airframe, Electrical, Radio, and Instrument systems.

It provides a lead for a similar research for other helicopters of the organization.

2. On the basis of this research, the operators can review their General Maintenance Practices (GMPs). It can assist in accurate and swift defect isolation, prohibit "No Fault Found (NFF)" situations and render planning for the scheduled or One Time Inspections (OTIs).

3. In the overall period of 3 years, Electrical system is found to be the most dependable system with less failing tendency.

4. In these 3 years, defects of Airframe system remained at number 1 position and were closely followed by Instrument system defects.

Figure 17. Availabilty of components

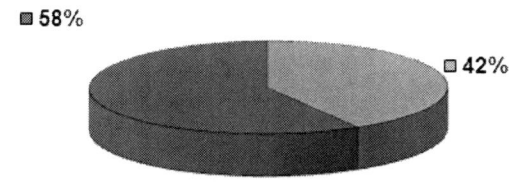

Figure 18. Availabilty of technical manuals

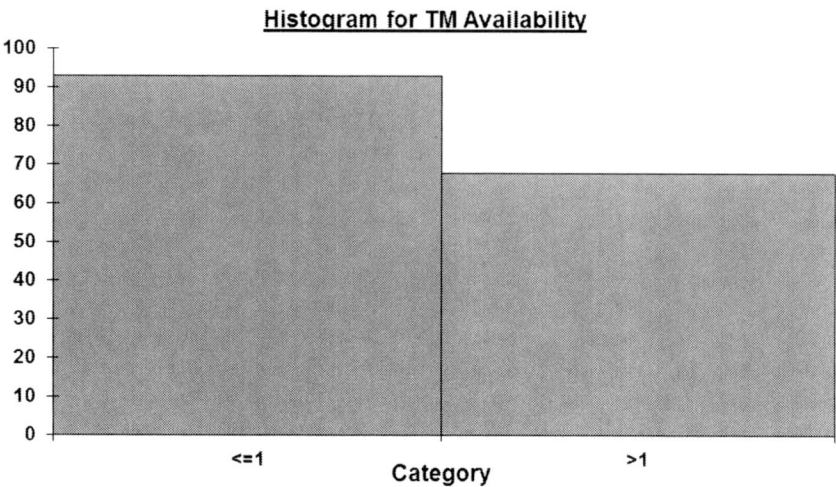

5. Increased defects of Airframe system are owing to certain occurrences; otherwise the system is defiant to frequent failures.

6. With every passing year, a defect growth tendency exists for the MI-172 helicopters. Similarly, with passage of time, a positive upward inclination is also noted in the flying hours.

7. Uphill travel of defects, with passage of time, is largely dependent on the increase in aircraft operating hours and ageing aspects.

8. There is a corresponding upward shift in the trend lines for defects and flying hours. It indicates that these variables are closely interlinked. It is therefore established that a direct and positive correlation exists between them.

9. A comparative analysis suggests that there are 41% defects, which can occur on MI-172 helicopters only (not on MI-17/171 helicopters). Thus, a separate maintenance setup is required for enhanced serviceability of these diverse machines.

10. There are 24% defects, which involve replacement of components. On the overall scale of defects, only 42% components are available (MI-17/171 helicopters components cannot be used). Thus, it necessitates a separate and comprehensive logistics backup to maintain the inventory level of requisite components for the recurring faults.

11. There are 42% defects for which no technical publications are available. It is a huge impediment to the maintenance setup and requires an immediate resolve.

Figure 19. Availabilty of technical manuals

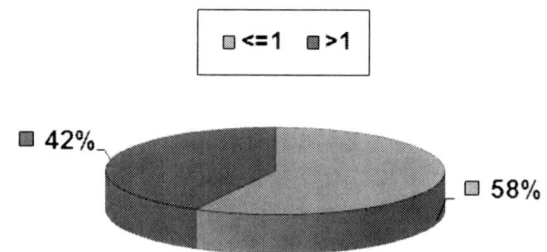

Figure 20. Availabilty of technical expertise

Histogram for Technical Training

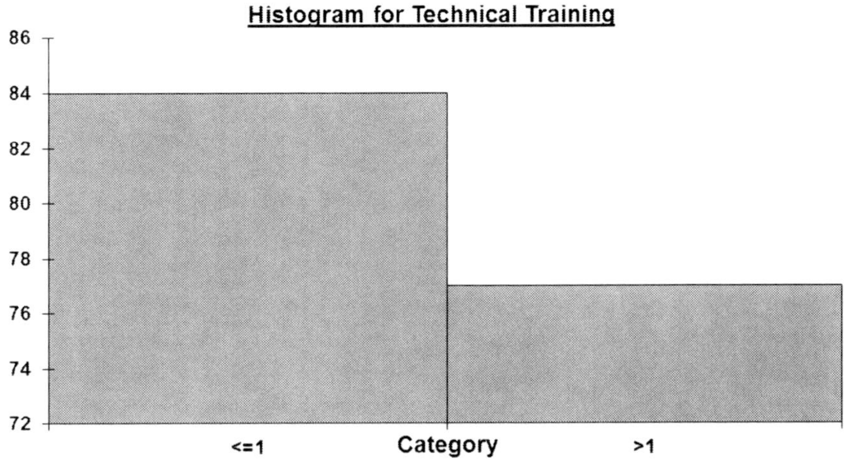

12. There are 48% defects, which cannot be resolved through the held expertise. For increased maintenance efficiency, a dedicated technical training is a foremost requirement.

It is expected that findings of this research chapter can be extremely helpful to eradicate the system flaws and improve it in totality. This study can be further extended to make it more practical for accurate futuristic predictions regarding impending defects and components' failures.

CONCLUSION

Statistical results, logical deductions, and analytical findings of this research chapter provide a realistic defect trend analysis of 5 major trades of MI-172 helicopters. An insight to the principal issues of the maintenance setup is also presented through numerical results. The foremost identified impediments include non-availability of components, TMs and technical expertise. The chapter suggests that a separate maintenance setup is re-

quired to address the identified issues. The research emphasizes that an unplanned new induction can heavily burden an aviation maintenance setup; resulting in irresolvable issues. This research can be extended to identify the impact of sub systems and components over the defect trend analysis of major systems. Moreover, it can be segregated on the basis of tiers of maintenance setup. Better statistical/decision making techniques; like Palisade software; can be used by the researchers to establish deeper links between the systems, sub systems, associated maintenance requirements and even calculation of maintenance costs as well.

Figure 21. Availabilty of technical expertise

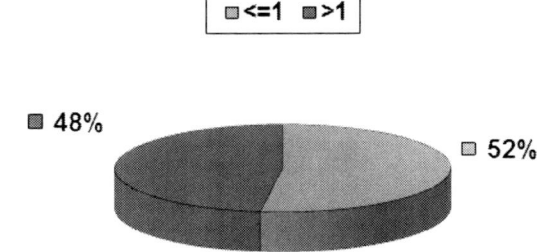

REFERENCES

Manarvi, I. A., & Umer, W. (2010). Analyzing the defects of C-130 Aircraft through maintenance history. In *Proceedings of 2010 IEEE Aerospace Conference (AERO 2010),* (pp 1-7). Big Sky, MT: IEEE Press.

Osmanbhoy, M. Z., Runo, S., & Mallasch, P. (2010). Development of fault detection and reporting for non-central maintenance aircraft. In *Proceedings of 2010 IEEE Aerospace Conference (AERO 2010),* (pp. 1-7). Big Sky, MT: IEEE Press.

Owotoki, P., & Mayer-Lindenberg, F. (2005). Comprehensible hierarchical intelligent (CHI) framework for monitoring and preventive maintenance of aircraft systems. In *Proceedings of Third International Workshop on Intelligent Solutions in Embedded Systems,* (pp. 175-184). IEEE.

Qi, Y., Lu, Z., & Song, B. (2002). New concept for aircraft maintenance management: New cognition for aircraft maintenance study in R&M field of China. In *Proceedings of Reliability and Maintainability Symposium,* (pp. 401–405). IEEE.

Qi, Y., Lu, Z., & Song, B. (2005). New concept for aircraft maintenance management: The "dolphin curve life cycle model" of a typical repairable system. In *Proceedings of Reliability and Maintainability Symposium*, (pp. 533-538). IEEE.

Tariq, A., & Manarvi, I. A. (2011). Defect tend analysis of F-7P aircraft through maintenance history. In *Proceedings of 2011 IEEE Aerospace Conference (AERO 2011),* (pp. 1-8). Washington, DC: IEEE Computer Society.

Wu, H., Liu, Y., Ding, Y., & Liu, J. (2004). Methods to reduce direct maintenance costs for commercial aircraft. *Aircraft Engineering and Aerospace Technology, 1*(1), 15–18. doi:10.1108/00022660410514964

Wylie, R., Orchard, R., Halasz, M., & Dube, F. (1997). IDS: Improving aircraft fleet maintenance. In *Proceedings of 14th National Conference on Artificial Intelligence and Innovative Applications of Artificial Intelligence,* (pp. 1-8). IEEE.

Section 2
Project Management

Chapter 8
Component Failure Analysis of J69–T–25A Engine

Muhammad Asim Qazi
Center for Advanced Studies in Engineering, Pakistan

Irfan Manarvi
HITEC University Taxila, Pakistan

Assad Iqbal
Bahria University Islamabad, Pakistan

ABSTRACT

Reliability and serviceability of jet engines in the aviation industry is of paramount importance and is directly related to flight safety. Tight maintenance programs, including scheduled and preventive inspection are in place worldwide for jet engines to ensure air worthiness of aircraft. Old age provides maintenance maturity to the system, but on other hand, it requires focused efforts to ensure reliability due to aging factor. J69-T-25A falls in the same category, as it has been in service for the last six decades. Despite all maintenance efforts, a variety of defects are being faced on J69 engines. The major defects include RPM fluctuation, noise, oil gain, vibration, and smoke. The troubleshooting process identifies a number of components that cause these problems. this chapter is based on statistical analyses of component failure in terms of frequency and fault isolation. The top ten components were selected based upon failure rates and were compared against reported problems to establish a relationship between defects and failed components. Based upon the result, various remedial measures are suggested to reduce defects in the future and increase engine reliability.

INTRODUCTION

J69-T-25A engine was designed by Continental Aviation Engineering (CAE). This is a robust designed jet engine which has been in service

for more than six decades. The engine has been utilized on many air vehicles, but most prominent and famous installation is on T-37 aircraft. The aircraft has been in service with numerous Air Forces around the world to meet the basic flying training needs. The aircraft has been called "Tweety Bird" because of engine sound.

DOI: 10.4018/978-1-4666-3658-3.ch008

The engine has served to its best performance during its six decades of service. Noise is the known weak area for this engine, which requires extra efforts to overcome. RPM fluctuation and vibration are also the known weak areas. Maintenance of known and unknown defects requires a healthy amount of finance in terms of man hours and spares support. This makes an important part of Life Cycle Cost (LCC) (Khan & Manarvi, 2010).

Statistical and Trend Analysis of the regularly collected defects data is of foremost importance in aircraft and its components management system, and in estimating Mean Time Between Failure (MTBF), Mean Time To Repair (MTTR), and production Delays (Khan & Manarvi, 2010; Manarvi & Umer, 2009). Bath Tub Curve Model and Spoon Shaped Curve Model have provided the different phases of component life. These have also provided relationship between product life and maintenance from angle of its maintainability (Manarvi & Umer, 2009; Qi, Lu, & Song, 2005). The cost of scheduled maintenance is less than unscheduled maintenance cost (Kumar, Croker, & Knezrvic, 1999). Efforts can be made to improve traditional maintenance policies and practices to overcome unscheduled removals and to enhance components life, which will result in more engine operating hours.

The concept of Defect Prevention (DP) can be used effectively to reduce unscheduled engine removals. In this data of present defects is analyzed to suggest preventive measure for future (Jalote & Agarwal, 2012). Scheduled inspections are planned inspections; therefore, these do not affect MTBF or MTTR whereas unscheduled removals or premature failures do have direct impact in MTBF or MTTR (Younus & Manarvi, 2010). Unscheduled removals or premature failures play an important role in enhancing or reducing reliability.

METHODOLOGY

The defects were collected from Log Books of engines for last two and half year (30 months). This data selection was done intentionally, with the aim to forecast and suggest remedial measures for remaining six months of Year 2011 through data analysis. Hours from last inspection and total operating life of top ten failed components was also collected from Log Books. The collected data was based upon following information:

1. Date on which defect was reported.
2. Defect category and nature.
3. Root cause identifying defective component.
4. Reason for component failure.

Total failures for each year were calculated. All scheduled and unscheduled engine removals were taken into consideration and in total 32 different components were identified. Top ten failed components were selected for analysis. The factors, affecting performance of components or contributing towards failure of component, like operating hours and operating conditions were used for failure analysis. During component failure analysis flight hours, schedule, and unscheduled removals cannot be over looked. Any failure during schedule inspection is expected that is the reason for schedule inspection after specified interval, whereas unscheduled removal is the major stakeholder for component failure analysis. The unscheduled removals help to decide component life, remedial measures, redefining inspection intervals, One Time Inspections (OTIs) and other steps to arrest the trend.

IDENTIFICATION OF DEFECTIVE COMPONENTS

The Log Books of the engine were reviewed. The year wise data for scheduled removals and unscheduled removals was segregated. Figure 1

Figure 1. Year wise unsch, sch, and total removals

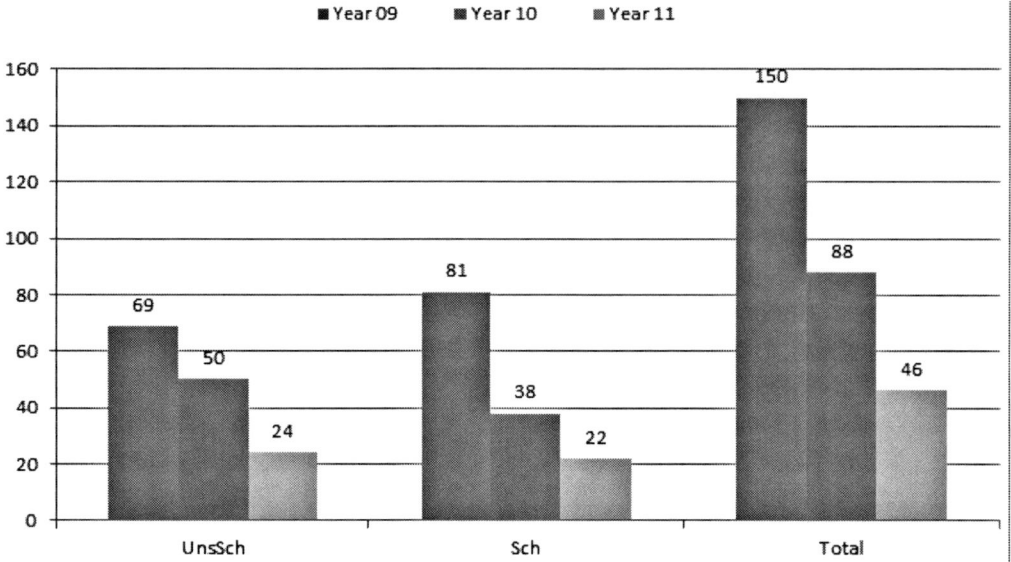

describes the year wise schedule and unscheduled removals, whereas Figure 2 describes the total flying for the year during which failures occur.

Analyzing Figures 1 and 2 in parallel following conclusion can be drawn:

1. Total engine removals, including schedule and unscheduled removals, in year 2009 are more as compared to other two years. At the same time flight hours for Year 2009 are also more as compared to other two years.
2. The correlation between flight hours and unscheduled removals is 0.98. This approximately shows a linear behavior between both variables. Therefore, it can be concluded that increase in flight hours will increase unscheduled engine removals.
3. The correlation between flight hours and schedule removals is 0.854. This is not a linear relationship between both variables. Increased flight hours do affect the schedule engine removals but engine removals may fall in next flight year.
4. On six monthly basis comparison, Year 2011 is the best because with second highest

flight hours it has less number of scheduled and unscheduled removals. This will subsequently result in less component failures.

The failed components were identified and then total failures per year were calculated. Total numbers of failures were plotted for each year as shown in Figure 3.

The following inference can be drawn from all above three graphs:

1. Fuel Control Unit (FCU) has the highest failure rate. Even in the 2011, where the data is only for first six months and flight hours are less as compared to 2009-10, it has the highest failure rate.
2. The failure rate of all the components is high in year 2009 because of more flight hours and more number of inspections.
3. Component failure rate for complete Year 2009 is high because in total 150 (81 schedule and 69 unscheduled) engine were inspected.
4. Year 2009 remains at the high even if we compare removals per 1000 flight hours. Table 1 provides the removal per 1000 hour.

Figure 2. Year wise 6 monthly and complete flight hours

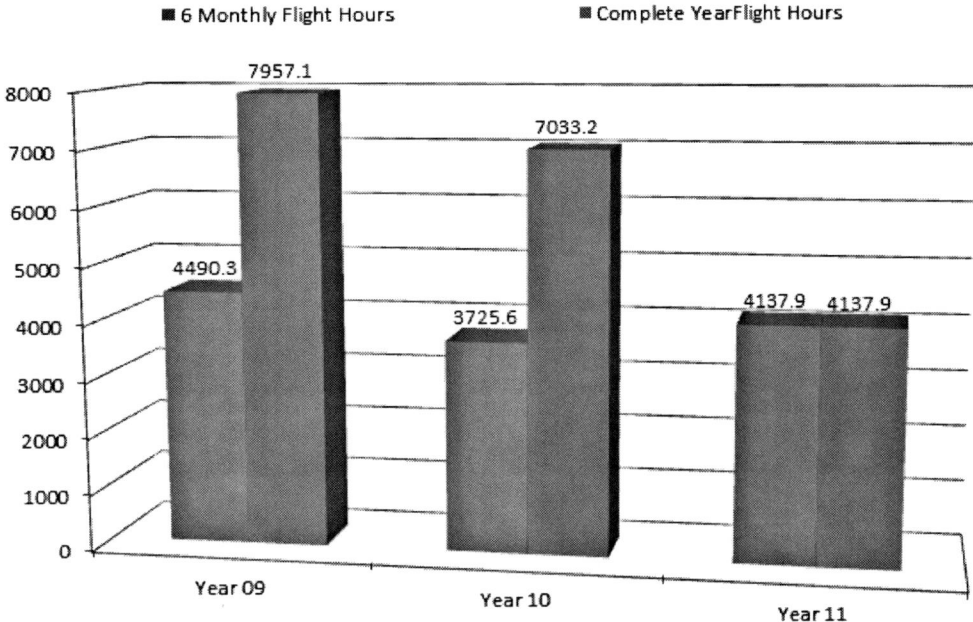

5. As schedule engine removals of one year can be carried forward to next year, therefore unscheduled removals per 1000 flight hours can provide a better picture for failure analysis.

6. Year 2011 is the best year for unscheduled removals as it has only 5.80 unscheduled engine removals per 1000 flight hours. Table 2 shows the unscheduled removals for all years.

7. For year 2011, FCU is the weakest area as 68% of total removals are because of FCU.

FCU FAILURES VS. OPERATING HOURS

An important outcome of safe aircraft operation is engine performance and reliability. On engine, Fuel Control Unit (FCU) plays a pivotal role, as it controls fuel supply and fuel scheduling for engine performance. FCU being the heart of the engine is getting special attention during engine operation and maintenance.

The following inferences can be drawn from FCU data analysis, Figure 3 and Figure 4:

1. During 30 months' time 17.58% FCU failed at zero operating hours, whereas 30% failed in less than 100 operating hours.

Table 1. Removals per 1000 hours

Year	Removals Per 1000 Hrs
2009	18.85
2010	12.51
2011	11.12

Table 2. Unscheduled removals

Year	Unsch Removals Per 1000 Hrs
2009	8.67
2010	7.11
2011	5.80

Figure 3. Top ten failed components

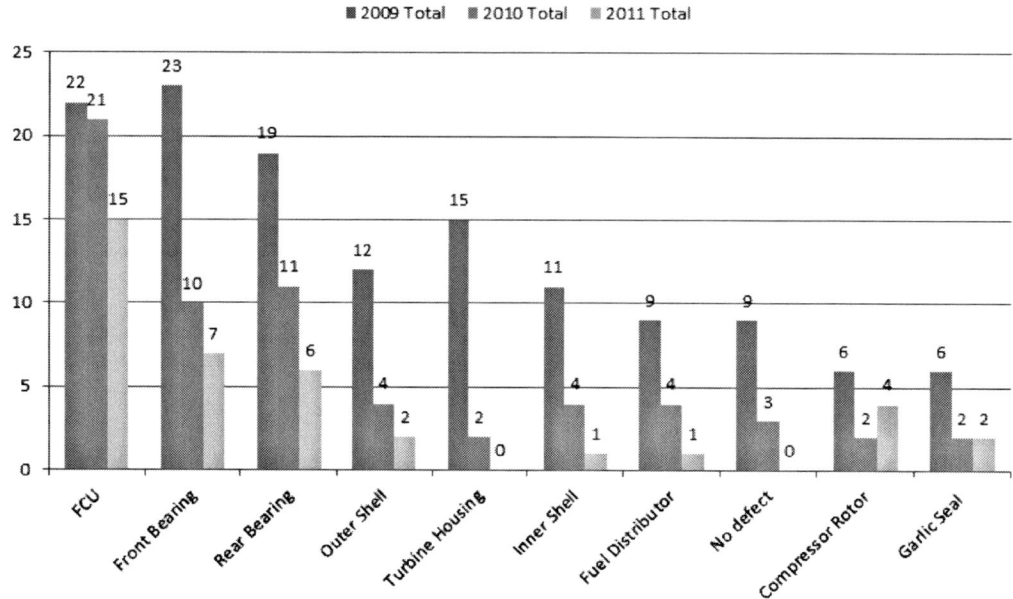

Figure 4. FCU failure trend

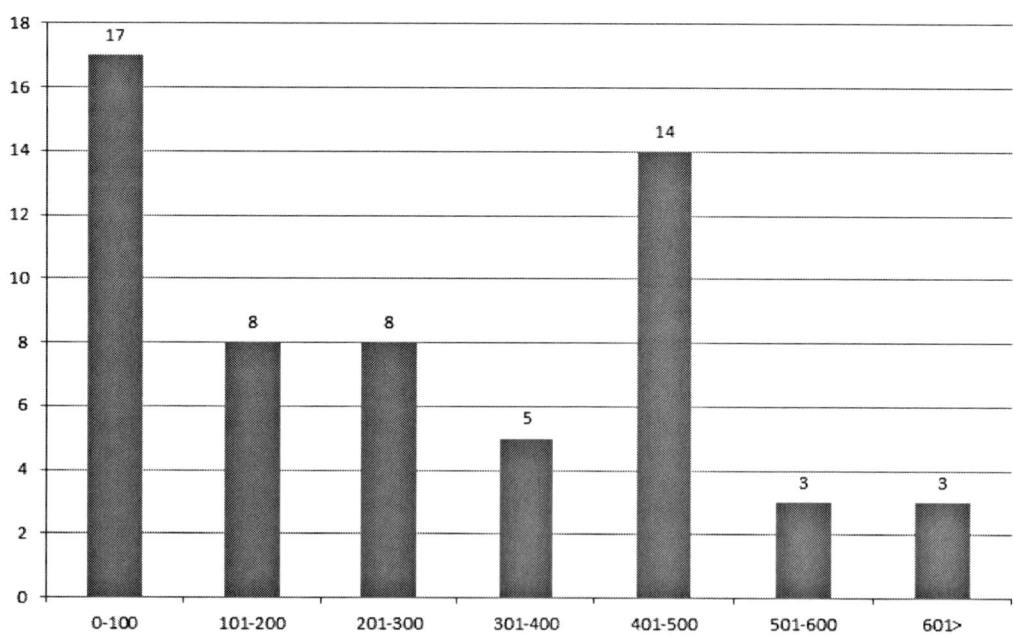

2.   FCU failure came out to be the weakest area on J69 engine.

3.   Maximum unscheduled removals are because of FCU failure.

4.   MTBF of FCU is 839.41 hrs.

5.   Reliability of FCU for a life cycle of 500 hrs. is 54.97%, whereas probability of failure during this time period is 45.03%.

6. The data mean is 260.5, Skewness is 0.155 and Kurtosis is -1.286. This shows that distribution is flatter than a normal distribution curve. Moreover, tails of distribution are heavy it is clear from graph, one tail is having 17 numbers whereas other has 20 numbers.

FRONT AND REAR BEARING FAILURES

42.5% of total failed front bearings, failed after schedule inspection, whereas 47.22% of rear bearings showed the same result. Figures 5 and 6 graphically represent the failure trend of both type bearings:

The following inferences can be drawn from Figures 5 and 6:

1. 57.5% Front Bearings crossed the operating life of 1000 hrs.
2. The graph depicts 19 failures between 501 to 1000 hrs. whereas in actual 10 out of 19 Front Bearings failed after completing 1000 hrs. operating life.
3. Only 10% out of total failed Front Bearings were able to cross 2000 hrs. operating life.
4. The average life of Front Bearing is 1033.675 operating hrs. Skewness is 0.87 and Kurtosis is 0.986. This means distribution curve of this data is sharper than normal distribution curve. Tails of the curve are not too heavy.
5. 63.88% Rear Bearings successfully crossed the operating life of 1000 hrs.
6. The graph shows 22 failures between 501 to 1000 hrs. Whereas in actual 11 Rear Bearings failed after completing 1000 hrs. operating life.
7. Only 2.8% out of total failed Rear Bearings were able to cross 2000 hrs. operating life.
8. The average life of Rear Bearing is 1012.05 operating hrs. Skewness is 0.218 and Kurtosis is -0.757. This means distribution curve for this data is flatter than normal distribution curve and tails are heavy. More failures are expected in early or late life cycle of Rear Bearing.

Figure 5. Front bearings failure trend

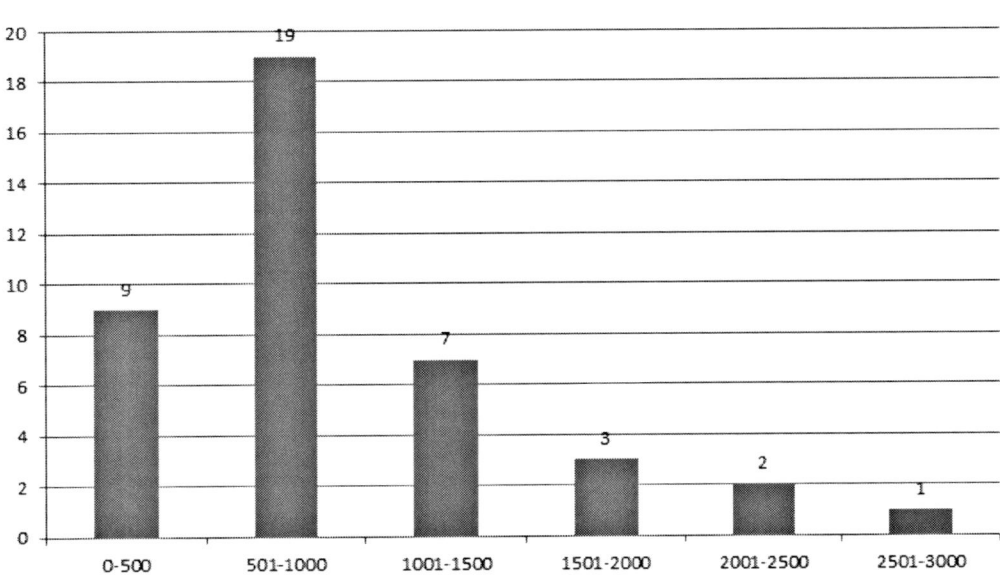

Figure 6. Rear bearings failure trend

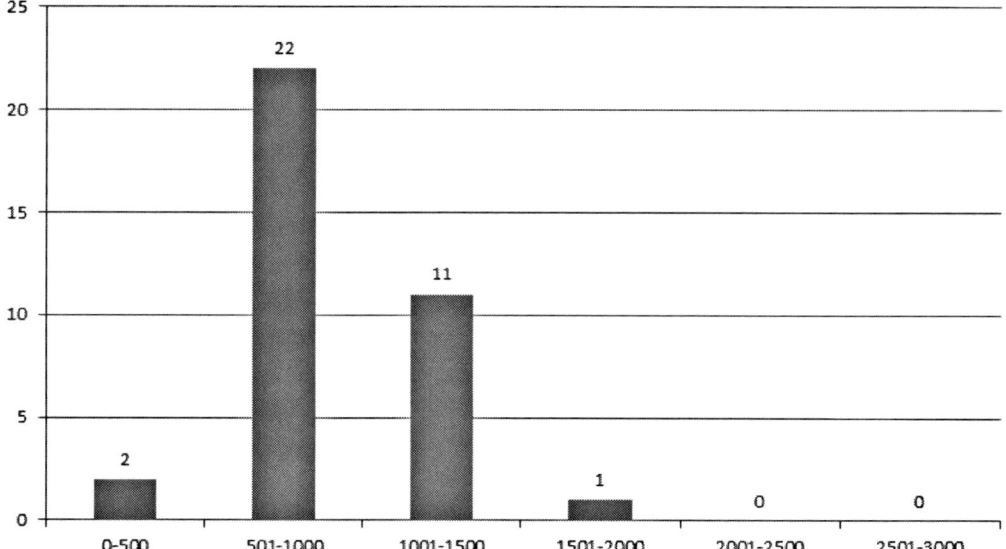

OUTER AND INNER SHELL FAILURES

Outer shell is one of the most important part of hot section area. The average life of outer shell is 3271.58 operating hours (see Figures 7 and 8).

The data analysis revealed that 66.66% outer shells failed between 2500 to 3000 operating hours, whereas in graph this failure is shown between 2001 to 3000 hours:

1. The safe life cycle for outer shell is 2500 operating hours.
2. 44.5% failures of outer shell are unscheduled. At the same time 33.33% outer shells crossed the life cycle of 3000 operating hours.
3. Outer shell failure and flight hours have positive correlation of 0.812 which mean increase or decrease of flight hours will have positive effect on outer shell failures.
4. More number of scheduled inspections in year 2010 had direct effect on number of failures.

5. Inner shell is also the one of the most important part of hot section area. The average life of inner shell is 2616.775 operating hours (see Figures 9 and 10).
6. The graph represents that 50% failure are between 1001 to 2000 operating hours whereas 37.5% failures are between 2001 to 3000 operating hours.
7. The data analysis revealed that 50% failures are between 1750 to 2000 operating hours.
8. The safe life cycle for inner shell is 2000 operating hours.
9. 6.25% inner shell failures are unscheduled and only 12.5% inner shells crossed the life cycle of 3000 operating hours.
10. Inner shell failure and flight hours have positive correlation of 0.870 which mean increase or decrease of flight hours will have positive effect on inner shell failures.
11. More number of scheduled inspections in year 2010 caused more number of failures.

Figure 7. Total outer shell failure trend

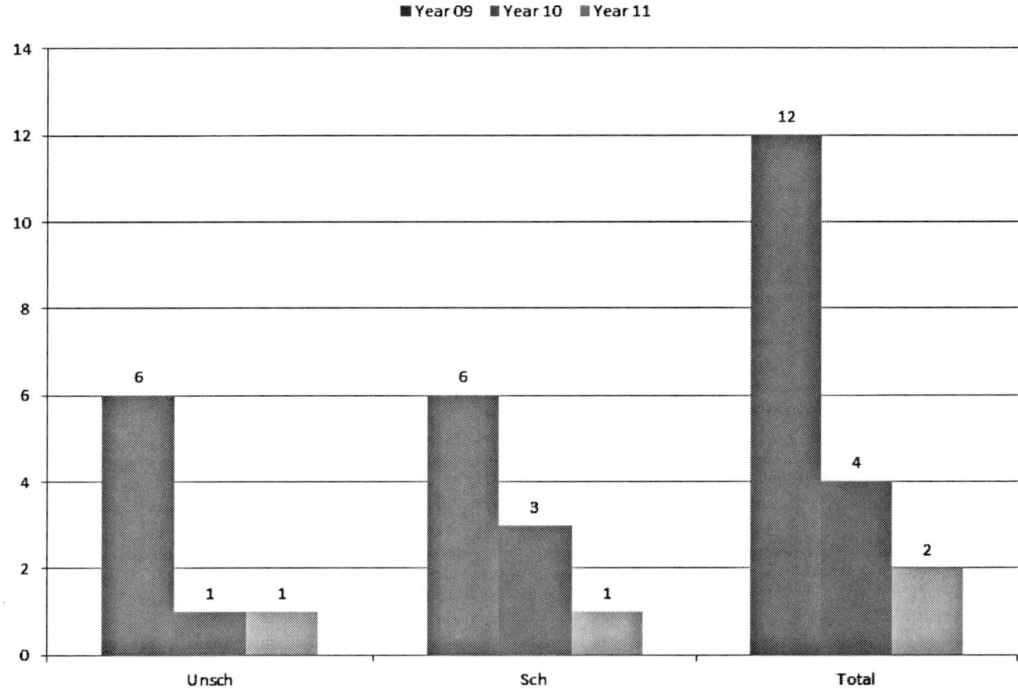

COMPRESSOR AND TURBINE HOUSING

The Compressor of J69 engine is single stage centrifugal. It consists of two parts first is compressor inducer and second is compressor rotor. The failure or rejection of this is because of material failure or grinding of ring for its balancing. The material failure factor is very minute whereas other is approximately 99%.

The following inferences can be noted from Figure 11.

The data mean is 6209.5 operating hours.

Minimum life of Compressor is above 3600 operating hours and maximum life is approximately 11000 operating hours, whereas maximum failures or rejection are between 5000 to 6000 operating hours.

Skewness is 1.72 and Kurtosis is 4.065. This shows that curve tails are lighter and data curve is sharper than Bell Curve. There is probability that maximum failures or rejection will occur in one standard distribution. Turbine Housing houses the hot section parts and its internal surface bears the thermal stresses due to combustion. Normal rejection is because of cracks in boss area or on the surface (see Figure 12):

1. The average life of Turbine Housing is 2616 operating hours.
2. Maximum failures are between 1000 to 3000 operating hours.
3. Skewness is 3.62 and Kurtosis is 13.84. The data is left skewed from its mean and a very high value of Kurtosis makes the curve shape sharper and tails lighter.

FUEL DISTRIBUTOR DEFECTS

Fuel Distributor (FD) plays a vital role in J69 combustion system because fuel is supplied and

Figure 8. Outer shell failure trend

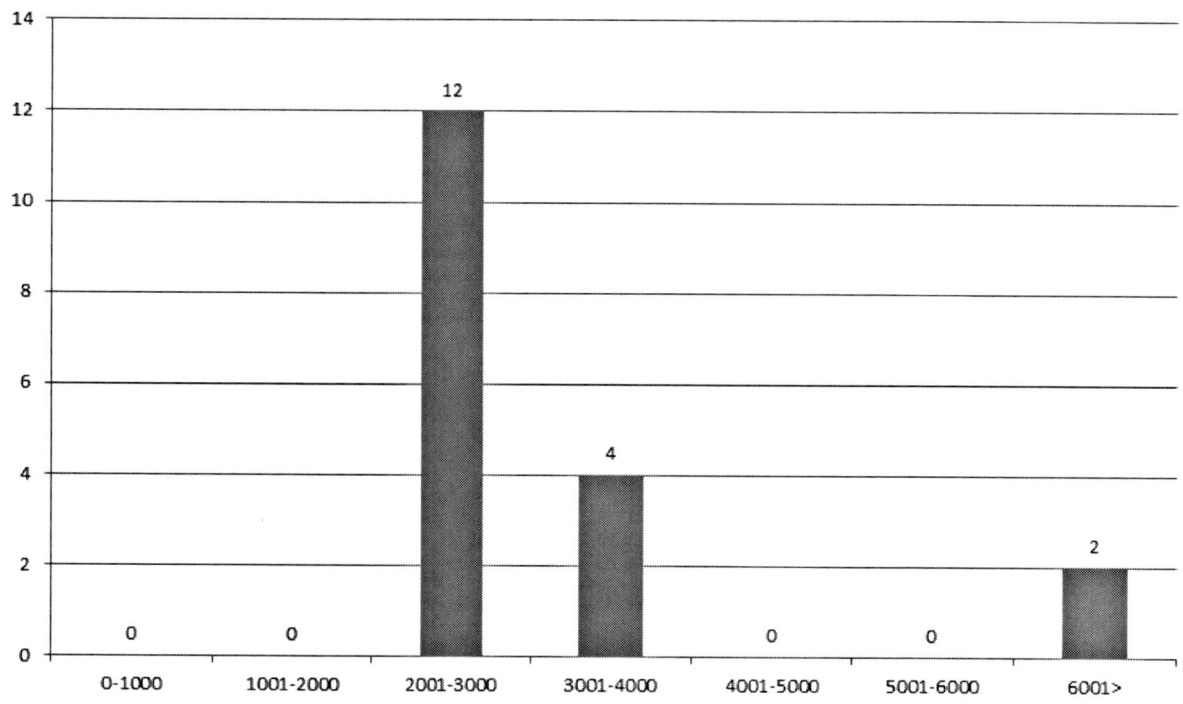

Figure 9. Total inner shell failure trend

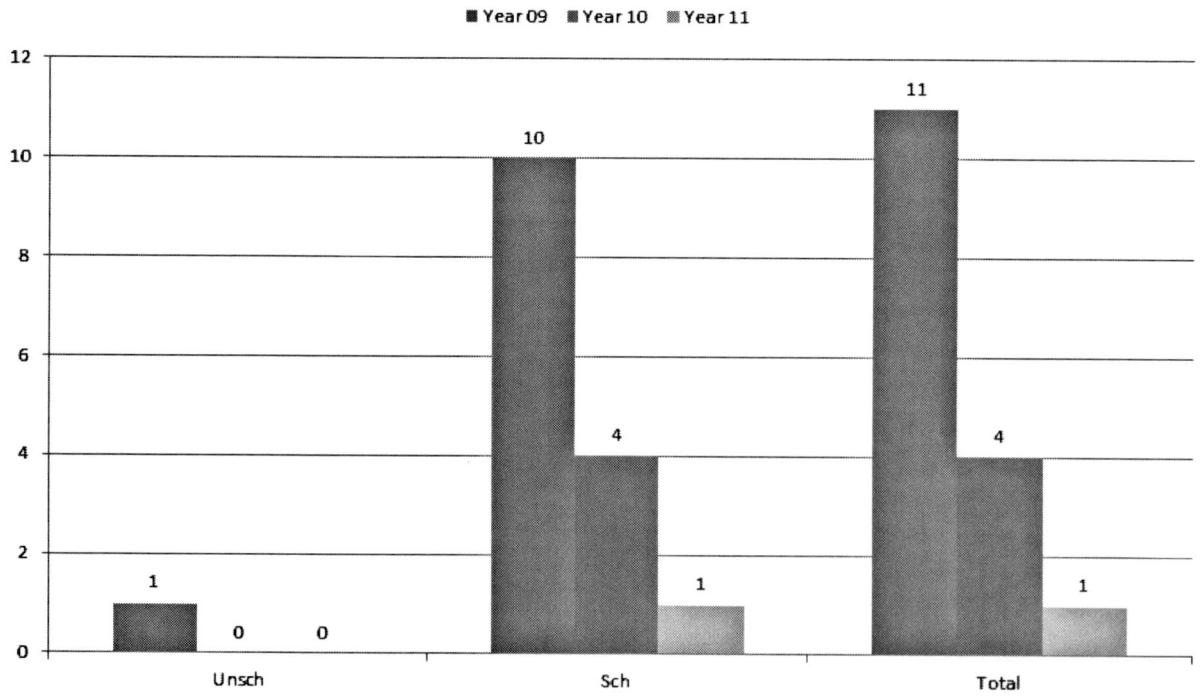

Figure 10. Inner shell failure trend

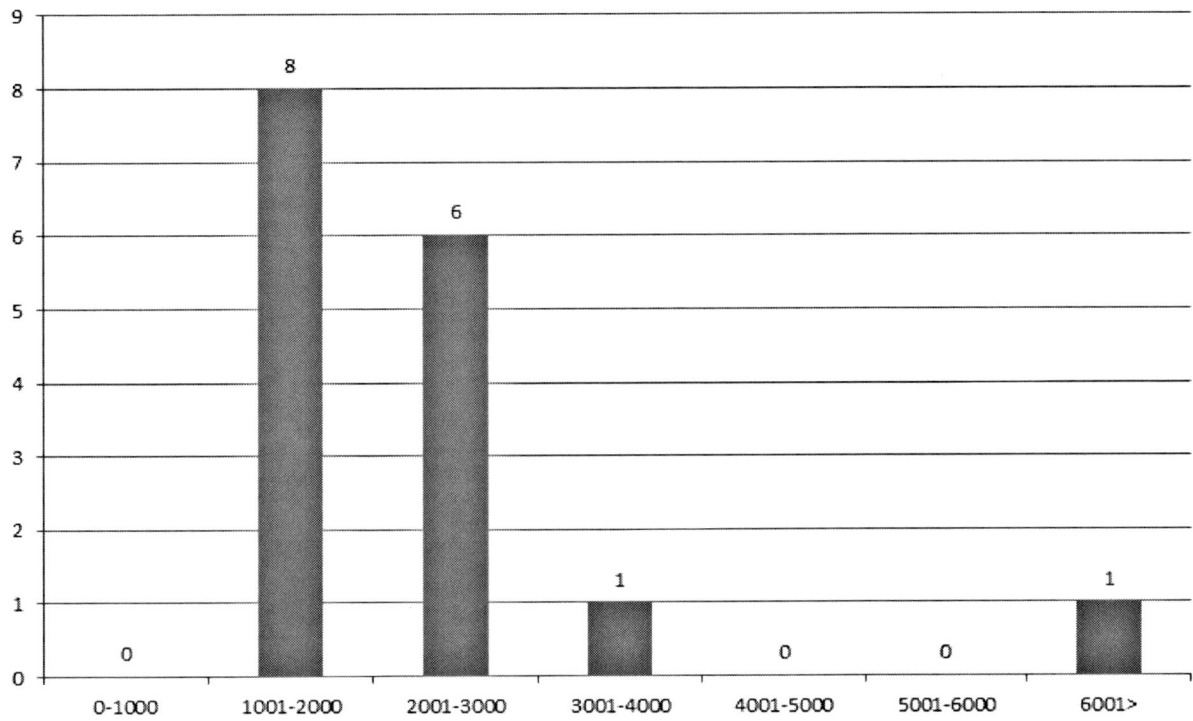

Figure 11. Compressor failure trend

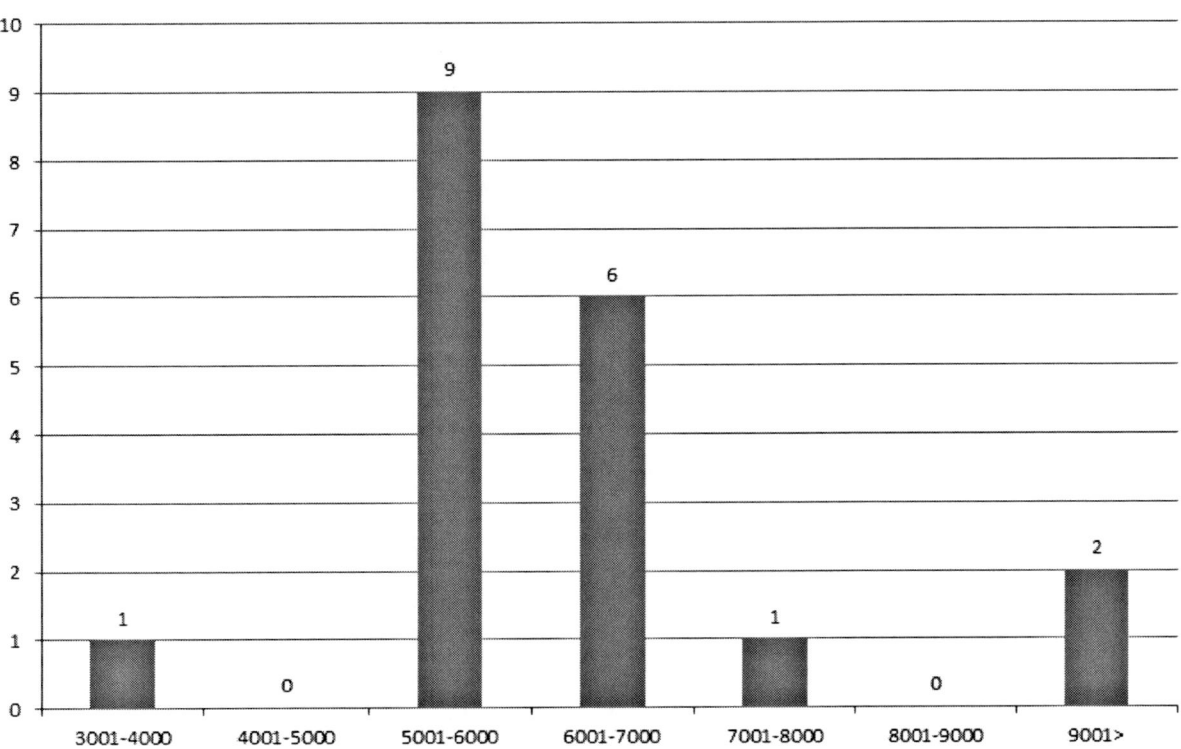

Figure 12. Turbine housing failure trends

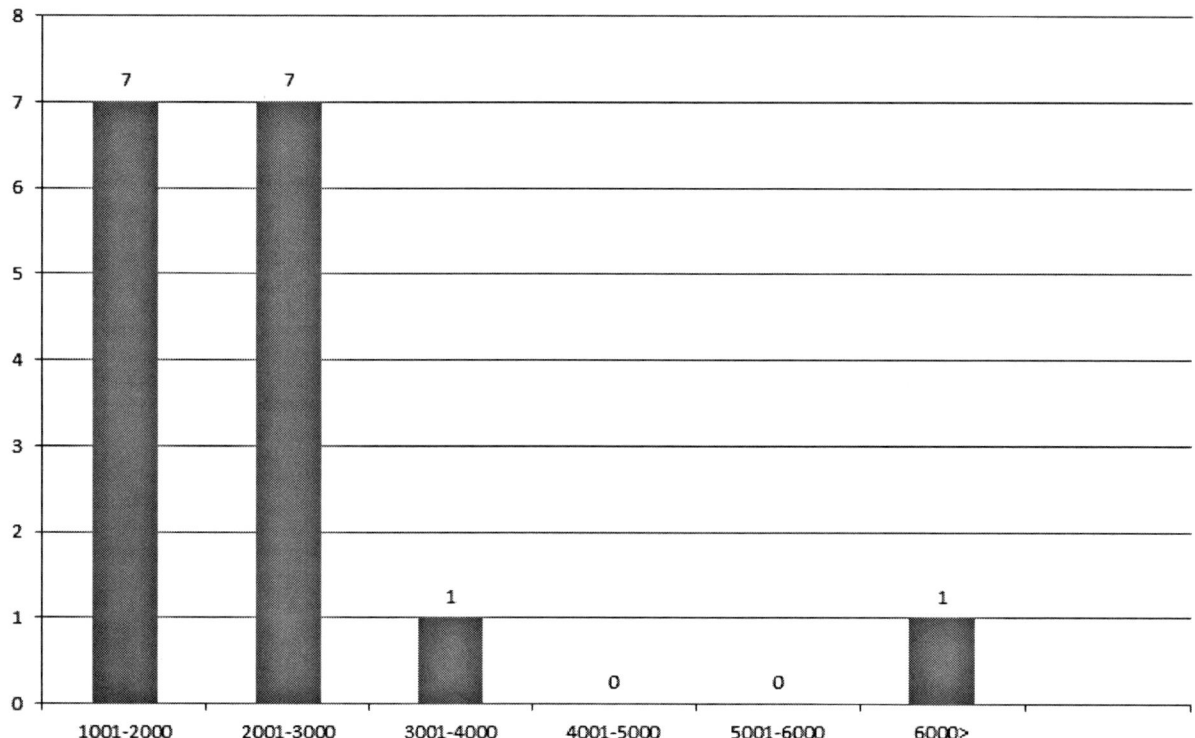

sprayed through its holes. A defective FD causes improper ignition and mixing of fuel with oil.

The inferences drawn from Figure 13 and data are:

1. The average life of FD is 3487.3 operating life.
2. The Skewness is -1.997 and Kurtosis is 4.616. This means that data is skewed towards right of the mean and curve is sharper than normal and tails are lighter.
3. The data analysis revealed that maximum failures are after 3500 operating hours.

NO DEFECT AND GARLIC SEAL FAILURE

Due to robust and old design and presence of Centrifugal Compressor makes the noise another weak area. Rumbling and howling noises are com-

mon on J69 engine. Figure 3 and data reveals that in total 12 engines have been removed till date but no defect have been found on these engines. It is pertinent to note that these all engine were removed unscheduled. These are 8.39% of total unscheduled engines whereas these are 4.22% of total engine removals.

A total of 6.99% of unscheduled removals were because of Garlic Seal whereas 3.41% of total removals were because of Garlic Seal. Now specific trend was monitored. This failure may be due to poor material of Seal of result of poor workman ship.

Table 3. Correlation values

	2009 Total	2010 Total	2011 Total
2009 Total	1.000		
2010 Total	0.808	1.000	
2011 Total	0.703	0.947	1.000

Figure 13. Defect trend of fuel distributor

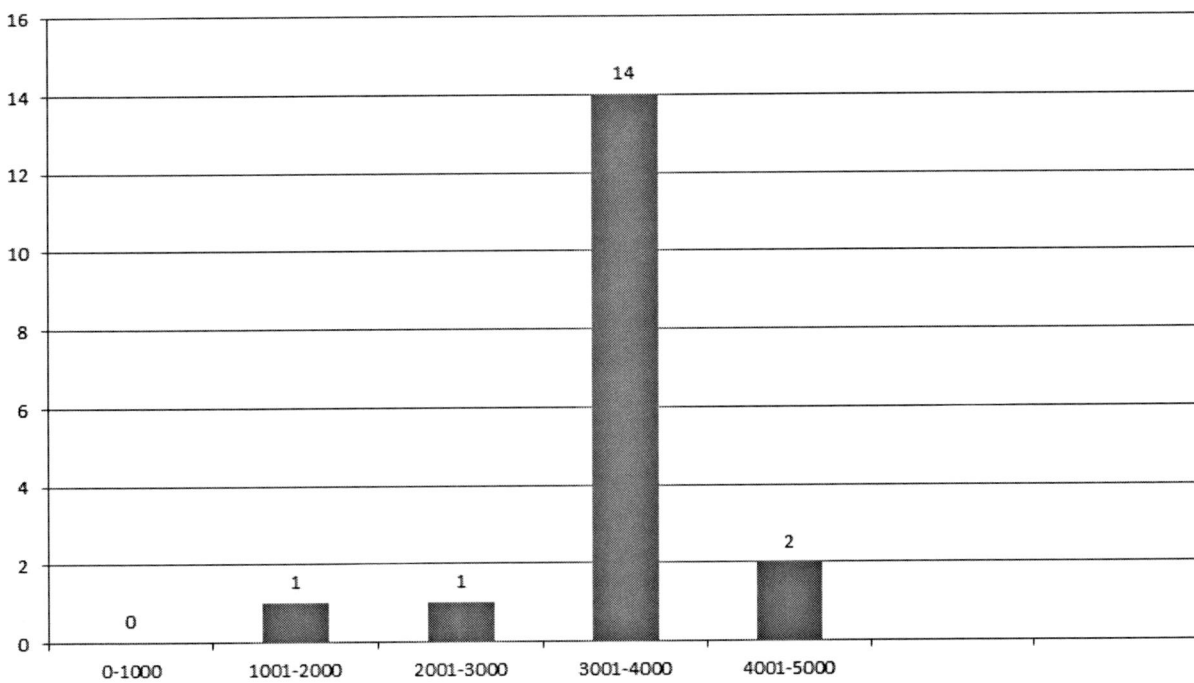

Figure 14. Defect categories

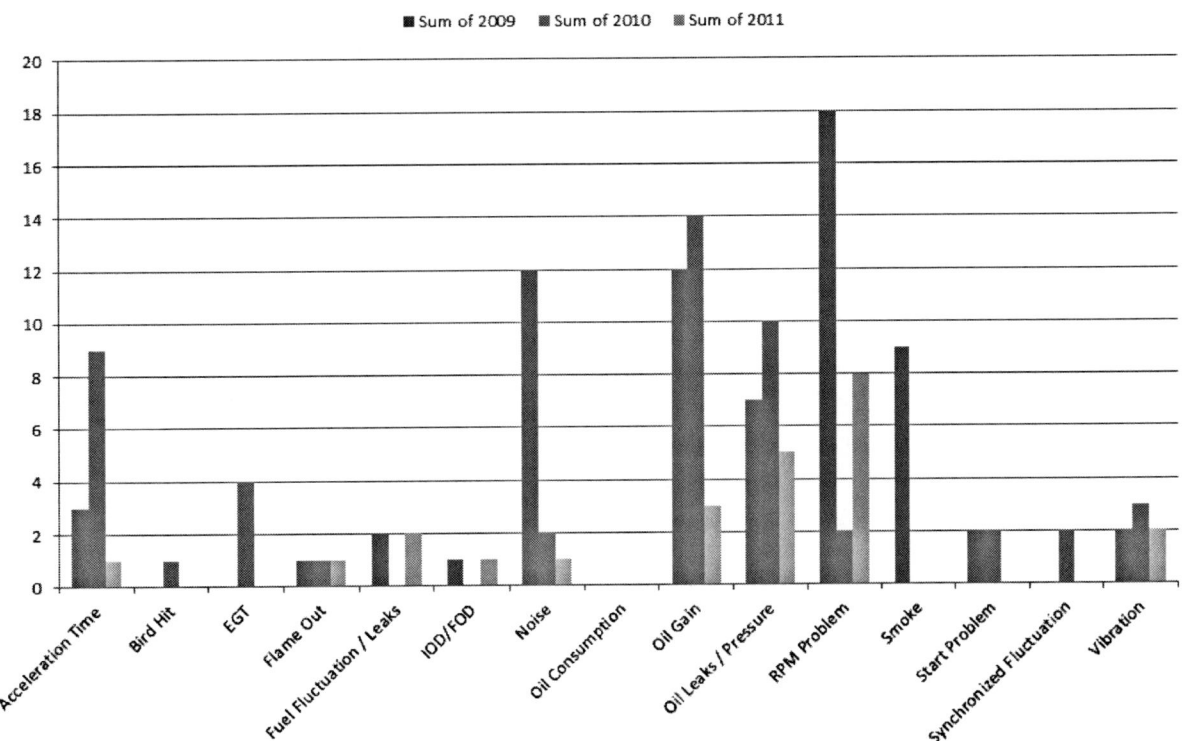

DEFECTS CATEGORIZATION

The unscheduled removals defects can be categorized in 15 broad categories. The categories are based upon nature of defects. Figure 14 shows the different categories.

CORRELATIONS OF YEARLY DEFECTS

The influence of yearly reported defects was checked on each other and correlations were evaluated. Table 3 shows the correlation values.

The correlations can be evaluated as:

1. The total defects of year 2009 have a positive strong correlation of 0.808 with year 2010 defects whereas on comparison this has a weak correlation with year 2011.
2. Very strong correlation of 0.947 is present between defects of year 2010 and 2011.
3. From this relationship it can be concluded that looking at defect trend of second half of year 2010, remedial measures can be taken to arrest defects of second half of year 2011.

FINDINGS

Following are the findings of this research through data analysis:

1. Fuel Control Unit is the weakest area of J69 engine.
2. The failure data of components does not follow normal distribution. The data is skewed either left or right. Moreover, data curve is either sharper or flatter.
3. Maximum unscheduled removals are due to Fuel Control Units.

4. Front and Rear Bearing has tendency to fail during scheduled inspection.
5. The unscheduled removals can be controlled by assigning life to components. This will reduce after inspection failures.
6. Operating hours have direct effect on components life.
7. Operating conditions and practices can enhance life of hot section parts.

CONCLUSION

J69 is an old engine but still serving and fulfilling the requirements. On J69 on two components are lifed by OEM whereas this study has revealed that assigning life to other components will increase reliability by decreasing unscheduled removals. Other operators may carry out their data analysis and suggest life to different components or reduction in inspection cycle. This all will help the operators to maintain reliable J69 engines.

REFERENCES

Jalote, P., & Agarwal, N. (2006). Using defect analysis feedback for improving quality and productivity in iterative software development. In *Proceedings of ITI 3rd International Conference on Information and Communications Technology: Enabling Technologies for the New Knowledge Society,* (pp. 703-713). ITI.

Khan, N., & Manarvi, I. A. (2011). Identification of delay factors in C-130 aircraft overhaul and finding solutions through data analysis. In *Proceedings of 2011 IEEE Aerospace Conference,* (pp. 1-8). Big Sky, MT: IEEE.

Kumar, U. D., Croker, J., & Knezrvic, J. (1999). Evolutionary maintenance for aircraft engines. In *Proceedings of Reliability and Maintainability Symposium,* (pp. 62-68). IEEE.

Manarvi, I. A., & Umer, W. (2010). Analyzing the defects of C-130 aircraft through maintenance history. In *Proceedings of 2010 IEEE Aerospace Conference (AERO 2010),* (pp 1-7). Big Sky, MT: IEEE.

Qi, Y., Lu, Z., & Song, B. (2005). New concept for aircraft maintenance management: The "dolphin curve life cycle model" of a typical repairable system. In *Proceedings of Reliability and Maintainability Symposium*, (pp. 533-538). IEEE.

Younus, B., & Manarvi, I. A. (2011). Defect trend analysis of airborne fire control radar using maintenance history. In *Proceedings of 2011 IEEE Aerospace Conference,* (pp 1-5). Big Sky, MT: IEEE.

Chapter 9
Lean Development:
A Tool for Knowledge Management in Software Development Process

Saqib Saeed
Bahria University Islamabad, Pakistan

Izzat Alsmadi
Yarmouk University, Jordan

Farrukh Masood Khawaja
Ericsson Telekommunikation GmbH & Co. KG, Germany

ABSTRACT

Software development is a complex activity, which is human intensive in nature. In order to build quality software systems, organizations need to follow mature software development practices, which are continually improved. As a result, the concept of software development process emerged, which highlighted a systematic set of activities required to develop a software system. Recently, agile development methodologies have provided a rich set of innovative software development approaches, aiming to optimize the software process. In order to be successful in adopting these approaches, a thorough understanding of their implementation procedures is required. In this chapter, we took a look at the lean development approach to understand how its principles pave the way in fostering knowledge management initiatives in software process development.

INTRODUCTION

Evolution of personal computers meant that computer systems are no more used by the programmer only but users from other domains could also benefit from the computing capabilities. As

DOI: 10.4018/978-1-4666-3658-3.ch009

a result, the evolution of high-level languages emerged, and software systems were designed to support users in government, business and other day-to-day tasks. The enhanced dependence on computing systems made the underlying software development quite complex. Since there was no standardized software development methodology in place, so the success of the project was mostly dependent on the skills and experiences of software projects and managers. As a result there were many software project failures and need was felt to define a standardized software process model. Software process is a set of activities required to realize a software product. The most basic definition of software process is presented by Humphrey (1989). He states "software process as a set of tools, methods, and practices used to produce a software product." While looking closely at software process the underlying activities could be classified into core and supporting activities. Core activities are mandatory activities without which software artifacts could not be finished whereas supporting activities improve the quality of the end product. A list of core and supporting software process activities is shown in Table 1.

With the passage of time, many software process models emerged (see Table 2). Waterfall/linear sequential model is the oldest software process model which advocates for carrying out all software development activities in sequence (cf. Royce, 1970; Pressman, 1996). Later different other models emerged to improve the weaknesses of this model. Among them prototyping (cf. Bischofberger & Pomberger, 1992), spiral (cf. Boehm, 1988), incremental model (cf. Larman & Basili, 2003), V model (cf. Pressman, 1996), Rapid application development (cf. McConnell, 1995), and object-oriented paradigm (cf. Booch, et al., 2007).

Recently agile methodologies having principles of limited documentation, active user involvement and short iterative cycles are getting quite popular (cf. Boehm, 2003; Schuh, 2005). Adoption to agile methodologies has helped in increasing project success rates (Coram & Bohner, 2005). Table 3 highlights 12 principles behind agile methodologies.

Agile methodologies is an umbrella notation of many development methodologies such as Scrum (cf. Schwaber & Beedle, 2002), Feature driven development (cf. Palmer and Felsing, 2002), Crystal methodologies (cf. Cockburn, 2008),

Table 1. Software process activities

Core Activities	Supporting Activities
• Project Initiation • Project Planning • Requirements Engineering • System Design • Software Coding • Testing • Maintenance	• Software Configuration management • Software Quality Assurance • Project Management

Table 2. Software process models

Traditional Process Models	Agile Process Models
• Waterfall/Linear Sequential Model • Prototyping • Spiral Model • Iterative Model • V Model • Rapid Application Development • Object Oriented Paradigm	• Scrum • Feature Driven Development • Crystal Methodologies • Adaptive Software Development • Dynamic System Development Methodology • Extreme Programming • Lean Software Development

Adaptive software development (cf. Highsmith, 2000), Dynamic System development methodology, Extreme programming (cf. Beck, 2000; Paulk, 2001), Lean software development (cf. Poppendieck, 2001).

Knowledge Management (KM) is an important methodology to improve organizations performance. There have been different research efforts to implement KM methodologies in different fields of life (cf. Fagrell, et al., 1999; Fagrell & Ljungberg, 2000; Fitzpatrick, 2002; Pipek, et al., 2003; Saeed, et al., 2008, 2010). However, there is a sparse body of knowledge on KM initiatives in software process. In this contribution, we highlight how lean approach helps in fostering an effective knowledge management methodology in software development process.

The remaining of this chapter is structured as follows: next section describes lean methodology principles in detail. Section 3 highlights the knowledge management channels fostered by adopting lean development approach and is followed by a conclusion.

Table 3. Agile principles (extracted from Agile Manifesto, 2012)

- Our highest priority is to satisfy the customer through early and continuous delivery of valuable software.
- Welcome changing requirements, even late in development. Agile processes harness change for the customer's competitive advantage.
- Deliver working software frequently, from a couple of weeks to a couple of months, with a preference to the shorter timescale.
- Business people and developers must work together daily throughout the project.
- Build projects around motivated individuals. Give them the environment and support they need, and trust them to get the job done.
- The most efficient and effective method of conveying information to and within a development team is face-to-face conversation.
- Working software is the primary measure of progress.
- Agile processes promote sustainable development. The sponsors, developers, and users should be able to maintain a constant pace indefinitely.
- Continuous attention to technical excellence and good design enhances agility.
- Simplicity--the art of maximizing the amount of work not done--is essential.
- The best architectures, requirements, and designs emerge from self-organizing teams.
- At regular intervals, the team reflects on how to become more effective, then tunes and adjusts its behavior accordingly.

UNDERSTANDING LEAN METHODOLOGY

After World War-II, the Japanese manufacturing industry was facing shortage of human, material and financial resources and it required an innovative low cost manufacturing methodology, to satisfy customers' demands of diverse high quality products. In order to survive in this situation Eiji Toyoda and Taiichi Ohno proposed guidelines to improve production system based on their analysis of Ford's Rouge plant in Detroit. They deployed this methodology in Toyota in 1960 which is now termed as lean production or Toyota production system. The focus of this approch is on utilizing fewer resources like time, space, human efforts, materials, and machinery to realize customers' products. The focus of this approach is on reducing cost to enhance profits but it should be done without destroying team member, reducing maintenance budgets, and weakening the company in the long run. The concepts of standardization, team empowerment, reducing Muda, decreasing defects are all interrelated activities. If these activities could be channelized effectively the manufacturing process would be improved significantly. Lean methodology requires sufficient training and skilled manpower to execute activities, and if this whole process is not properly planned and executed the stated benefits could not be achieved (Gurumurthy & Kodali, 2008).

Table 4 highlights key principles of lean methodology and in the following paragraphs we discuss them in details. A salient feature of lean approch is to optimally utilize available human resources. Maximum utilization of materials may result in over production, so lean approch focuses on maximum utilization of human resources. Its focus is on introducing human flexibility, so that people centered processes are developed that flow smoothly and safely (Shah & Ward, 2007). Another important feature of the lean approach is teamwork. Pascal (2002) describes safety first, employment security, uniforms, absence of executive offices, no executive dining and parking facilities, and Genchi genbutsu as the key features of team work. He describes safety is as important, as production and quality, the continuous mixing of managers and team workers helps free communication among employees and management and could avoid the problematic situations early on thus enhancing the productivity (Holweg, 2008).

Pascal (2002) argues that the human motion can be categorized into three categories actual work, auxiliary work, and Muda. Muda is a Japanese word describing the work, which does not have value and is not paid by customer. If the efforts are invested on a work, which is not worthwhile to customer, it will increase cost and decrease productivity. There are eight different categories of Muda which should be identified and reduced to minimum possible level (Hicks, 2007; Shah & Ward, 2007).

Table 4. Lean principles (adopted from Liker, 2004)

• Continuous Process Flow	• Error Prevention
• Promote Leadership	• Visual Control Mechanisms
• Team Development	• Reducing Muda
• Improve Supplier/Partner Quality	• Long Term Planning
• Level Out Work	• Decision Making by Consensus and Rapid Implementation
• Use Reliable Technology and Equipment	• Continuous Improvement (Kaizen)
• Understand Problems by Yourself (Genchi Genbustu)	• Standardization and Employee Involvement

Another important characteristic of lean production contributing to the productivity is stability in Man/Woman, Machine, Material, and Methods. Enhanced stability in these four key aspects leads to standardization, which ultimately reduces defects and problems, and as a result, productivity will be enhanced. Standardization introduces process stability, training of employees, Kaizen, employee involvement, audit and problem solving, organizational learning and performance of process. Once the standardization takes place employee becomes aware of whole process and equipment which makes his/her interaction with the equipment safer and also the optimization in the process helps in better productivity. In lean, visual management and 5S system are basis for standardization. 5S system supports Total Productive Maintenance (TPM), and Just In Time (JIT) production (Hopp & Spearman, 2004).

Error prevention helps is reducing the cost of the system and also helps in saving the effort and time spent on rectifying those errors which in turn improves the delivery time. Lean term Poka Yoke is used for error prevention which means adopting simple low cost devices that detect abnormal situations before they occur or if they occur stop the line to prevent defects. Lean methodology advocates for a comprehensive inspection system to control these errors. The continuous inspection of the equipment makes it sure that it is safe and compliant to the standards (Pascal, 2002). The focus of involvement activities is to improve productivity, quality, cost, delivery time, safety, environment, and morale by solving specific problems, reducing hassles and risks and by improving employees' capability. In order to ensure involvement Kaizen circle activity is proposed by lean approach (Womack, 2003).

The lean production cannot be achieved overnight and there are specific steps, which need to be followed for a successful implementation of lean approach. These include motivating people by orientation towards continuous improvement, empowering employees with required skills and by creating awareness. There is also a need for extensive information sharing. Another important factor is employee's involvement in decision making at lower level. An atmosphere of experimentation should be adopted by tolerating mistakes and willingness to take risks. A realistic reward and evaluation process should be in place so that employees could benefit from this. Then there is a need that the changes in organizational settings are not done at once, instead pilot projects should be carried out and after the success of pilot projects these activities should be extended to whole organization. Continuous improvement requires incremental improvements of products, processes, and services. In the process there are always some mistakes but these mistakes should not be repeated. When the continuous improvement is pursued, this also helps in rectifying problematic processes and improving the process so that the productivity could be increased (Tapping, et al., 2002).

Polanyi (1966) categorized knowledge into tacit and explicit. Tacit knowledge normally resides in human brains whereas explicit knowledge is rational. As a result, transfer of explicit knowledge is relatively easier than tacit knowledge. However, tacit and explicit knowledge are interchangeable and subject to cognition and externalization processes. Spender (1996) categorized organizational knowledge into four classes, which are conscious, objectified, automatic, and collective. Rowley (2002) described Knowledge management as a methodology to optimally utilize organizational implicit and explicit knowledge resources. As a result recent software development models have built-in processes for knowledge management.

KNOWLEDGE MANAGEMENT AND LEAN METHODOLOGY

Since activities performed during software development process are knowledge intense so in order to perform them optimally knowledge produced during entire development process is quite vital (Robillard, 1999). Knowledge management could be defined as "process of sharing, distributing, creating, capturing, and understanding of a company's knowledge" (cf. Cohen & Prusak, 2001). Knowledge management is a contested term (cf. Attewell & Rule, 1984; Cohen & Prusak, 2001; Huysman & Wulf, 2004, 2006; Chou, 2005; Joia, 2007) but effective knowledge management methodologies should be in place to store, codify and reuse this knowledge for continuous improvement.

Successful knowledge management at organizational level requires a considerable amount of resources and planning. In this section, we look at how lean approach helps in implementing KM initiative in software development process. The basic notion of agile approach advocating on extensive communication among team members to reduce the problems is in line with the findings of Alavi and Leidner (2001), who advocate for facilitation of communication to improve knowledge transfer. Similarly low focus on documentation helps to keep the focus on software itself. The customer orientation focus of lean approach helps in involving the customers in every development phase which increases the chances of product

acceptability and lower the chances of product failure, as diversions could be detected early on.

The optimal utilization of human resources advocates for human flexibility in carrying out different tasks. The continuous changeover of duties helps the team members to grasp variety of expertise, and as a result, the experts increase in the team paving a way for knowledge transfer among work force. The teamwork aspect which advocates for continuous mingling of all team members also results in effective knowledge transfer among them. Furthermore, as Poppendieck and Poppendieck (2006) describe that training of team supervisors, distribution of responsibility and a culture of pride in workmanship are important measures to get better productivity from employees.

In order to be market leaders there is always a need to produce products with advanced features but software developers should also keep in mind that these features are really needed by users. A major reason of project failure is the identification and adherence to exact software requirements. An important lean attribute focusing on reduction of "Muda" helps in keeping the software requirements free from unnecessary extra features. This phenomenon which is known as gold plating consumes considerable resources but the required benefits are not proportional. The continuous refinement of requirements to adhere to project needs at every software development helps team members to learn from the previous mistakes and considerable lessons are learned for future projects. Enhanced focus on users to accommodate their needs during requirements and system design highlighted by user centered approaches (cf. Galer, et al., 1992; Vredenburg, et al., 2001) is quite effective in reducing the redundant requirements. Redundant requirements will lead to resource wastage in system design, coding, testing and maintenance stages.

Although software development is human intensive activity but standardization feature is widely used in every software development step.

The evolution of software standards such CMMI, ISO 9000 provide best practices and standard guidelines for software development process. Every organization can develop its own set of practices based on those standards or their own experience. The focus of lean approach on standardization helps to adhere to these guidelines. Similarly lean advocates for continuous auditing, employee training and application of visual tools for status reporting. These better auditing and reporting mechanisms help to keep the control of the software development process as well as an important tool for knowledge transfer about the status of software process among stakeholders.

The quality of the software product is vital and the error prevention feature of lean approach helps in reduction of errors and delivery of a reliable product. An organization wide inspection system helps in evaluating the performance of software process as well. The knowledge about deficiencies and strong points of software process help the organizations to better learn and improve their practices to realize mature products. It is recommended that software project is divided in smaller batches and cycles so that each iteration is very small and the process is not stuck for a long time. Furthermore, product packages should be loosely coupled to have minimum dependency. This would help managers and developers different options to schedule activities without disturbing the overall project timeline. The continuous monitoring and status reporting helps in fostering a knowledge management environment where every team member actively knows the project status.

The cost of problems and errors is minimized if they are detected earlier on in the lifecycle. The communication barrier among team members discourage the reporting of problems and diversions earlier on. Poppendieck and Poppendieck (2006) suggest that developers should be quality conscious even before code writing and avoid any code duplications. Similarly, routine activities should be automated so that they do not cause problems for other team members. Lean develop-ment encourages employees to be actively involved in the software process activities to establish an effective knowledge sharing culture among teams. Lean's focus on continuous improvement fosters a knowledge management culture where continuous data of the performance of software process is collected and results are again verified after improving the work practices. This helps in setting up a quality culture and results in setting up best practices and guidelines for optimally carrying out tasks. Furthermore, this helps in setting up a research based culture where everyone is responsible for identifying the causes of failures and committed to improve the work by carrying out continuous improvement.

CONCLUSION

Agile methodologies are an important tool for enhancing software development activities. Lean approach is one of the famous agile techniques. In this chapter, we have discussed the important principles of lean manufacturing approach. We highlighted how lean principles help in setting up knowledge management culture during the software development process. This chapter will help students and practitioners to better plan the implementation methodologies in software development process by understanding KM perspective of the lean approach.

REFERENCES

Agile Manifesto. (2012) *Website*. Retrieved from http://www.agilemanifesto.org

Alavi, A., & Leidner, D. (2001). Review: Knowledge management and knowledge management systems: Conceptual foundations and research issues. *Management Information Systems Quarterly*, 25(1), 107–136. doi:10.2307/3250961

Attewell, P., & Rule, J. (1984). Computing and organizations: What we know and what we don't know. *Communications of the ACM, 27,* 1184–1192. doi:10.1145/2135.2136

Beck, K. (2000). *Extreme programming explained: Embrace change.* Reading, MA: Addison-Wesley.

Bischofberger, W. R., & Pomberger, G. (1992). *Prototyping- Oriented software development– Concepts and tools.* Berlin, Germany: Springer-Verlag. doi:10.1007/978-3-642-84760-8

Boehm, B., & Turner, R. (2003). *Balancing agility and discipline: A guide for the perplexed.* Boston, MA: Addison-Wesley.

Boehm, B. W. (1988). A spiral model of software development and enhancement. *IEEE Computer, 21*(5), 61–72. doi:10.1109/2.59

Booch, G., Maksimchuk, R., Engle, M., Young, B., Conallen, J., & Houston, K. (2007). *Object-oriented analysis and design with applications* (3rd ed.). Reading, MA: Addison-Wesley Professional.

Chou, S.-W. (2005). Knowledge creation: Absorptive capacity, organizational mechanisms, and knowledge storage/retrieval capabilities. *Journal of Information Science, 31*(6), 453–465. doi:10.1177/0165551505057005

Cockburn, A. (2008). *Crystal methodologies.* Retrieved from: http://alistair.cockburn.us/Crystal+methodologies

Cohen, D., & Prusak, L. (2001). *In good company: How social capital makes organizations work.* Boston, MA: Harvard Business School Press. doi:10.1145/358974.358979

Coram, M., & Bohner, S. (2005). *The impact of agile methods on software project management.* Washington, DC: IEEE Press.

Dennis, P. (2002). *Lean production simplified: A plain-language guide to the world's most powerful production system.* New York, NY: Productivity Press.

Fagrell, H., Kristoffersen, H. S., & Ljungberg, F. (1999). Exploring support for knowledge management in mobile work. [Berlin, Germany: Springer.]. *Proceedings of ECSCW, 1999,* 259–275. doi:10.1007/978-94-011-4441-4_14

Fagrell, H., & Ljungberg, F. (2000). A field study of news journalism: Implications for knowledge management systems. In *Proceedings of PDC 2000.* PDC.

Fitzpatrick, G. (2002). Bootstrapping expertise sharing . In Ackerman, M., Pipek, V., & Wulf, V. (Eds.), *Sharing Expertise: Beyond Knowledge Management* (pp. 81–110). Cambridge, MA: MIT Press.

Galer, M., Harker, S., & Ziegler, J. (1992). *Methods and tools in user-centered design for information technology (human factors in information technology).* London, UK: Elsevier Science Ltd.

Gurumurthy, A., & Kodali, R. (2008). A multi-criteria decision-making model for the justification of lean manufacturing systems. *International Journal of Management Science and Engineering Management, 3,* 100–118.

Hicks, B. J. (2007). Lean information management: Understanding and eliminating waste. *International Journal of Information Management, 27,* 233–249. doi:10.1016/j.ijinfomgt.2006.12.001

Highsmith, J. (2000). *Adaptive software development.* New York, NY: Dorset House.

Holweg, M. (2007). The genealogy of lean production. *Journal of Operations Management, 25,* 420–437. doi:10.1016/j.jom.2006.04.001

Hopp, W. J., & Spearman, M. L. (2004). To pull or not to pull: What is the question? *Manufacturing and Service Operations Management, 6,* 133–148. doi:10.1287/msom.1030.0028

Humphrey, W. (1989). *Managing the software process.* Reading, MA: Addison-Wesley.

Huysman, M., & Wulf, V. (2004). *Social capital and information technology.* Cambridge, MA: MIT-Press.

Huysman, M., & Wulf, V. (2006). IT to support knowledge sharing in communities: Towards a social capital analysis. *Journal of Information Technology, 1*(21), 40–51. doi:10.1057/palgrave. jit.2000053

Joia, A. L. (2007). Knowledge management strategies: Creating and testing a measurement scale. *International Journal of Learning and Intellectual Capital, 4*(3), 203–221. doi:10.1504/ IJLIC.2007.015607

Larman, C., & Basili, R. V. (2003). Iterative and incremental development: A brief history. *IEEE Computer, 36*(6), 47–56. doi:10.1109/ MC.2003.1204375

Liker, J. K. (2004). The Toyota way: 14 management principles from the world`s greatest manufacturer. *Business Book Review, 12*, 1–11.

McConnell, S. (1995). *Rapid development.* Redmond, WA: Microsoft Press.

Palmer, S. R., & Felsing, J. M. (2002). *A practical guide to feature-driven development.* Upper Saddle River, NJ: Prentice Hall International.

Pascal, D. (2002). *Lean production simplified: A plain-language guide to the world's most powerful production system.* New York, NY: Productivity Press.

Paulk, M. (2001, November-December). Extreme programming from a CMM perspective. *IEEE Software*, 19–26. doi:10.1109/52.965798

Pipek, V., Hinrichs, J., & Wulf, V. (2003). Sharing expertise: Challenges for technical support . In Ackerman, M., Pipek, V., & Wulf, V. (Eds.), *Sharing Expertise: Beyond Knowledge Management* (pp. 111–136). Cambridge, MA: MIT Press.

Polanyi, M. (1966). *The tacit dimension.* London, UK: Routledge and Kegan.

Poppendieck, M. (2001, June). Lean programming: Part 2. *Software Development*, 71-75.

Poppendieck, M., & Poppendieck, T. (2006). *Implementing lean software development - From concept to cash.* Reading, MA: Addison-Wesley Professional.

Pressman, R. S. (1996). *A manager's guide to software engineering.* New York, NY: McGraw-Hill.

Robillard, P. N. (1999). The role of knowledge in software development. *Communications of the ACM, 42*(1), 87–92. doi:10.1145/291469.291476

Rowley, J. (2002). Reflections on customer knowledge management in e-business. *Qualitative Market Research, 5*(4), 268–280. doi:10.1108/13522750210443227

Royce, W. (1970). Managing the development of large software systems. *Proceedings of the IEEE, 26*, 1–9.

Saeed, S., Pipek, V., Rohde, M., & Wulf, V. (2010). Managing nomadic knowledge: A case study of the European social forum. In *Proceedings of the 28th International Conference on Human Factors in Computing Systems.* Atlanta, GA: IEEE.

Saeed, S., Reichling, T., & Wulf, V. (2008). Applying knowledge management to support networking among NGOs and donors. In *Proceedings of IADIS International Conference on E-Society.* Algarve, Portugal: IADIS.

Schuh, P. (2005). *Integrating agile development in the real world.* Hingham, MA: Charles River Media.

Schwaber, K., & Beedle, M. (2002). *Agile software development with scrum.* Upper Saddle River, NJ: Prentice-Hall.

Shah, R., & Ward, P. T. (2007). Defining and developing measures of lean production. *Journal of Operations Management, 25*, 785–805. doi:10.1016/j.jom.2007.01.019

Spear, S., & Bowen, H. K. (1999). Decoding the DNA of the Toyota production system. *Harvard Business Review*. Retrieved from http://hbr.org/1999/09/decoding-the-dna-of-the-toyota-production-system/ar/1

Spender, J. C. (1996). Making knowledge the basis of a dynamic theory of the firm. *Strategic Management Journal, 17,* 45–62.

Tapping, D., Luyster, T., & Shuker, T. (2002). *Value stream management: Eight steps to planning, mapping, and sustaining lean improvements.* New York, NY: Productivity Press.

Vredenburg, K., Isensee, S., & Righi, C. (2001). *User-centered design: An integrated approach.* Upper Saddle River, NJ: Prentice Hall.

Womack, J., & Jones, D. (2003). *Lean thinking.* New York, NY: Free Press.

Chapter 10
Management Practices in Exploration and Production Industry

Kashif Saeed
Wintershall Holding GmbH – Kassel, Germany

Georg Ziegler
Wintershall Holding GmbH – Kassel, Germany

Muhammad Kashif Yaqoob
Mubadala Petroleum – Abu Dhabi, UAE

ABSTRACT

This chapter is divided into three main sections; project management, HSE management, and quality management. A focus description of the different elements of exploration and production industry along with implementation of management practices on each of these elements including asset/portfolio, resources, time, project planning and scheduling, and proactive risk management are presented. Health safety and environment and quality management are dealt with as separate sections.

INTRODUCTION

Maximizing the value on investment responsibly is the ultimate goal of any engineering management project. Addressing exploration and production business needs is always challenging due to several reasons. First of all there are huge sums of money at stake. It takes enormous amount of capital expenditure to initiate any project, which can be up

to billions of euros for the surveying, processing, technical studies, drilling activities, completions, and work-overs. Additional infrastructure and community development is a necessity. Sometimes, the operational expenditure exceeds initial capital. Apart from that, extraordinary long duration of exploration and production projects makes it challenging to manage projects professionally. The average time-line of hydrocarbon projects can be spread over tens of years, starting from regional surveillance and reconnaissance surveys

DOI: 10.4018/978-1-4666-3658-3.ch010

to depletion and abandonment of hydrocarbon reservoirs. Additional factors can be technology applications and blending with experienced professionals within organizations, protection of environment and eco-system, personnel and resource management.

The synergistic approach helps most to successfully execute project plan. Interdisciplinary integration of technical, human resources and economic teams is very essential for successful accomplishment of projects. The geologists, geophysicists, petrophysicists, reservoir engineers, production engineers, facility, and process engineers should all focus together on the definitive project goal and support each member of the team. It will not only boost the morale of the team as a whole but will also incorporate the important values like openness, flexibility, communication, and coordination among the team. Therefore, it is recommended to make project based teams instead of department based teams.

The application of management on early stage of a project can yield enhanced results and augment profits for the organization. In addition, dividing a mega-project into small tasks helps effective implementation of the project plan leading to successful achievement of the ultimate objectives. To maximize the return on investment it is significant to deplete the hydrocarbon reservoir intelligently. Thus, strategic, technical, and planning capabilities are essential at every stage of the project life cycle.

This study is divided into three main sections; project management, HSE management and quality management. A focus description of the different elements of exploration and production industry along with implementation of management practices on each of these elements including asset/portfolio, resources, time, project planning and scheduling, proactive risk management which are presented as section 'project management,' health safety and environment and quality management have been dealt as a separate sections.

PROJECT MANAGEMENT

Project management is the concept which has taken significant importance in all scopes of businesses and especially in the exploration and production industry, where the projects are extra-long and huge sums of money is involved. Fifty years ago, projects were executed and completed without even considering any idea like project management which is not only the subject of how to manage projects within given time and budget but also coping with different targets, environments and people with understanding the success key skill set essential for a certain project. Along with 'know-how,' the concept of 'do-how' can help greatly to accomplish the hefty accomplishments.

Dealing with a wide variety of projects including exploration, appraisal, production, disposal, sequestration, and storage, etc. is the hallmark of the exploration and production industry. This makes it a very diverse process, and at the same time, each type of project needs a specific set of expertise to execute successfully. However, over the years, some management tools, techniques, and problem solving approaches have proved to be essential for almost any type of project. Thus, at least such a basic skill set is considered to be a must for execution of such projects. Any additional expertise required for the specific projects are generally provided by the organizations in the form of trainings, seminars, and/or lectures.

The project management practices are very essential for the effective utilization of resources and thus to achieve the ultimate goals. Clear organization and scope setting at initial stage of project helps in great deal to define a project plan, which should be followed throughout the project lifecycle. In addition, identifying key success skills and performance indicators with the development of scenarios allow the efficient mitigation of any bottlenecks and risks. These are properly communicated to each level of hierarchy to continuously improve the progress until the project closure.

The key attributes of the project management to achieve specific goals including planning, organization, managing, leading, and controlling resources with reference to the exploration and production industry projects are discussed in this section. In addition, their interaction with primary and secondary challenges (achieving goals and objectives) and constraints (scope, time, and budget) are elaborated.

Introduction to Project Management

Project Management (PM) can be defined very simply as a body of knowledge, which is concerned with principles, techniques and tools used in planning, controlling, monitoring and reviewing of projects. The structure of project management includes several elements like process groups and control systems, which aligns all resources to achieve the goal efficiently. Every project is significantly unique. However, different project development phases of any project are initiation (project plan and design), execution, monitoring and controlling (value addition) and finally closure of the project.

The nature and scope of project is determined in the initiation phase. It is very essential for successful execution of project to perform well the initiation phase and develop/design a plan to execute project. This is primarily dependent on the understanding of business environment and making sure that all necessary controls are incorporated in the project definition. Any deficiencies and risks are identified and reported at this early stage with recommendations to mitigate them. The essential task set to be performed at this phase includes:

- Understanding global business needs and requirements
- Current company operations and key focus
- Financial analysis (costs and benefits) for the project
- Stakeholder analysis

- Project character (tasks, deliverables, schedule and costs)

Three of the key attributes in project deliverables are time, cost and scope. To achieve these targets, the need of time, budget, and quality planning can be directly analysed. A detailed project plan with schedule is very helpful at the initial stage of the project. The main purpose of such a plan is to estimate the work needed and to effectively manage risks during project execution. The attributes of such a plan include developing scope, level of detail, selecting project team, identifying deliverables, creating breakdown structure, identifying activities to achieve deliverables, estimating resources (time and cost) required for these activities, developing schedule and budget, risk planning and obtaining approval for project kick off.

After successful initiation phase comes the execution phase, which consist processes used to complete the work defined in the project plan to accomplish the project deliverables. The execution process is attributed by coordinating people, resources, and essentially performing activities as part of the project management plan. The deliverables are produced as output of the processes and activities performed in accordance with the project management plan. The constant value addition is very helpful at each stage by monitoring and controlling. The final project closure phase comes at the end to finish the process. The lessons learnt should be reported irrespective of whether the defined goals are achieved or not.

Conceptual Workflow

The exploration and production project management workflow (see Figure 1) describes the set of activities typically performed during a project. Different identifiable phases of the project have a unique set of challenges to be managed. These basic project phases represent the major factors influencing a project success or failure. The proj-

Figure 1. Generalized exploration and production project management workflow

ect goals and deliverables are strongly dependent upon how management practices are followed in each phase as explained by Wit (1988).

Project organization includes a structured set of project executers and deliverables. Different types of projects are carried out in the exploration and production industry, which yield to a wide variety of integrating teams to perform a task and achieve the common goal. As bigger the project more professionals are involved and thus a clearer organization structure is very essential and yields desired objectives and goals. The project organization should harmonize essentially each member of team to execute specific task. Not always specialists in their fields can better perform the job, mostly integrated correspondence in a team yields better results.

The project scope involves the setting of clear deliverables as primary or secondary goals. Questions of 'what to do' are predominantly addressed and should be made clear to each member of the team. Any questions, queries, suggestions, and

recommendations are discussed before starting the planning phase. This is where the project work actually starts and the team begins to take shape.

Project planning is the answer to 'how to achieve set targets.' It is the project scope of further development to detailed level as possible. Scheduling and intermediate project outcomes and milestones are identified and strategies of how to achieve them are established. Definition of required elements of work tasks and feasible sequences of executing them is set in this phase. Estimations of money and time required (deadlines) are finalized.

Key success skills are sometimes made as a part of the project planning where the right direction to achieve goals is figured out. In this step, areas of strength for the project goals are defined which are of utmost importance throughout project lifecycle.

Progress and performance marks the project execution work, which is performed under a watchful eye of a project management team. Progress and performance is continuously monitored and appropriate adjustments are made according to the strictly regulated change in management procedures. These changes are adequately recorded and reported at the end. This phase is considered as core of the process and the project team concentrates on meeting objectives like time, budget, and resources agreed upon at the beginning of the project.

Risk management is an essential phase and needs to be visible in the workflow right from the start. Inadequate risk management can result into project disaster, especially in the exploration and production industry, where the economic situation is unclear. This can give a significant jolt to the organization by relinquishing key assets due to malicious or non-risk management. The key items in proper risk management include progress tracking of outstanding action plans, describing responsibilities for actions and setting the expected timeframe for resolutions.

Communication and continuous improvements include the feedback and progress of the tasks being performed. It is very essential for the project management team to know about any activities of all team members if something goes in a wrong direction there will be enough time to get back in order before it is too late.

Audit, documentation, and closure mark the end of project. Internal and external audits during and after the projects are helpful for the time, budget and benchmarking of efficiency deliverables. It is clearly reported about the set of primary and secondary objectives and if deliverables are achieved fully or partially. Apart from that a concise reporting of lessons learned during project is also performed.

Project Organization/Hierarchy

Project organization includes setting-up the team and harmonizing it with intended project scope. The solution of continuous development is more complex and organizational problem-solving strategies are not trivial and thus needs excep-

tional cooperation between corporate divisions and specialists. Traditional management and organizational concepts are mainly characterized by splitting-up competence and management (which focuses on an efficient and effective job processing). The process involves the best project organization, configuration of the project team and decision powers. Since the projects are always time-constraint, tolerance for adoption or failure correction is hardly available. The set-up of the project organization is the first of many obstacles to project success.

The project organization decision normally considers a maximum degree of freedom for involved departments and project members avoiding havoc caused by unclear job description or accumulation of coordination requirements. The idea boosts the concept of no single best option for setting up a project organization. The specific organizational structure should reflect the requirements of the project. The project management has to identify prior to project start, internal and external requirements for best possible recommendation for a successful project.

Project Organizational Models

Project management practicing over the years, various project organizational models have been developed. Two extreme ways of handling projects are 'projects as part of the functional organization' and 'project as a free standing part of the parent organization.' A hybrid technique of 'matrix organization' also exists:

- Line organization/functional organization:
  - Does not have a specific position for project manager
  - Is divided into partial tasks and delegated to responsible departments
  - Team members continue to report to line directors and upper management
- Project organization:

- Project manager is fully responsible for a group of specialists
- Specialists temporarily dedicate their entire workforce for the project
- Matrix organization:
  - Combination of a functional and a pure project organization
  - Organizational structure allows participation on multiple projects
  - Greater integration of expertise and project requirements can be achieved

Selection of Project Organization

The decision of formation, preparation, and initiation of the project organization is the first and most important step to the project success. Therefore, a great deal of time should be spent in these considerations.

A step-wise approach is recommended for the goals and implications of the project within the organizational structure. However, following indications are considered as reconnaissance overview of what should be considered:

- Project definition:
  - Size of the project
  - Corporate profile of project
  - Innovation needed
  - Inter-department integration requirements
  - Number of external interfaces (complexity)
  - Budget and time constraints
  - Resource level
  - Stability of resource requirement
- Key tasks determination
- Breaking down project into work packages
- Special characteristics or hampering factors
- Choice of structural organization

Project Structural Organization and Roles

The internal distribution of tasks to perform different operations and relationship between individuals and departments is controlled by project structural organization. In contrast to structural organization which defines who has to perform a particular job, operational structure is more dynamic and defines when, where and how often something has to be done.

Two main tasks of the structural organization in exploration are position of the project in the parent organizational model and the internal structure of the project team.

The general structure of an internal project organization is displayed in Figure 2.

Different elements of an internal structural organization are:

- Steering Committee
- Project Sponsor
- Advisory Committee
- Project Manager
- Team Leaders and Team Members

Steering Committee:

- Represents a group of senior managers responsible for business issues effecting project
- Have budget approval authority and can make decisions about changes in goods/scope
- Are highest authority to resolve issues and disputes
- Assist with resolving strategic level issues and risks
- Can approve/reject changes to project keeping in view time and budget
- Has to access project progress and report to senior management and higher authorities

Figure 2. The hierarchy of project management team from steering committee to team members

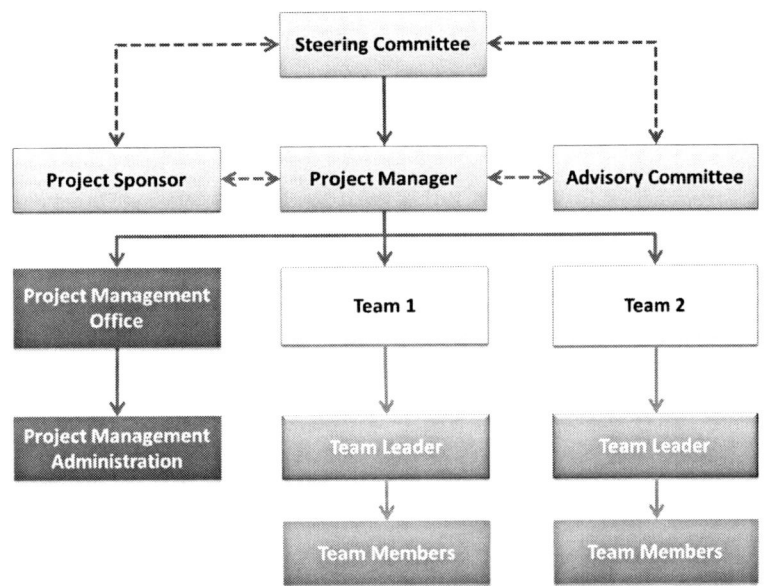

Project Sponsor:

- Is a manager/executive within organization, not directly involved in operational work but has ability to oversee a project
- Can provide support as trainer/coach to project manager
- Has sufficient authority/influence to direct all staff involved in a project and cooperation of key stakeholders
- Ensures that project is aligned with organizational strategy and compliant with policy
- Has ability to monitor the effectiveness of project manager in larger projects

Advisory Committee:

- Consists of a group of people representing key project stakeholders
- Generally recruited from senior management
- Cannot make decisions regarding the project

- Enhances insight of the team regarding stakeholder interest, technical advice and relevant initiatives
- Work along with steering committee to assist resolving issues and risks

Project Manager:

- Key person within the project organization having overall responsibility for meeting project requirements within agreed time, cost, scope, and quality constraints
- Reports to the steering committee and generally performs following task sets:
  ○ Supervision and guidance of the project team
  ○ Regular project status reports to steering committee/project sponsor
  ○ Chair risk and change control committee
  ○ Attend steering committee meetings
  ○ Execute project management processes: risk, issues, change, quality and documentation

○ Ensure project plan, schedule and budget are up-to-date

○ Detect and manage variances

Team Leader:

- Is responsible for managing one part of the project 'subproject'
- Only exists for large projects which are subdivided in different sections
- Ideally has project management skills (human resource management)
- Is usually junior project manager
- Reviews all sub-team deliverables, holds regular sub-team status meetings and provide regular status reports to project manager

Team Members:

- Are professionals assigned to perform specific tasks
- May report directly or indirectly to project manager

Project Plan and Scope

The most important measurement of project efficiency is time and project work is always being carried out under the constraint pressure of time. A lot of changes occur with the passage of time therefore it is very important to have an updated project plan. It is often the biggest source of information for stakeholders. An updated project plan is dynamic and changes all the time with project execution.

An example of a project plan for a reservoir engineering project is shown in Figure 3. Managing reservoirs is a comprehensive and consistent strategy for intelligent depletion of reservoirs. Representing reservoir management, four main factors are purpose, environmental considerations, reservoir description, and implementation of technology.

The reservoir engineering project definition initiates with the application of project management where operations are done properly through professionals on time and within budget. The 'purpose' of the reservoir engineering project starts with the background about resource production, where production optimizations including well potential evaluations, maximizing economic production rates, minimizing production downtime, identifying production restrictions and production planning outlook are observed. Along with the resource production, the economic value of the end product and key business indicators are also monitored with their effect on corporate policy of the organization. Resource conservation and storage are also important factors to consider.

With emerging technologies and new sophisticated techniques, the discussion on environment has also heated up. Governmental, semi-governmental, and non-governmental organizations are very active with reference to the environment and responsibility. This makes it an effective practice in modern day reservoir management. Recent studies have shown that disposal of brine in reservoirs cause more earthquakes than fracturing. Issues like this have to be taken care of and well addressed early in the project life. The government and local regulatory authorities should be on-board and well informed about the activities being carried out. Economic factors and their relation with capital and operation expenditures should be properly foreseen. There should also be public perception workflows initiated by the companies in order to let common people know about the latest technology and how the system works. This will bring clarity in the process allowing more space for the companies to operate.

A through insight of the reservoir is very essential for the effective reservoir management. Initial exploration phase contains uncertainties and less knowledge of the reservoir. With availability of production data and more wells being drilled in the reservoir, more information is available which makes our reservoir knowledge less uncertain. The

Figure 3. Workflow of exploration and production project planning and scope (concept modified after Walkup Jr., et al., 2006)

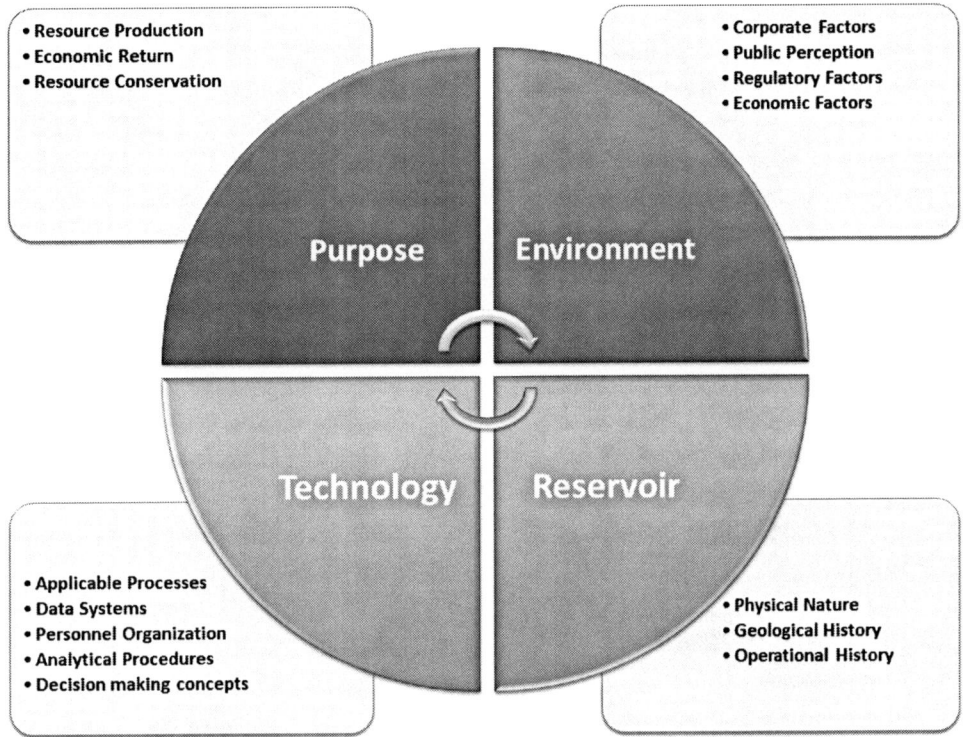

physical nature of reservoirs includes porosity, permeability, net-to-gross, lithology, and rock strength properties. A complete depositional history of basin imparts the geological knowledge of reservoir, which in combination with production data gives us almost a complete picture of reservoir behaviour.

The application of new technology on exploratory as well as mature fields yields economically efficient results. Organizations are very keen and focused on the incorporation of new techniques and methods to company assets (Yeo, 1993). This adds value to the assets producing beautiful blend of experienced professional with new technology implementation.

Keeping in view these basic elements, the project plan and scope is defined in reservoir management projects. Project scope outlines the precise explanation of expected project results or product for a customer, on external as well as internal point of view in specific, tangible and measurable way. The project scope will be fixed in the document and is agreed upon with the customer. Scope describes what is expected to be delivered to the customer and other participants for planning and measuring project success. It is a practice to make a checklist of all elements in order to complete the scope definition. The project scope generally contains following elements:

- Project objectives
- Deliverables
- Milestones
- Technical requirements
- Limits and exclusions
- Reviews with customer

Over time, it has been observed that many project scopes tend to expand and this has been termed as scope creep. Project character or scope statement help avoiding scope creep. In addition, setting clear project priorities help avoiding scope creep.

The management and/or customers define the order of priorities in accordance with their needs, which can be set arbitrary and vary from project to project. For the success and meeting or exceeding the expectation of customers or management of a project, three important factors are cost (budget), time (schedule) and performance (scope), as described in Figure 4. These factors are inter-related and depend heavily on each other. For example, when some resource is not available, it would cost extra time and money for execution of a particular project step.

Project Quality Improvement

Project quality improvement is ensured by progress and performance measurements. It is very essential to keep the whole project on track on time and within the given budget, specifically for large projects to evaluate and control the project at least at each benchmark or milestone. Project control is either simple or complex depending on the size of the project. Small projects can be completed without formal control whereas it is very essential for large projects. Control keeps projects on track by holding people accountable and preventing small problems to get big. Continuous quality improvement is ensured by effective feedback and efficient project control structure. It is very helpful right from the execution of activities for the goal and objectives. Appropriate level of control is very essential for the project success; too much control is time consuming whereas too little control is very risky. For effective control, the project should have a single information system where all up-to-date information is available. By building such a system, three key points are what data to collect, analysis of data and current status report.

Basic steps for the project control structure include:

- Service excellence
- Minimum baseline plan
- Measuring progress and performance feedback
- Maturity practice
- Comparing plan against actual
- Reviews and lessons learnt
- Corrective action
- Good and complementary practice

Figure 4. The triangle for project scope management with key elements

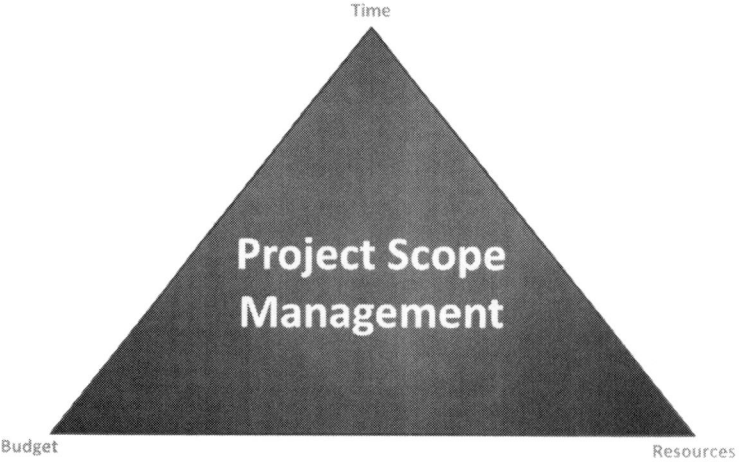

The control cycle (see Figure 5) consists of three main areas: monitoring, evaluation and controlling. The monitoring phase consists of performance comparison to predetermined set standards in the initial project kick-off meeting. Apart from standards, plans and objects are also compared. At each benchmark there is a comparison of how close (or far) the project is to set standards, plans, and objectives. The second area of focus in project control is evaluation in which the performance is evaluated with reference to strong areas and loopholes/pitfalls. Finally is the controlling area itself which is attributed by detection and rectifying the errors and prevention plans for their reoccurrences. It is very essential to figure out errors at an early stage to have sufficient reaction time.

Risk Management

Risk management is an essential part of the project lifecycle and adequate risk management can bring desired results without time or money loss. It is one of the key factors for the success or failure of the project. Since exploration and production projects are dependent heavily on the future expectations, there are always a lot of uncertainties involved. Uncertainties can result in an outcome that can be positive or negative evolving threats and opportunities. Threats are events having negative outcome whereas opportunities are events with positive impact on results. Uncertainty encompasses complete range of positive and negative impacts. It is very important to define a difference between uncertainties and risks. Risk is often defined as 'possibility of suffering harm, loss, or danger.' This is essentially having a negative impact on the project outcome whereas uncertainties can bring positive or negative outcomes.

Risk management controls occurrence of possible problems by planning proactive countermeasures. Thus, project team is directly avoiding prolonging time or exceeding planned budget. In risk management, an attempt is made to minimize adverse effects of possible loss by early warning systems. This can be achieved by identifying potential source of loss, measuring financial consequences and using control to minimize ac-

Figure 5. Project control cycle with different stages

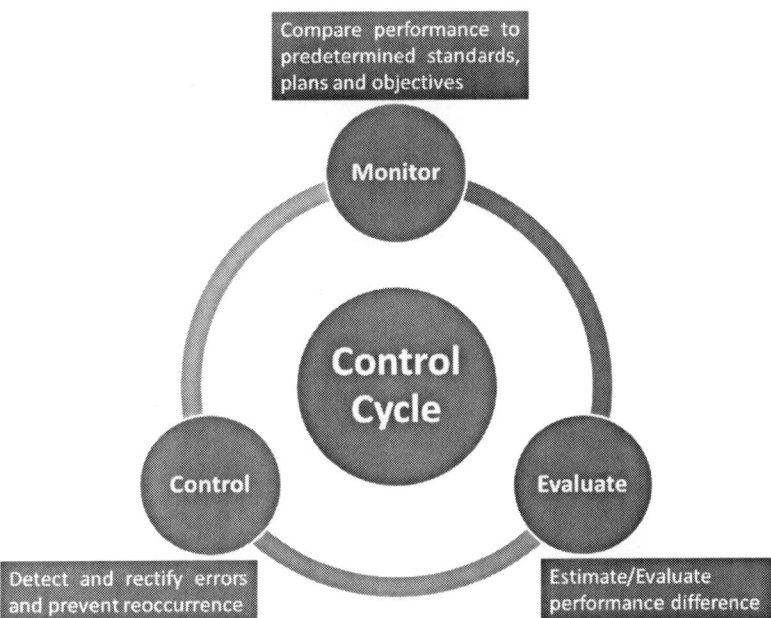

tual loss and their consequences. The purpose of monitoring all project risks is to increase value of each activity within the project.

Figure 6 explains the workflow of risk management process. Risk identification is the first stage of the process and is done by all stakeholders of the project and generally a watchful eye is kept on all project risks. The risk identification session mostly includes participants from following:

- Project team
- Risk management team
- Subject matter experts
- Customers and end-users
- Other project managers and stakeholders
- Outside experts

Participants may vary but the project team is always involved because they are dealing with the project day by day and therefore needs up-to-date information. External stakeholders and experts can provide an independent and unbiased opinion.

There are internal and external factors involved while dealing with risk identification. External factors include economic, social, legal, political, demand fluctuation and security situation whereas internal factors can be staff culture, capabilities, capacities, systems and technologies, procedures, leadership effectiveness and communication effectiveness.

Information gathering process starts with activities of prior projects and experience, developments, hints, failures and risks. 'Lessons learnt' from old projects is the first step of gathering information. With emergence of new tools and techniques, it is now possible to document, review, and analyse information right from the start of the project. The project plan and planning documents include project character, project scope, work breakdown structure, project schedule, cost estimates, resource plan, procurement plan, assumptions list, and constraints list.

Risk reports from previous projects explain the list of identified risks, list of potential responses,

Figure 6. Risk management elements and workflow

root cause of risks and updated risk categories. This information can be very helpful for risk identification for new project.

Risk identification is done in a sequential way to ensure all the important activities and potential consequences. The possibility of outsourcing the risk management process was also tried in some exploration and production companies but an in-house approach has been proved more effective as experienced over the years.

Risk analysis is based on the risk identification and covers a complete continuous evaluation, which should be realized qualitatively as well as quantitatively to analyse all possible risks. Analysing risk helps in prioritizing and helps identifying consequences and organizational goals. The demands of risk evaluation include:

- Objectivity
- Comparability
- Quantification
- Consideration of interdependencies

There are many techniques (matrix mode, failure mode, effect analysis, etc.) used for analysing risk but the most common used of them is 'scenario analysis' which simply consists of probability of events and impact it has on the project. It should be decided which level will be used for risk evaluation before starting a risk analysis. Traditionally these can be marked with very low, low, moderate, high, and very high. Quantitative numbers can be decided depending upon the parameters and projects. After risk evaluation, its effect on time, cost scope, and quality is measured along with the probability of risk.

Risk response refers to decision of reaction after a risk has occurred. There can be a choice between mitigate, avoid, transfer, share or retain the risk. The project manager chooses the appropriate options. Mitigation means a reduction of the impact and possibility of risk occurrence. In such case, testing and prototyping can be done which

means testing the project at a smaller scale. This can be very helpful for preparation of the project team and foresee the possible risks. However, cost and time cannot be mitigated because money is already spent and time is gone. The risk management however, can solve the issue by keeping extra (backup) time and budget for unseen events. This can be done before starting the project. Avoiding risk is often proved as more drastic approach since whole project schedule can effect by avoiding a particular risk. An example of avoiding a risk could be using well-known technology in a large project instead of experimenting new technology. Risk transfer is another technique in which risk is not completely removed, eliminated or dampened but transferred. A common example of that is outsourcing which is often done but risk transfer in this way also costs extra money.

Risk control is the very last step in risk management including executing the risk response strategy, monitoring and triggering events, initiating contingency plans and continuously watching for new risks. In risk control, a change management system is an important part to adjust for the changes in scope, budget, and schedule, which has to be dealt after start of the project:

- *Change in Scope* can be dealt with when the client or manager wants to implement an extra feature or change in design/execution which really represents a big change
- *Implementation of contingency plan* is applied when a risk has really occurred and counteractive measures has to be taken. These need resources in terms of cost and schedule and so represent a change to baseline
- *Improvement changes by project team members* is initiated when a change is associated with supplier and a new supplier can deliver the same quality goods at cheaper rate

All changes usually represent a big challenge to the team as a whole and to the project manager. In exploration and production projects, changes are unavoidable and are taken care of in project planning and scope definition.

Time Management

In the exploration and production business world, time is today a premium. Concentrating on work and just not staying busy is a key value, which is implemented in the project team, and comes when each member of the team has a good idea of the concept. For the projects, logs are maintained which help tracking the time and outcome and value of time is transparent and accessible for everyone. Project teams are made focused by goal setting practices and this practice is implemented on all levels of the project team. While setting goals with reference to time, the attributes like clarity, challenge, commitment, feedback, and complexity are taken care of. A very famous practice while setting goals is called SMART goals where SMART is the abbreviation of specific, measurable, attainable, realistic and timely. When goals are set they need to be prioritize. This is the time where distinction between importance and urgent is made. For a successful time management, interruptions should also be managed and proper backup time should be reserved ahead of the project start.

Scheduling is an important phase and refers to a vital tool in time management. It represents a form of organization in which important tasks are recognized and performed at priority. The key factors in scheduling include:

- Allotting time for all necessary tasks and functions
- Cutting off unproductive interruptions
- Limiting meetings to the scheduled time
- Allocating time for breaks

It is practice in exploration and production projects to develop a master schedule at start and split it up into medium and short-term (hourly or daily) sub-schedules.

Time management adds value to the project outcome and is an essential process to finish the project on time. All the activities in the project should have a deadline for an up-to-date time plan, a record of activities performed and activities pending. Different processes involved in project time management are displayed in Figure 7. All of these processes interact with each other and with processes in other organizational competence. Each process involves effort from one person or group of persons from the team. Each process has different tools and techniques and occur at-least once, although some processes can occur more than once as well.

Activity definition involves the definition of operations to be performed for any project. It is an essential phase and consistently used throughout the project lifecycle as the basis of all further steps. Defining operations means determination and documentation of operations and transections are developed such that they are structured and defined in each individual work package. This is done because each activity definition phase needs time and effort by work breakdown structure specified and partial delivery items are developed.

Activity schedule includes definition of operational sequences with documentation and determination of relationship between them. At the end of this practice, a final process flow chart is available defining a realistic and practical time plan. For short-term projects this activity can be done manually but for large projects with multiphases, a sophisticated computerized activity schedule is made because it is easier to handle and update the activity plan.

Duration estimation involves the number of work periods for activities and generates it when information is required on:

- Scope of work
- Required resource type
- Estimated resource quantity
- Resource availabilities

The input for the activity duration estimation originates from the person or group on the project team who is most familiar with the nature of work content in the specified schedule activity. The process is highly dependent upon quality and quantity of input data and the resources available for the project. All data and assumptions made for the duration estimation are documented for each activity duration.

Schedule development includes the set of instructions for a whole project activity from start to end. If the project does not finish in time, additional or backup time has to be planned before start and there should be a schedule development. Mostly time schedule development is repeated many times before the project time schedule is finished.

Schedule control is concerned with three basic principles:

- Influencing the factors which create schedule changes, determining the current status of project schedule
- Determining the changes in project schedule
- Managing the actual changes as they occur

Schedule control is connected with other control procedures.

Audit, Documentation, and Closure

Audit

An audit may be defined as evaluation of a person, organization, system, process, project or product. Audits are performed to ascertain validity and reliability of the information. It also provides an assessment of a system's internal control. The ultimate goal of an audit is to check if the project is according to the expectations, for example is it within time, budget, etc. In general, an audit is a tool to improve quality.

Figure 7. Sequential time management workflow

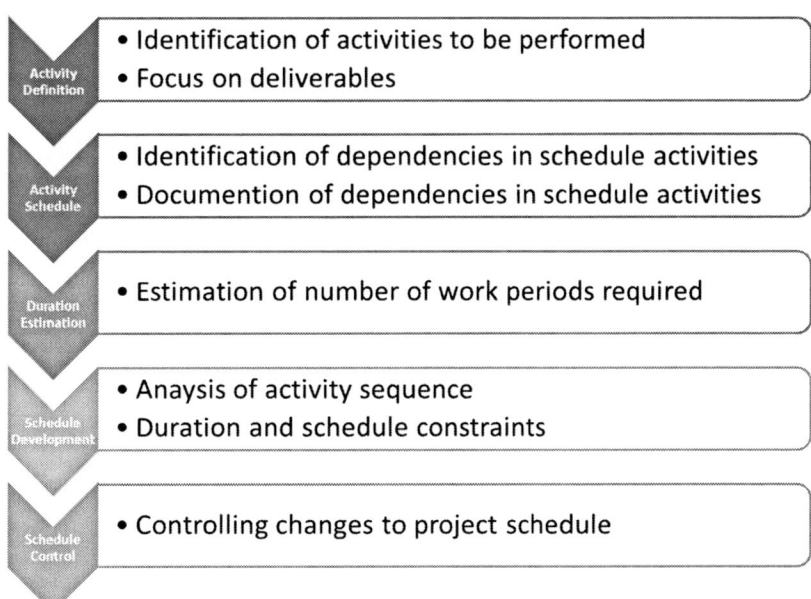

Activity Definition	• Identification of activities to be performed • Focus on deliverables
Activity Schedule	• Identification of dependencies in schedule activities • Documention of dependencies in schedule activities
Duration Estimation	• Estimation of number of work periods required
Schedule Development	• Anaysis of activity sequence • Duration and schedule constraints
Schedule Control	• Controlling changes to project schedule

There can be internal or external independent audits, also single or multiple audits depending on the needs of the project. The current status can be analysed and ways to move forward are defined depending upon the result of an audit. Audits play an important role during installation, certification and running of management systems as well as for the evaluation of projects.

Main types of audits include:

- The product audit
- The process audit
- The system audit
- Internal and external audits

Motives of the audit are clearly defined and fitness for use, conformance to requirements and pursuit to excellence can be some examples of motives of audits. Thus, audits are part of quality assurance services and enhance project capability a great deal. Recently developed idea of audit cycle is very efficient in the exploration and production industry where the audit is being performed continuously as the projects moves from one phase to another and actually is a part of the project internal deliverables.

Documentation

The project documentation system is a very helpful tool for the systematic documentation and in case when one of the documentation personnel is not available. It enhances internal and external communication and facilitates project management. Too much of documentation is a problem as too little. It is essential to have a balanced, precise and to the point documentation.

Best practices in documentation have been proved very helpful yielding better project transparency. It is also very essential for leading, defining, planning, controlling, organizing and closing a project. It provides order and structure to the project by giving direction and setting parameters

increasing efficiency and effectiveness. Project documentation may consist of following items:

- Procedures
- Project Manual
- Project Library
- Flowcharts
- Forms and Reports
- Memos and Newsletter
- History Files

Project Closure

Project closure is the officially last stage of any project and with reference to project management. It is something more than just packing and move to the next project plan. The closing process mainly includes a separate closing contract and administrative workflow at the end of each project. A successful project end has to be defined in the scope of the project in the kick-off meeting. Traditionally, projects are closed when they meet or exceed the stakeholders' expectations. Another way can be the end of process before completion due to unavoidable circumstances. Different types of project closures include:

Normal closure refers to the common circumstances where the project goals and objectives are accomplished up to the stakeholders' satisfaction. This is the ideal close and is wished before start of any project.

Premature closure refers to the finishing of project while some parts of the project have been eliminated. This type of closure can occur by the pressure of managers or steering committee because of some reason. An example of such a reason can be declining market situation. The associated risk and consequences are reviewed carefully and assessed by all stakeholders and management.

Perpetual closure is implemented on types of projects, which are never coming to an end because of many reasons. Most of the times the reasons are constant add-ons rather delays. The managers are constantly implementing small changes for quality improvement and as a consequence the prolonging of project closure. In such a case, it is advised to refer back to project scope statement. Such projects can be handled by limiting time, budget, or redefined project scopes.

Some projects simply fail because of circumstances. In practice, it is possible that planned projects have a developed prototype for new product and technologies, so that something can be extracted out of the project at the end.

Contract closeout refers to the formal closing of the projects and the completion is done according to terms set in the contract. The described work is completed to satisfaction and with a required level of accuracy. These levels are agreed before start of the project between steering committee, project manager, and stakeholders. Contract closure also includes the step of recording and archiving for the future references. Some contracts have special conditions for the closeout like there may be formal written information about the project completion and the stakeholders have to get results and quality checks. If any result or services are not according to the agreed standards, these have to be upgraded until the time it is up to standard mutually agreed on.

Administrative closure includes the steps of:

- Performance measurement
- Product documentation
- Archiving
- Formal acceptance
- Lessons learnt

Conclusions

The section of project management is focused on set of activities being performed in the exploration and production industry and the sequence starts from project organization, covering all elements of project lifecycle down to project audit, documentation, and closure.

Project organization is the first stage and can easily determine success or failure of projects. Setting clear roles and responsibilities is very essential for the project success right from the beginning of project. In addition, it is very necessary to communicate professional that everyone know their roles and responsibilities. The centralizing role of project managers is to harmonize the team achieving final goals/objectives. Project managers act as backbone and play a key role by interacting with different levels of hierarchy understanding their needs and demands. Along with project manager, steering committee and advisory committee plays significantly important role and of course, the key assets for any project team are the people working for the project.

Project plan and scope definition is also very important as clear statements about what to achieve and a mature rundown sequence of activities play a key role in project success focusing on deliverables like time, budget, and quality. The continuous quality improvement process also helps achievement of set goals. Such a process includes control system on the basis of monitoring, evaluating and controlling.

The process of risk management plays a key role in identifying and developing early plans in order to foresee and react at earliest stage with any possible risks in the project. Proper analysis, response and control criteria development helps in minimizing the damage from potential risks in the project.

HEALTH, SAFETY AND ENVIRONMENT (HSE) MANAGEMENT

Energy by far is the most essential commodity in almost every aspect of our life. It also plays a significant role in defining economic success or failure. According to International Energy Agency (IEA), more than 80% of total consumed energy is obtained from fossil fuel, which includes coal, oil, and gas. Two long-term energy challenges are always in focus by using fossil fuels, which are: tackling climate changes by reducing carbon dioxide (and other greenhouse gases) emissions and ensuring secure, clean and responsible energy. The term "responsible energy" covers a whole wide spectrum of Health, Safety, and Environment (HSE) issues. HSE ensures health and safe operating environment and our lusting energy behaviour is not fabricating any possible climate risks for our rational and sensitive future generations.

Today, proficient HSE practices are undoubtedly one of the main focuses in exploration and production industry. The idea of HSE was implemented belligerently in last three decades. New and improved techniques are being adopted to make workplaces healthier, safer and environment friendly. As a result, rates of deaths, injuries, and work-related illness have declined. Carrying on with this positive trend and to achieve the landmark of excellence, companies conduct HSE trainings and provide appropriate facilities to ensure safety for people and environment. In addition, information of HSE related accidents, near misses, and pitfalls have to be reported to administrative authorities, adding value to efficient energy production.

As a company, huge sums of money are invested on trainings, facilities and ensuring safer and environment friendly workplace in difficult operations. The HSE is given same priority and managed in the same way as production and cost control. Activities are only undertaken when an agreed HSE compliant plan has been developed. Every effort is made to minimize the occurrence of accidents and occupational diseases by preventive measure. However, in the event of an accident, proper contingency plans and emergency measures are initiated to limit the consequences as far as possible. The responsibilities are enhanced manifold where there are joint ventures and operator/consultant interactions are involved. The exploration and production projects on average involve 6-10 companies onsite, responsible for different types

of tasks, depending on the scope of project. There is always a foolproof HSE management system for all companies in operation. In addition, there is a combined field HSE plan for overall site, ensuring the safe workplace. With offshore operations, it becomes even more challenging because there are limited spaces and escape routes and the artificial exploration or production islands have to be out in the ocean with extremely dangerous weather conditions. It must be ensured that risks in changing workplaces are properly monitored and controlled.

Policy Guidelines and Regulations

Internal Policies and Regulations

It is very important to develop a genuine policy and realistic guidelines for the company in general and also before the start of any operation, keeping in view the specific operating environment. Today, almost all companies have developed policy and guidelines keeping in view the safe operations and responsibility to environment. In such a policy, the company commits to conduct activities in such a way that the foremost account of employees and other people's health and safety and to give proper regard to the conservation of the environment. In addition, the terms Health, Safety, and Environment (HSE) are defined and stance of the company on these is put in black and white (Asrilhant, et al, 2004).

The common HSE aspects put into policy and guidelines can include:

- The company should conduct a way to avoid harm to the health of their employees and others; and promote the health of their employees.
- The company should work on the principles that all injuries should be prevented and promote the high standards of safety consciousness and disciple.

- The company should pursue in their operations progressive reduction of emissions, effluents, and discharge of waste materials that are known to have negative impact on the environment.
- The company should assess health, safety, and environmental matters before entering into new activities and reassess them in case of significant changes depending on the circumstances.
- The HSE policy of the contractors should be fully in compliance with the company. Any changes should be known, discussed, and addressed before the start of operations.
- The company should recognize the concerns of employees, shareholders and society on HSE matters; provide relevant information and discuss company policies and practices.
- The company should develop and maintain contingency procedures in cooperation with authorities and emergency services to minimize harm from any accidents.
- The company should work with the government in the development of improved regulations and industry standards which relation to HSE matters.
- The company should conduct or support research towards the improvement of HSE aspects of their products, processes, and operations.
- The company should facilitate the transfer to others, freely or on a commercial basis, on developed knowhow in these fields.

Policy development phase should include a very thorough and detailed study of the operational conditions. Appropriate knowledge and experience in similar conditions is significantly helpful. The overall workflow of policy development can be summarized as:

- Proper inspection and potential hazards identification

- Investigation of accidents and incidents in similar operational conditions
- Appropriate trainings and facilities required
- Assessment of safety cases
- Information management and guidance
- Implementation of technical standards

The International Regulatory Authorities

Over the years, governments have become increasing aware of their responsibilities to protect their citizens and environment against undue hazard and pollution. This has resulted in the continuous change in the regulatory field. With the production in oil and gas industry and implementation of proper managerial skills, it has been established that proper management of HSE requires special attention of top management of companies in order to other disciplines like engineering, operations, finance, planning, etc. The company top hierarchies consider the HSE as a top priority, which yields to a special improvement and dedicated considerations for HSE.

The petroleum industry has created several agencies to formulate and represent the industry position in international developments. Some of these are CONCAWE (The Oil Companies European Organization for Environmental and Health Protection, Brussels), E & P Forum (Oil Companies International Exploration and Production Forum, London. In addition, IEA (International Energy Agency), and EIA (Environmental Impact Assessment) keep a close look on HSE issues in exploration and production sector. The industry produces many documented guidelines which are made available to the governments, non-governmental organizations and common public. This information plays an important role in government attitudes and regulatory activities.

Internationally, IMO (Inter-Governmental Maritime Organization, London), ILO (International Labour Office, Geneva), and UNEP (United Nations Environmental Program, Geneva) are the leading world-wide agencies with regulatory impact. These organizations are working effectively, closely together with the bodies like EEC and NW European Conference in a regional structure.

International Maritime Organization

The International Maritime Organization (IMO) is a UN legislative body for maritime matters. It is the United Nations specialized agency with responsibility for the safety and security of shipping and the prevention of marine pollution by ships. It is located in London and has around 120 governments memberships. The main activities of the IMO include:

- **Maritime Safety Committee:** Safety of Life at Sea Convention (SOLAS), Safety Code for Mobile Offshore Drilling Units (MODU), safety zones, emergency preparedness, survival training, life-saving appliances certification for crews, etc.
- **Maritime Environmental Protection Committee:** Oil/water monitors, ballast-water reception facilities, use of dispersants, discharges from offshore platforms, etc.

The International Labour Organization

The International Labour Organization (ILO) in Geneva contributes mainly to the health and safety regulatory fields. ILO was emerged with the League of Nations from the treaty of Versailles in 1919. It was founded to give expression to growing concern for social reform after the World War I and the conviction that any reform had to be conducted at an international level. It has around 183 member states. It consists of tripartite structure where representatives from employees, workers and governments have an equal voice. This promotes the social dialogue.

The main focus of ILO is on safe conditions at the workplace, safety codes, reporting and ac-

cident statistics. Its influence is towards providing natural labour unions with goals, standards, codes of practice, etc. ILO has four principal strategic objectives:

- To promote and realize standards, and fundamental principles and rights at work.
- To create greater opportunities for women and men to secure decent employment.
- To enhance the coverage and effectiveness of social protection for all.
- To strengthen tripartite structure and social dialogue.

United Nations Environmental Program

United Nations Environmental Program (UNEP) is the United Nations agency to promote worldwide environmental awareness and protection. Regional action plans to unite nations in the protection of a common sea area have been initiated from Geneva office, e.g., the Mediterranean, the Gulf area, the South China Sea and the Caribbean. The mission of UNEP is to provide leadership and encourage partnership in caring the environment by inspiring, informing, and enabling nations and peoples to improve their quality of life without compromising future generations.

The UNEP Paris office for the 'Industry and Environment' provides, in close cooperation with the exploration and production industry, active guidance to governments on issues of environmental concerns, such as the environmental effects of drilling mud discharges, the use of dispersants in oil spill combat and acid rain.

Occupational Safety and Health Administration

The Occupational Safety and Health Administration (OSHA) is part of the United States Labour Department and is based on occupational safety and health act 1970. Its main objective is to assure safe and healthful working conditions for working men and women by setting and enforcing standards and by providing training, outreach, education, and assistance.

OSHA and its partners have currently more than 2400 inspectors, 550 state consultants, and complaint discrimination investigators. Apart from that, they also have engineers, scientists, and state facilitators. These staffs establish protective standards, enforce those standards, and reach out to employees and employers through technical assistance and consultation programs. The mission of OSHA is to ensure the safety and health of American workers. This is achieved by setting employer's responsibilities, employee rights and standards (see Table 1).

European Agency for Safety and Health at Work

The European Agency for Safety and Health at Work (European OSHA) was set up in 1996. It collects, analyses and communicates OSHA-related information across the European Union. The primary mission of the European OSHA is to make Europe's workplaces safer, healthier, and more productive, by promoting a culture of risk prevention. It is located in Bilbao, Spain.

European OSHA:

- Raises awareness and disseminates information on the importance of worker's health and safety for European social and economic stability and growth.
- Designs and develops hands-on instruments for micro, small and medium-sized enterprises to help them assess their workplace risks, share knowledge and good practices on safety and health within their reach and beyond.
- Works side-by-side with governments, employers' and workers' organizations, EU bodies and networks, and private companies. The voice is multiplied by occupational safety and health network represented by a dedicated focal point in all EU Member States, EFTA countries and candidate and potential candidate countries.

Table 1. The responsibilities and rights of employer and employees (after OSHA homepage)

Employer must	Employees have the right to
Follow all relevant OSHA safety and health standards	Working conditions that do not pose a risk of serious harm
Find and correct safety and health hazards	Receive information and training about chemical and other hazards, methods to prevent harm and OSHA standards that apply to their workplace
Inform employees about chemical hazards through training, labels, alarms, colour coded systems, chemical information sheets etc.	Review records of work related injuries and illnesses
Notify OSHA within 8 hours of a workplace fatality or when three or more workers are hospitalized	Get copies of test results done to find and measure hazards in the workplace
Provide required personal protective equipment at no cost to workers	File a complaint asking OSHA to inspect their workplace if they believe there is a serious hazard or that their employer is not following OSHA rules.
Keep accurate records of work-related injuries and illnesses	Use their rights under the law without retaliation or discrimination. If an employee is fired, demoted, transferred or discriminated against in any way for using their rights under the law, they can file a complaint with OSHA.
Post OSHA citations, injury and illness summary data and the OSHA Job Safety and Health – It's the Law poster in the workplace where workers will see them	
Not discriminate or retaliate against any worker for using their rights under the law	

- Identifies and assesses new and emerging risks at work and mainstream occupational safety and health into other policy areas such as education, public health and research.

At regional level, in several areas of world, governments have come together to jointly agree on measures that will improve the quality of life in their area of concern. Examples are codes of safe working practice in industry, training and certification requirements, emission limits, requirements for environmental impact assessment, identification of blacklisted substances.

Where possible, the petroleum industry endeavours to provide timely technical recommendations. This is normally done through contributions for instance from E&P Forum, OCIMF or CONCAWE. Sometimes, however, there is also dedicated regional industry co-operation, which can address specific concerns such as the North Sea Operators' Clean Seas Committee. Whatever the international/regional arrangements may be, it remains the prerogative of the national government to set its own standards and requirements for industry operations within its jurisdiction. Major accidents, such as 'Piper A,' 'Alexander Kielland,' 'Exxon Valdez,' and 'Ocean Ranger' often trigger specific national precautionary measures such as increased stringency in certification requirements or emergency measures, special shipping lanes and requirements for unlimited compensation.

The Cullen Inquiry into the Piper A disaster had a major impact on the regulation of the petroleum industry as a whole, particularly in the UK and Western Europe. The principles formulated by the Inquiry also affected the legislation for the petroleum industry in many other parts of the world. These principles include:

1. **The industry is responsible for its own safety management:** This requires competent management in general, not technological solutions or specific problems.

2. **The regulator should set the objectives to be achieved, not prescribe the means to achieve them:** This approach requires the operator to identify the means to achieve the objectives. It promotes a constructive dialogue and ensures visibility of the objectives. It encourages 'fit for purpose' designs and procedures and a constant pressure for improvement.

3. **The operator will implement Safety Management Systems:** The operator will be required to demonstrate systems in place to manage the safety aspects of its business effectively, particularly with respect to safe design and operation of its installations, and to verify effective working of systems. The regulator will audit both processes: the system and the verification (internal audit). Confidential-Property and Copyright: SIPM, 1991 2-9 Health, Safety and Environmental Conservation.

4. **All installations will be subject to Formal Safety Assessment:** The operator will be required to demonstrate for each installation that hazards have been systematically identified and eliminated where possible and that the means to control the remaining hazards, including contingency plans, are in place.

Codes and Standards

Codes and standards for safe design and practice are issued (and sometimes enforced) by a number of bodies. The intent is to aid the designer and to assure that installations and operating practices minimise the risk of mishaps.

Design and operational standards and specifications are intended to embody the entire engineering knowledge and experience, within the framework of codes and regulations. They provide the basis for a uniform minimum level of engineering quality and serve as a reference level for the necessary audits and Quality Management. The following types of code should be part of local design and practice considerations:

1. **International codes:** Codes prepared and issued by international bodies such as ILO, UNEP, ISO, IMO, are widely used. A typical example is the Mobile Offshore Drilling Unit (MODU) 'design code,' issued by IMO.
2. **National codes and regulations:** These are issued by governments and are legally enforceable, e.g. Health and Safety at Work Act, UK; Norwegian Petroleum Directives (NPD), Norway; Mining Regulations, Netherlands; etc. Frequently a well-considered national code is also used in other countries for local regulation.
3. **Codes and regulations issued by institutions:** These are prepared and issued by recognised national and industry institutions, such as BSI, ANSI, DIN and IP, API, NFPA, ASME, and are frequently adopted by industry as the principal 'design code.'
4. **Company guides and practices, standards and specifications:** Exploration and production companies have developed their own standard guidelines and practices, which are in compliant with the local and international standards of designs and standards.

Safety in Design

Safety in design should start very early in the planning stage of a project. It should be clear and sequence of steps should be followed for setting up a design for new installation or for modifications of existing facilities. The generalized hazard related aspects are highlighted in Figure 8.

Figure 8. Elements of safety in design with associated potential hazards

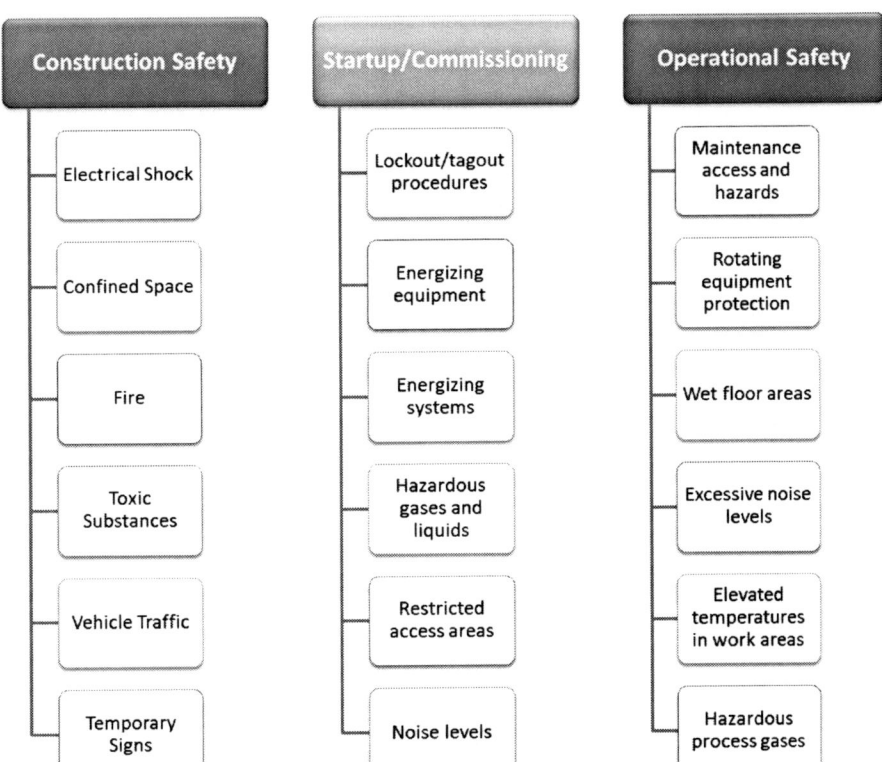

It is important to develop design suggestions to minimize/eliminate hazard.

A workflow of potential sequence of events is very helpful at the very start of the project. A typical workflow can include following components (see Figure 9):

- Establishment of a safety philosophy
- Determination of standards, which will be used
- Identification of the hazards at the earliest possible stage
- Designing out hazards, develop procedures and train personnel to control all operations
- Limiting the effects of possible unwanted events, e.g., fire, by:
  ○ Early detection
  ○ Protection (fire-resistant wall, etc.)
  ○ Adequate fire-fighting facilities
- Formally reviewing the design at specific stages, a review of layout aspects, Quantitative Risk Assessment for evaluation of alternatives and a pre start-up audit

Project Organization

The communication between different departments in operation is very essential for successful risk free operations. Personnel responsibility for project design, operations, maintenance, and safety must integrate and cooperate interdisciplinary right from the beginning of the project. It is very essential for the safety, efficiently operated, environmentally acceptable, and well-maintained installation.

A structured approach to hazard identification and control will facilitate communication and ensure that the key hazards are properly managed at every phase from initiation through production and abandonment.

Hazard Identification in Facilities Design

Early identification of the hazard is significantly helpful in successful mitigation and minimum hazard aftermaths. A workflow consisting of sequential checks is the normal routine adopted

Figure 9. The process lifecycle of safety in design

in exploration and production industry. A list of checks and a short description is:

1. Design and operating standards
2. Flow scheme checking
3. Site layout drawings
4. Hazardous area classification
5. Equipment specifications
6. Shutdown/Blow down philosophy
7. Structural layout design

Design and Operating Standards

The most important aspects in the facilities design are the safety implications and ensuring that standards are consistent with company and international requirements. With reference to that, up-to-date and flawless design and operating workflows should be developed. Emergency response and manning policies are in particular very important.

Flow Scheme Checking

The process flow diagrams and process-instrument diagrams should be thoroughly discussed and reviewed after they have been finalized by the team of engineers. This step is important in order to keep the track of any unseen incident or dangerous situation that may arise during the operations.

Special attention should be given to the modifications (change control) and updating of original documentation (drawings, manuals etc.) after modifications have been made and re-issuing them to all the stakeholders of original documents.

Site Layout Drawings

Site layout drawings should be checked carefully with reference to the distance to neighboring installations, operational aspects, ergonomics, topographic data (levels), access for fire-fighting, compartmentalization of fire hazards and emergency evacuation and escape routes.

Hazardous Area Classification

It is significantly helpful to classify the hazardous areas, especially where flammable materials may be present. The corresponding drawings are mandatory. This should be carried out after the review of flow schemes and before the layout is finalized.

Equipment Specifications

Ordering or purchasing equipment is a normal practice. It is very essential to include all safety aspects such as safety requirements for the design of vents and drains, details of special connections frequently opened and closed, seals and packing bushes, relief valves, purging facilities, ignition systems, noise and vibration limits. Critical control equipment may have to be fire resistant e.g., valves, actuators, cables, cable trays etc.

Shutdown and Blow Down Philosophy

Process upset and emergency situations can arise any time. Therefore, it is very important to develop a strategy for the shutdown in case of an emergency situation. These action plans must be reviewed in consultation with the operations and safety departments using aid and effect diagrams. Fire protection requires isolation and depressing of vessels.

Safety standards must be kept in mind where the blow down is needed. Approximately 15 minutes is normally allowed as blow down time to achieve half of the operating pressure or 7 bar whichever is less, as per API 521. However, the adequacy of code requirements should be checked against the specific fire and explosion scenarios to affect the facilities. Where high pressure gas power occurs, the survival times of vessels and pipework can be considerably less than 15 minutes.

Structural Layout Drawing

Structural layout drawings should show fire barriers, structural fire proofing and fire walls; each bounded area requires separate valve drainage. On offshore platforms and installations, the fire walls should also be gas tight and explosion resistant.

It is the responsibility of the project engineer to follow up all recommendations arising from hazard identification at the drainage stage. A model or CAD (Computer Aided Design) is very helpful to assist hazard identification as a part of design procedure for very complex models.

Reviewing Safety in Design

Safety plans often need a backup or review and certain techniques are used as standard practice in exploration and production industry. These techniques should be applied in a structured and consistent manner for hazard identification and assessment and safety reviews.

The industry standards for such environments are the Hazard and Operability Studies, which provide a systematic analysis of the system compartments for their behaviour/failure under circumstances deviating from normal, a so called 'what-if' analysis. Similarly, the failure modes and effects approach, analyses the effects of failure of system components on the performance of the entire system.

The Hazard and Operability Studies should be carried out for all designs and modifications, especially when proven standards are not available. It is recommended to develop a fire protection analysis as a systematic approach to review the fire protection of new and/or existing installations. It is very important to have experienced professionals working in coordination with young engineers.

Quantitative Risk Analysis is normally adopted in the exploration and production industry for the identification of potentially hazardous events and estimation of likelihood and consequences to people, environment, and resources of accidents developing from these events. The entire process of risk analysis, interpretation of results and recommendation of corrective actions is usually called "risk assessment."

Quantitative Risk Analysis is most useful in project definition and conceptual design stages to compare the risks inherent in alternative schemes. In the project definition stage, only a coarse quantitative risk analysis is possible which should be followed by a detailed analysis later in the detailed design stage.

Examples of quantitative risk analysis include comparison of manned versus unmanned or temporarily manned platforms, the need for subsea safety valves in pipelines, platforms with integral accommodation or separate living quarter platforms, etc.

Special Operative Conditions in Design

The most important considered hazards in the exploration and production industry involve fires, explosions or the dispersion of gas clouds. The physical phenomenon and the results can be modeled mathematically and it is very handy to maintain a suite of computer models for fires, radiations, explosions and dispersions, etc.

It is pretty complex to model the explosions in semi-confined areas such as offshore modules or compressor houses.

The special operative conditions which must be considered in design phase include:

- H_2S Hazards
- Other Toxic Components
- Dispersion of Flammable and Toxic Gases
- Heat Radiation
- Explosion Hazards
- Noise Abatement
- Ionising Radiation
- Design of Specific Process Safety Equipment

H_2S Hazards

Hydrogen sulphide which is an extremely toxic gas may be present with oil and gas. This is because when source rock is matured to yield oil and gas, it needs specifically reducing environment. Sulphur rich reducing environments in some fields cause H_2S production.

At low concentrations, H_2S has a smell of rotten egg. At high concentrations, sense of smell is anaesthetised and a victim can lose consciousness and die if not rescued quickly.

The H_2S requires extensive precautionary measures, in particular at exploratory drilling sites but also in facility design and operations. Special sour gas design standards should be strictly followed. H_2S may also form at the surface of stagnant oily water (as a result of sulphate reducing bacteria activity) in storage tanks, etc. Strict operating procedures must be enforced and emergency preparedness maintained in all situations where H_2S may occur.

Whenever it is necessary to work in the areas where the concentration of H_2S in the atmosphere exceeds 10 ppm by volume, protective equipment must be worn (10 ppm is termed as the Threshold Limit Value or TLV and is a time-weighted concentration for an eight hour period to which a person can be exposed without harmful effect).

Other Toxic Components

Crude oil and natural gas may contain components which are toxic and/or carcinogenic at low concentrations, such as benzene (TLV=10ppm) and toluene (TLV=100 ppm). Toluene and to some degree benzene can also be absorbed through skin following skin contact with the liquid. Proper protective measures must be taken to ensure that exposure limits are not exceeded.

Displacement of Flammable and Toxic Gases

When venting significant gas volumes, the extension of the gas cloud should be calculated with regard to lower flammable limits or Threshold Limit Values (TLV) under various weather conditions. The design and citing of the vents should be crosschecked with this aspect.

Heat Radiation

Heat Radiation produced by flares or by accidently ignited vents should be calculated and the recommended practices and guidelines must be followed.

Siting, design and operation of vents and flares should take into account the expected radiation levels. The maximum exposure limit for personnel is 6.3 kW/m^2 in freely accessible areas from which easy escape is possible.

Explosion Hazards

The design should take account of the possible effects of a confined or partially confined explosion. Buildings should be designed and sited to withstand the likely blast effects. Techniques are available for predicting overpressures.

Noise Abatement

Noise levels of equipment, flares or rotating equipment should specifically be specified by the designer to keep noise levels at the installation within required limits. Every company have developed their guidelines which specify that persons should not be exposed to steady noise levels of more than 120 dB or impulse noise levels above 135 dB with or without hearing protection, or to a personal noise doze above 85 dB per shift. The distance from the noise source should also be taken care of.

Ionising Radiation

Radioactive materials are sometimes present in gases and liquids which may cause enhanced ionising radiation. Typical sites can include condensers, separators, slug catchers, etc. Onsite investigations are required to assess the need of removal or neutering action. Some level detectors use radioactive sources which must be safely disposed when replaced.

Design of Specific Process Safety Equipment

In consultation with the equipment and safety engineers, special attention should also be given to those items of plant where non-routine operations are carried out and/or which are required to isolate differently pressure-rated or classified sections of the facility from each other. These include:

- Pressure relief valves and systems
- Vents, vent systems
- Blow-down systems
- Flares, flare systems
- Flame arrestors
- Shutdown valves
- Drains, drain systems
- Equipment isolation procedure and hardware
- Heating, ventilating and air conditioning systems

Planning and Cost Estimating Equipment

All installations should be reviewed with regard to application of safety equipment and safety equipment should be considered in the cost estimate of an installation. There should be guides for available safety equipment and suppliers complied by each company (Carrillo, 2004).

Special attention should be given to maintain the operability of safe equipment particularly after long periods of non-use and/or adverse climate storage conditions.

Detection and Alarm System

Detection and alarm systems should aim at providing prompt warning for loss of containment and/or fire and automatic initiation of appropriate action.

Permanently Installed Instrumentation

There are placed in the field where a fixed instrumentation is required. These can be used for gas detection (toxic gases, H_2S etc.) and fire detection (heat, flame, and smoke). Fixed detectors require a reliable electric power supply and often signals are relayed to a panel located in a central control room.

Suitable measures to allow testing of such systems should be included in the design. Measures to ensure the detector operability during an electric power failure should be considered.

Portable Detection Instruments

In addition to the fixed detectors, portable (hand or belt carried) instruments will be required in areas where the presence of flammable and/or toxic gases can occur or where an abnormal amount of inert gases can cause oxygen deficiency e.g., nitrogen, carbon dioxide etc.

Fire-Fighting Facilities and Equipment

All facilities which contain flammable material will require fire protection to be considered at an early stage in the design. The facilities selected should be based on the risk of fire, its consequences and the probability of successful control by the system proposed.

Facilities can include:

- Fixed Facilities
- Portable Facilities
- Fire Water Pumps
- Fire Trucks (Fire Stations)
- Fire-Fighting Vessels
- Passive Facilities

Safety in Operation

The operations cover both technical operations (procedures and housekeeping in drilling, production, maintenance, etc.) and supporting operations (transport, fire/gas protection, emergency contingency planning, survival, search and rescue, etc.)

Safe Operating Practices

Several safety concerns apply equally as to design e.g., H_2S, noise; others apply specifically to operating activities such as handling of explosives, welding and cutting, diving, transport, chemical handling and hazardous waste disposal, etc.

Adequate safe operating practices require the following:

- Visible management support of the safety regime
- Clearly defined responsibilities and accountabilities for all levels of the line
- Clear definition and strict enforcement of procedures and regulations
- Dedicated training and exercising
- Continuous awareness and alertness

Detailed procedures, guidelines, and checklists should be described in specific documents, originating from the company's concerns.

Operational Activities for Special Safety Concerns

Transport

Transport on roads, waters and air needs special attention in exploration and production sector, as it may involve accidents resulting nonreturnable loss. The rule of thumb for most cost effective way of reducing transport accidents is to minimize the need of transport. Some of the factors that can be reviewed in this regard include work patterns, site visit frequency, on/off schedule, adequate telecommunication.

Road Transport is most dangerous as there is a high exposure and fatal accidents can occur. The exploration and production industry follows strict rules and provide necessary training and facilities for their employees with reference to traffic regulations, seatbelts and rollover bars, defensive driving courses, speed limits etc.

Air Transport is mostly used for the contractor crew and equipment. Local management satisfaction is guaranteed to maintain the safe air traffic. This is achieved by inspection of quality at regular interval of time. In addition, training and experience of crew, conditioning, repair and maintenance of equipment is ensure well in-time in accordance to technical and safety procedures.

Marine Transport is mostly used for the passenger transfer in sea, coastal or river and it is also monitored under strict safety control and procedures. Education and facilities like life jackets, discipline on-board, rescue facilities and other safety measures are clearly visible. In some instances swimming certificates and man-over board exercises are mandatory.

Seismic

Seismic activities are often carried out in remote and far-flung areas and mostly in inhospitable conditions which present many kinds of challenges for the seismic teams. The commitment and responsibility to hire local labour adds responsibility because of being unskilled with no prior experience of seismic surveys.

Standards guidelines, procedures, and industry practices are adopted in the companies to ensure safe operational conditions for online and/or offshore seismic operations. These guidelines include the sequence of operations from beginning to end.

Drilling

Drilling operations are marked by extraordinary exposure of crew to potentially hazardous situations and the nature of work is labour intensive. These activities are physically demanding and require coordinated activity of each member. Therefore high safety standards are established in drilling and work over operations. It requires well designed and properly laid out equipment operated by professionally managed and well trained personnel.

H_2S

The H_2S is a design as well as operational hazardous. It requires strictly enforced and well exercised regime of awareness that the serious accidents during the H2S emergencies can be prevented. H_2S detection and alarm systems, breathing apparatus, emergency preparedness and evacuation plans should be in operation and tested frequently.

Disposal of Hazardous Wastes

Specific precaution and procedures must be followed in order to dispose the hazardous waste material. These can include toxic gases or poisonous vapours affecting portable water or edible matter (after dumping, burial and decomposition). Some of the examples of hazard wastes can be leftovers mud components, well completion chemicals, treatment bio-acids, transformed oil containing extremely poisonous, non-degradable and toxic polymers.

Personal Protective Equipment

Personal Protective Equipment (PPE) is not a substitute for effective engineering controls, safe working conditions and sound work practices. However, PPE does play an essential role in protection of employees working in difficult operational environment. PPE helps in providing a mean of controlling individual exposure to particular hazard.

Emergency Planning and Controlling

There is a proper workflow developed in each organization for the emergency planning and controlling of hazardous situations. The exploration and production industry is pretty responsible in the recent years and proper emphasis is given and proactive action plans are decided before any hazardous situation. Most common emergency situations in exploration and production can be well blow-out, fire/explosion, spill/leakage, helicopter operational emergencies, medical emergencies, structural failures, severe storms, and earthquakes, etc.

Safety Trainings

All exploration and production companies invest a huge sum of money on training employees. Training is a major part of the safety program, ensuring that all employees are aware of policy and guidelines set by the company. This makes them aware of their responsibilities (to themselves and towards others) and rights. Potential hazard identification and mitigation plans are also included in training programs.

Different elements of the safety trainings are explained in Figure 10.

Safety Induction is necessary for all workers working in field locations. Those who will be exposed to specific risks such as air or water transport or offshore environment, should receive appropriate training beforehand. All new recruits have to attend the specific trainings required for the job.

Safety Skills include the technical matters in which certain employees need to be trained. The skillset for the employee is normally designed by the company to access the needs of the employee aligning with type of operations of the company and the operating conditions in which company is active. Some basic training is essential for all employees like use of hand extinguishers and first-aid kit etc.

Technical Safety includes technical audits, design reviews, and fire protection engineering.

Supervising Safety is the concept, which covers the hierarchy in the organizations. The safety supervisor is the key to an effective safety program. In addition, basic and essential trainings are to be carried out often and for groups.

Managing Safety includes the safety marginal roles and technical staff interaction. The new manager should be informed about the rules, regulations, and guidelines of the company cover-

Figure 10. Safety trainings areas and their attributes

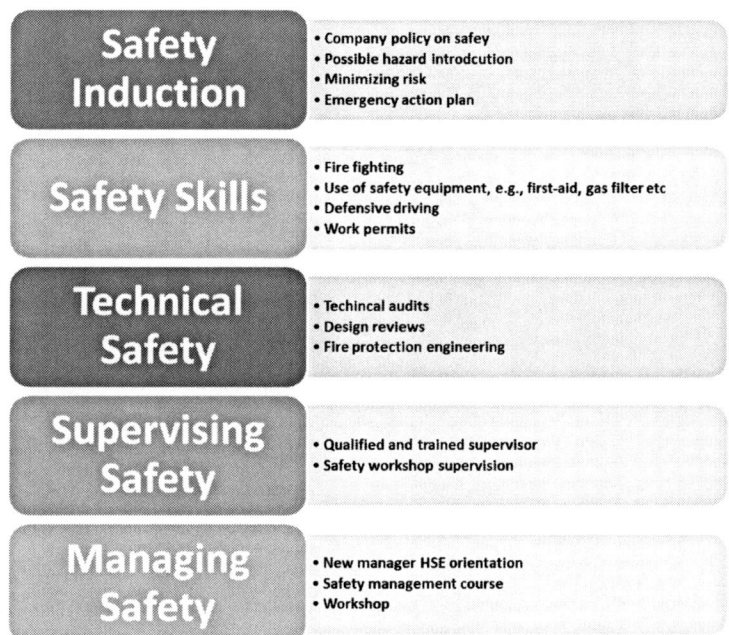

ing all the aspects of his role in safety program. There should be reviews of progress and direction of company staff effort.

Audits

Regular audits are very significant in tracking and verifying that the procedures are practices applied are being applied rigorously in the form specified and endorsed by management. The additional effect is the validity of the effectiveness of controls.

Audits should be part of the routine operation of the business and it is shortsighted to issue instructions, guidelines, policies or procedures without following up to see if they have been received, understood, and followed. There is always a regular check that the procedures stay up-to-date and are not obsolete.

Within organizations, there are audits at number of different levels:

- Self-Audit
- Cross-Audit
- Management Visits
- Internal Team Audit
- External Team Audit

Conclusions

Health safety and environment is the most important concept in exploration and production industry recently. Over the years, organizations have developed health, safety and environment best practices and procedures. The policy guidelines, regulations, and procedures are matured both internally and globally. International regulatory and administration authorities have developed codes and standards in order to meet the HSE criteria and put people as priority.

There are two main ways the safety can be ensured: in design and in operations. In design, safety is ensured by hazard identification, planning and cost estimation for the safety practices and facilities for special environment. On the other hand safety in operations is ensured by personal protective equipment, implementing warning and prohibition signs at work, emergency planning and controlling and regular safety inductions and skills.

QUALITY MANAGEMENT

The advent of new era has paced improvement of technology and thus refined the productivity efficiency. This idea was greatly accepted and rigorously adopted in exploration and production projects to enhance the productivity and profitability, which directly boosts the company assets. Alongside improvement of technology, the phenomenon of quality management has been developed very recently in modern industry. Although, the panoramic idea is very deep rooted and relates back to the civilizations supporting arts and crafts, allowing stakeholders to choose products and services meeting higher quality standards. The difference is that standards now are very well set, widely recognized, and accepted.

The quality management process is focused on the improvement of quality with better planning, controlling, and quality assurance throughout the project lifecycle. It is also directly related to the technical professionals, budget, and time specified for the project. Ensuring realistic manageable goals for quality standards today is a routine practice for the exploration and production industry. This adds value to the project by avoiding non-conformance, which saves significant project time and budget. Quality management is a slow evolutionary process, which makes it deep-rooted and well established in larger organizations. Mid-size to small organizations are committed to the concept and are continuously improving standard practices for different types of company assets and projects. Ultimately, this yields enormous time and cost efficiency, which is very essential for the growing organizations.

Introduction to Quality Management (QM)

Quality Management (QM) is the management of activities, which are directly related to the quality of products and services imparting value addition to role of business efficiency and competiveness. The concept of quality management is very significant for the exploration and production industry keeping in view the global energy requirements and interaction with fossil fuels.

The Figure 11 shows quality management is a function of quality planning, quality control and quality improvement. Quality assurance combines elements of quality planning and quality control. This is well established in exploration and production companies with emphasis on application in engineering projects.

Quality Planning

Quality Planning is the setting of corporate quality goals, strategies and transactions into objectives, controls, policies, plans, standards and working procedures, preventing deviations from the quality norms.

Quality Control

Quality Control is the detection of deviations when they occur via inspection and testing. It also includes the corrective measures to prevent their reoccurrence.

Quality Improvement

Quality Improvement is concerned with seeking a 'breakthrough' to new, more efficient operating levels by project-by-project improvement which seeks out areas of 'chronic waste' in business operations. It also identifies the root cause of

Figure 11. Quality management as integration of planning, control, and improvement

inefficiency and implements a permanent improvement.

The interaction between three elements of quality management is explained in Figure 12.

Quality Criteria in Exploration and Production Industry

The fundamental basis of quality improvement is the definition of quality itself. In exploration and production industry, quality can be defined as 'conformance to the agreed customer requirements.' Quality is the satisfaction of customer, which they receive from the performance of products or services in meeting their needs and the absence of dissatisfaction caused by deficiencies or non-conformances.

Within exploration and production companies, it is marked by doing things right the first time and therefore eliminating waste, scrap and repeated work.

The Essence of Quality Improvement

The quality improvement may result from three sources:

- Added value improvement (normal line management activities)
- Quality systems and their implementations (follow-up quality audits)
- As a result of specific quality improvement projects

The project-by-project approach to quality management has great potential to remove chronic waste and to solve problems of cooperation at organizational interfaces. The approach of quality management involves following aspects:

- Identify customers
- Agree requirements
- Full customer satisfaction
- Measurements of quality performance
- Cost of poor quality
- Quality improvement structure
- Top management commitments
- Continuance

The key benefits of a quality improvement project approach include structured short-term projects can compete on-going business activities at an equal level for resources and priority.

The Corrective Action Process

The individual quality improvement projects are undertaken by specifically performed project teams. The structure of corrective action teams include:

- The diagnostic journey:
  ○ Clear definition of the problem and its symptoms
  ○ Checking and confirmation of needs for the project
  ○ Symptoms analysis
  ○ Development of theories on causes
  ○ Measurement and collection of relevant data
  ○ Establishment of root causes
- The remedial journey:
  ○ Identification and development of remedies
  ○ Anticipation resistance to change
  ○ Selection of best remedy
  ○ Implementation of permanent solution
  ○ Development plans to hold the gains

  ○ Development plans for follow-up and review
  ○ Communication of results
  ○ Identification of areas for replication

Several problem-analysis and problem solving tools and techniques exist to give support to the activities involved in diagnostic and remedial journey action plans during exploration and production projects.

Quality Management Trainings

Quality management trainings are an essential part of quality improvement and well practiced for better execution of exploration and production industry projects. There are well-organized quality management training programs with the emphasis on quality improvement.

Essential modules for the training packages include:

- *Proof of need* to identify and confirm quality improvements in the organization.

Figure 12. Interaction between elements of quality management

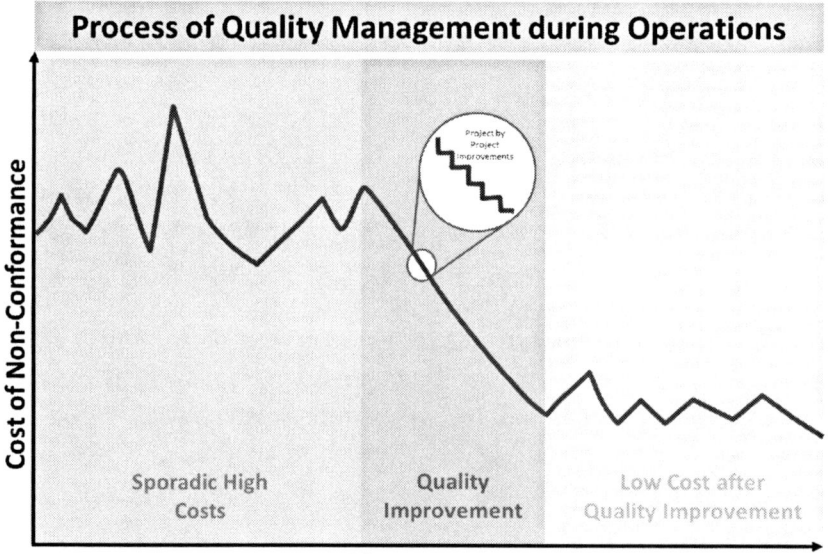

- *Quality management steering committee training* to enable top management to better implement quality improvement within the organization.
- *Quality management advisor training* to enable quality management advisors to give support, guidance and assistance to all levels in the organization.
- *Quality improvement team training* to translate the organization-wide directives of steering committee into firm action plans.
- *Corrective action team training* to enable corrective action teams to start and complete a specific improvement project.
- *Individual awareness training* to provide an introduction to quality improvement process to those members of the organization who have not been exposed to purpose specific training module.

Quality Management Support

There are specific quality management focal personnel assigned for specific quality improvements in the project. Exploration and production industry has a continuous quality improvement cycle. Figure 13 explains the workflow of how quality improvement in exploration and production industry is ensured by value-added quality management practices. The initiation workflow takes place from management responsibility, which interacts continuously with stakeholders involved in the individual project. In the next step, management of available resources is ensured and realizations results are generated. In generation of result realizations, requirements of stakeholders play a key part. Final results are developed in the next phase followed by measurement analysis and improvement which directly relates to the stakeholder's satisfaction. This specific step can be cyclic and long process until the time satisfaction is ensured and the set standards are met.

Figure 13. Value chain for the continuous quality improvement

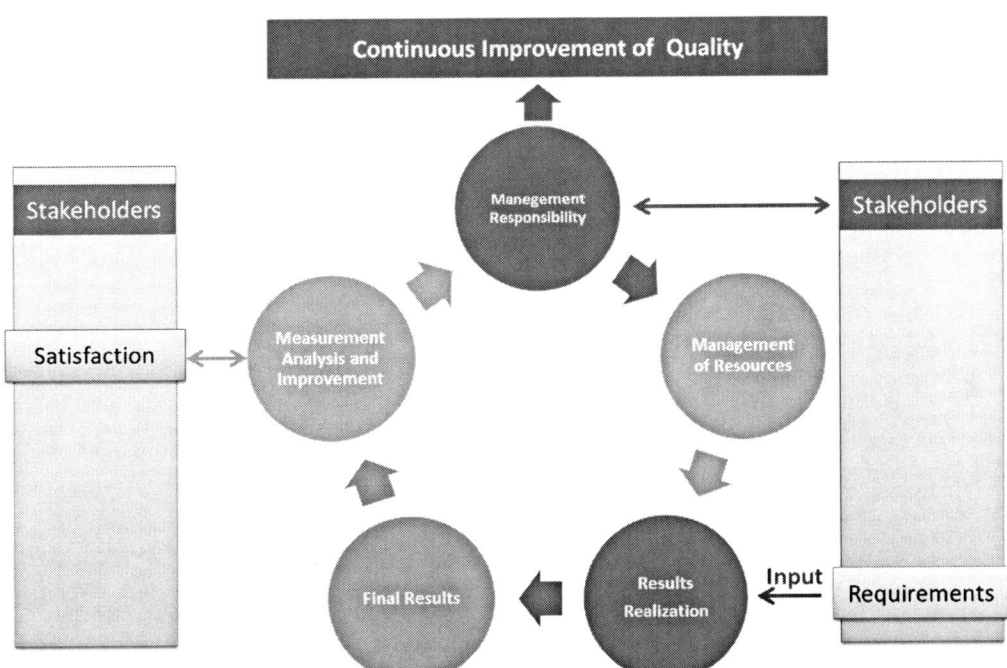

CONCLUSION

Quality management is a recent concept in engineering fields that ensures the product and service values and compares it with standard sets in scope of the project. It is essentially implemented in the oil and gas sector keeping in view the global demand of energy requirements from the exploration and production industry. However, it is a stepwise procedure and improves with every project. Initially in the extra project time and budget is kept in order to handle non-conformance. Modern practices and technology improvements help a lot to enhance quality. Additional technology acceptance is positive for exploration and production industry.

The proactive correction plan is routine practice in the exploration and production industry where diagnostic and remedial action plans are set. Along with that, the management trainings to improve quality management practices are organized by companies, which help great deal to improve the process. The overall lifecycle of quality improvement processes is very benefiting for exploration and production projects.

REFERENCES

Asrilhant, B., Meadows, M., & Dyson, R. G. (2004). Exploring decision support and strategic project management in the oil and gas sector. *European Management Journal, 22*(1).

Carrillo, R. (2004). Managing knowledge: Lessons from the oil and gas sector. *Construction Management and Economics, 22*(6).

de Wit, A. (1988). Measurement of project success. *International Journal of Project Management, 6*(3).

Walkup, G. W., Jr., & Ligon, J. R. (2006). *The good, bad, and ugly of stage-gate project management process as applied in oil and gas industry*. Paper presented at the SPE Annual Technical Conference and Exhibition. San Antonio, TX.

Yeo, K. T. (1993). System thinking and project management – Time to routine. *International Journal of Project Management, 11*(2). doi:10.1016/0263-7863(93)90019-J

KEY TERMS AND DEFINITIONS

Capital Expenditure (CAPEX): Is the initial amount of money creating future benefits. This is the first investment in any project and important part of cash flow statement.

Exploration and Production Industry: Includes the setup of facilities and operations in order to explore and extract the hydrocarbons (oil and gas) from the subsurface.

HSE Compliance: Refers to the efficient and realistic standards set on work places to keep in mind the health, safety, and environment by value added consideration for safer workplace and environment friendly production.

Key Success Skills: Refer to the important key factors essential for the project success. These are the primary focal points for the project manager to keep in mind in order to achieve the project goals.

Operating Expenditure (OPEX): Is the ongoing cost to maintain and operate properly the assets. It also includes the cost of workers, rents, utilities etc. A proper balance in CAPEX and OPEX is ensured in exploration and production industry.

Project Organization: Is the hierarchy defined as the first step of project initiation and includes the roles and description for each team member starting from steering committee to team members.

Quality Improvement: Is the step-wise process of value addition by seeking breakthrough to new, effective and efficient operating levels reducing chronic waste and inefficiency.

Synergism: Refers to the process of integrated performance and team-work by geologists, geophysicists, engineers and surface facility professionals in order to successfully achieve the ultimate project goal.

Chapter 11

Reasons Behind IT Project Failure:
The Case of Jordan

Emad Abu-Shanab
Yarmouk University, Jordan

Ashraf Al-Saggar
Irbid Electricity Company, Jordan

ABSTRACT

Information Technology (IT) projects have high failure and escalation rates because of the nature of domain and the rapid technology changes. It is important to understand the factors causing IT project success or failure. This chapter reviews the literature related to project failure and escalation and concludes with 17 important factors that cause IT projects to fail and 10 factors that contribute to the escalation of projects in time, cost, or scope. The concluded factors are utilized in an empirical study to explore the Jordanian environment and check the rank of these factors as perceived by Jordanian specialists. Conclusions and future work are stated at the end of this chapter.

INTRODUCTION

The Jordanian Information Technology (IT) sector is an important contributor to the national economy and an active generator of new projects, jobs, and investments. IT projects suffer from many problems resulting from the sophistication

involved and the changing nature of the domain. The rates of failure and escalation in IT projects are still high, and such failures and escalations are costing organizations their money and even their survival. Only 32% of IT projects were successful, 44% of IT projects were challenged, and 24% of these projects failed (Standish Group International, 2009). In addition, even when projects are executed successfully, they still fail to deliver business benefits.

DOI: 10.4018/978-1-4666-3658-3.ch011

Once a project is proposed and approved, the process of implementation starts with many reasons for deviations from the plan of execution. Projects in the IT domain starts to escalate in time and cost. Some research emphasize the danger of escalation more than failure as the decision to abort a project might be easier and more clear when a project is a total failure. Catastrophic situations occur when a project is not considered a total failure, but the project team still extends its execution period and cost. Such important phenomenon deserves some consideration to understand why IT projects fail and the reasons behind such failure. In addition, it is important to understand the reasons behind the escalation of projects as much as the reasons behind their failure.

The aim of this chapter is to review the reasons behind IT project failure or escalation. In addition, to explore the case of Jordanian IT sector by utilizing an empirical test conducted by utilizing a sample of Jordanian IT firms. This chapter will be divided into 6 sections. Following, a literature review will be conducted to understand the reasons behind project failure or escalation. The following section will illustrate an empirical test utilizing a Jordanian sample of firms. Finally, conclusions and future work are depicted at the end.

LITERATURE REVIEW

Information and Communication Technology (ICT) plays an essential part in our life and the way we react to external environmental factors. It is essential that information system projects be properly scoped and implemented successfully; many of national surveys showed that around 70-80% of all information technology and information systems projects fail totally or suffer from partial failure. Such rates of failure might be caused by lack of certain factors that are explored in the literature under the name of critical success factors (Yeo, 2002).

All projects, IT or otherwise, move through five phases in the project management lifecycle: initiating, planning, executing, monitoring and controlling, and closing. Each phase contains processes that move the project from idea to implementation.

Basic Project and Project Management Definitions

The Project Management Institute (PMI, 2004, p. 5) defines project as a "temporary endeavor undertaken to create a unique product or service. Temporary means that every project has a definite beginning and a definite end. Unique means that the product or service is different in some distinguishing way from all similar products or services." Yourdon (1997, p. 9) also defines project management as "the application of knowledge, skills, tools and techniques to project activities to meet project requirements." He indicates that the role of the project manager is to facilitate the entire process and meet the needs and expectations of involved personnel and project stakeholders affected by project activities.

The Importance of Project Management

Based on the previous definition that IT projects consist of a series of steps, points and techniques that guarantee and ensure the successful implementation of projects in an efficient way to meet specified objectives and goals, project management plays an essential role in the successful transition between the project lifecycle phases. The role of project management techniques is to implement projects in a successful manner taking into consideration the following areas: The planning and control of time, project cost, quality aspects, efficiency, and effectiveness, which all directly affect project success (Munns & Bjeirmi, 1996).

Projects are short-term activities that create a unique product or service, such as implementing new systems, removing old servers, deploying new architecture, developing a custom e-commerce site, creating new desktop images or merging databases. All projects are constrained by three factors: time, cost and scope. For a project to be successful, these three constraints must be balanced and managed in the most effective way. If any constraint is out of balance, the project is heading for a disaster and may result in a total failure.

Other definitions of project management indicated that it consists of a series of flexible and iterative steps through which you identify where you want to go, a reasonable way to get there, with specifics of who will do what and when (HHMI, 2006). A project defines the framework for the objectives, expected results and resources needed to reach the project objectives (Lubani & Qirjo, 2002).

Information Technology Projects

IT projects are defined as the application of modern project management techniques and systems, to the execution of an IT Project from start to finish, to achieve predetermined objectives of scope, quality, time and cost to the project sponsor satisfaction (Clark, 2010). Bakker, Boonstra, and Wortmann (2010) define IT projects as projects that aim at the development and implementation of computer software. Other definitions of IT project and IT project management focus on the set of tools, processes, and competencies utilized by people in order to enhance an organization's services and practices (Cai, Ghali, Giannelia, Hughes, Johnson, & Khoo, 2003).

There are certain features that differentiate IT projects from other types of projects (i.e. engineering projects), where they are characterized by an increase in the complexity and probability of project failure. Based on that, IT projects are considered as high risk projects from the perceptions of all parties involved (project stakeholders), and they necessitate careful development, planning and implementation process at the initiation stage

of the project to indicate the most important factors influencing project success (Repiso, Setchi, & Salmeron, 2007a, 2007b).

The previous discussion stresses the importance of IT project lifecycle, which plays an essential role in IT project development. IT project lifecycle includes the requirement analysis, design, testing, implementation, and the operation and maintenance; this cycle directly affects the implementation of IT project towards the achievement of intended success (Wong, Jingchun, & Ming, 2009). The authors also concluded that the absence of a scientific and rational life-cycle process in project management can lead to project failure. Another description of IT project lifecycle includes three major stages: Project preparation period, which includes: the decision-making phase that consists of the requirement identification, IT project definition, and IT project feasibility study. The second stage is the project construction period, which is divided into the design and implementation phases. The third stage is the project operation stage, which consists of the operation and maintenance activities (Wong, et al., 2009).

Project Success and Critical Success Factors

Project management plays a vital role in implementing IT projects successfully. Based on that, project success is a strategic management concept where the project efforts must be aligned with short and long term goals and objectives of the organization (Al-Tmeemy, Abdul-Rahman, & Harun, 2011). Assessing and evaluating project success can be done utilizing internal measures such as: meeting schedule and budget requirements to overcome escalations and overruns, and achieving performance measures (Chiong, 2008). Furthermore, the assessment of project success is integrated with top management role in project processes, so the measures should be set before project initiation (Huang, Poli, & Mithiborwala, 2009). Ding and Yang (2008) indicated that there are three main aspects that influence project success and they are the following: Project technology, project management, and project governance.

Research indicated that there are two basic elements for project success: project success factors and project success criteria (Jugdev & Muller, 2005; Muller & Turner, 2007). Project success factors include elements of a project that can influence project success (independent variables). On the other hand, project success criteria judge the successful outcomes of the projects (dependent variables).

Research in the area of IT project management focused on the factors that are considered critical to the success of projects (Critical Success Factors – CSF). Fan (2010) lists some of the factors that influence the project success directly:

- The in-depth understanding of modern project management methodologies and tools.
- The establishment and effective facilitation of project communication.
- Enhancing the collaboration and information sharing process between project team and within teams.
- Managing project scheduling related to the time allocated and spent on the process of initiating, planning, executing, controlling and ending of the project under the guidance of the project objectives and requirements.

On the other hand, Ding and Yang (2008) described some major CSFs for projects described in the following: A competent project manager, clear project objectives, well-organized project schedule and control, well-selected project team and human resource management, top management support and authorization, open and effective communication, closely controlled project plan, and finally user involvement. Meredith and Mantel (2003) concluded to some criteria for project success, which include among them: project's efficiency

in meeting the budget and schedule, customers' impact on the process, customers' satisfaction, business direct success, and future potential benefit for the business. Finally, project success relates directly to the quality of talents employed; this factor it is very important and management should effectively deploy needed talents in the project (Laplante, 2003). An efficient project manager must keep project members involved in the project and set clear expectations from the team and the project.

IT Project Success/Failure

According to the Standish Group International (2009), which tracks IT project success rates, only 32% of IT projects were completed successfully. 44% of IT projects were challenged and 24% of these projects failed. Challenged projects mean the late delivery of stages, cost overrun, shortcomings in the required features and functions to meet project objectives. In addition, when projects are executed successfully, many fail to deliver business benefits as expected.

Why many IT projects fail? What are the main factors that influence these failures? These questions and many more are very important to organizational top management, project management teams, and organization's stakeholders. These factors affect organization's success and in some cases impact organizational survival. Many IT projects fail because of many reasons such as: these projects may exceed their budget, time or don't deliver what they are expected to deliver after completion. Failures in IT projects are more common than failures in other sectors of modern businesses. In addition, IT projects escalate in budget and time of completion, but still large number of IT projects is considered total failures (Nulden, 1996a).

In many IT project reports, in the last few decades, data indicated that more than 30% of IT projects are cancelled before completion due dates. One of the highest reasons behind such failure rates are due to poor project management performance. Project management would face more complications if we consider that failure is not only a state of total malfunction or breakdown, failing to complete the project on time, and budget will increase the indicators that are considered as symptoms of failure. Based on that, research indicated that more and more IT projects are considered as failures than successes (Sundari, Barwal, Prakash, Yadav, Garg, & Jain, 2009).

When we talk about project failures, research puts some boundaries for such phenomenon that indicate the meaning of it, and the main indicators that present such situation. Based on that, project failure is defined as a failure to complete the project or the completion of a project in a state where: a) The schedule and the budget have been seriously overrun, b) the project does not satisfy the previously agreed requirements, c) and the value of the project has decreased significantly during the execution period (Storm & Savelsbergh, 2005). Managing IT and IS projects is a very difficult and challenging task, where most of these projects fail to achieve the aimed goals and objectives (Latendresse & Chen, 2003).

Research indicated several reasons for IT project failures mainly based of the nature of IT as a changing discipline. Other reasons include the usual project management challenges such as: committing to deadlines, budget constraints, and the shortage of personnel devoted to project execution. In addition to that, they also face unique technology challenges ranging from hardware, operating system, network or database, security risks, interoperability issues, and the changes in hardware and software configurations. Based on that, projects face failures or escalations that lead to wasted money, time, and other resources (Taylor, 2002). IT projects fail at early stages due to a lack of sufficient planning. An IT organization must take into consideration the needed resources to execute the project, the skills required to complete the execution, the stakeholders who need to be involved, and realistically consider the

time it will take to create, test and implement the project successfully. If such factors are not taken into consideration, the organization will never complete the project on time, according to budget, or with the required functionality (Nulden, 1996b).

Research indicates that IT and IS projects failures are associated with poor project definition and an extra emphasis on technology. The successful implementation of IT projects must meet three criteria: On time, on budget, and to specification. The successful IT project execution also requires a careful execution according to good project management standards, but still focusing on matching users' needs. Finally, when challenges in implementation occur, these are often handled by reductions in specifications or scope (Grant & Qureshi, 2006).

There are some numerous techniques and strategies that are designed to ensure IS project success such as: Information System (IS) development methodologies, risk and project management techniques, and software process improvement. Still, it is not possible to ensure project outcomes acceptance by all stakeholders of the project. Such strategies and techniques are recommended for achieving successful outcomes that focus on the IS project and the process itself; while ignoring the organizational and social settings (Winklhofer, 2001).

The accurate estimation of project duration and required resources are the most important issues in project success. Overestimating time, effort and budget due to the lack of resources or delay in project completion influence firms to refuse projects that might have a contribution to their business. On the other hand, underestimating such factors may cause projects to fail in delivering their outcomes within their available time and budget (Morgenshtern, Raz, & Dvir, 2006).

Literature indicated that sources of failure are: project team, suppliers, customers and other key stakeholders. Shou and Ying (2005) identified four major types of IT projects failures: Correspondence failure (when system design is not suitable to the main objectives), process failure (occurs when IS cannot be developed within an allocated budget and/or schedule), interaction failure (end-user usage levels as an alternate performance measure), and expectation failure (inability of a system to meet its stakeholders' requirements, expectations, or values). On the other hand, research summed the primary causes of project failure into the following: Poor planning, unclear goals and objectives, changing objective during the project execution, unrealistic time and resource estimation, lack of executive support, lack of user involvement, failure in communication between project team members, incorporate skills and knowledge, and lack of project leadership (Al-Neimat, 2005; Hartman & Sharafi, 2004; Thite, 1999).

Reich (2007) classified other factors that influence IT projects failures like: Failure to learn from past projects, competence levels among project members, incomplete integrating, and transferring knowledge, volatility in governance team, conflicts in ideas and opinions between team members, and exit of team members. On the other hand, the main reasons for failures, when related to managerial process, are summarized as the following: unclear/changing project requirements, lack of control, poor project management, poor resource management, and poor cost management (KPGMS, 2003).

Poor planning plays an essential role in project success, where many methods and techniques are developed to manage projects efficiently from initiation to completion. Some projects have a tendency to run over budget, take more time to complete than expected, or fail to deliver expectations in terms of quality, scope, safety or other key stakeholder expectation (Hartman & Sharafi, 2003). The failure in IT projects is directly influenced by the effort made to understand the responsibilities of all parties involved in these projects, recognize their roles, and understand where failures occur. This indicates that technological factors are rarely causing project failure, where

it is concluded that people and organizational factors are perceived as the major factors for IT project failure (Warne & Hart, 1996; Thite, 1999; Imamoglu & Gozlu, 2008).

Evans, Abela, and Beltz (2002) described typical characteristics for IT project failure as: Failure to apply essential project management practices, unrealistic management expectations and unwarranted optimism, effective software practices not implemented, lack of program management leadership, untimely Decision-making, and lack of pro-active risk management.

Projects do not fail for just one single reason they fail for multiple reasons; research indicated that the common problem related to project failure was inadequate requirements. Such issue leads to unrealistic expectations by customers, because in many cases customers do not meet developers to define their proper requirements and cause poor initial requirements (Cerpa & Verner, 2009). The same research stressed other factors such as: adding extra staff later in the project lifecycle to meet aggressive schedule, and unstable requirements that may lead to changes in scope during the project lifecycle. Hidding and Nicholas (2009) concluded to other factors for IT project failure like: Agreement on project goals, use of an inappropriate software development methodology, dissimilarity to previous projects, errors and mistakes during project implementation, requirements volatility, and inadequate technology base or infrastructure.

IT Project Escalation or Cancelation

Clients and senior management continue to invest in a project even though it has some problems in one of its constituents like its database, decision support, and integration technologies. People in the project realized such fact long before the project is canceled, but did not come forward with this information; They (i.e. project managers) were too committed and confident that problems would be solved with time (Nulden, 1996a). Similar to project failure/success factors, research indicated many reasons for project escalation (Keil & Mann, 1997; Keil, Rai, Mann, & Zhang, 2003): Underestimation of time for projects completion, senior management monitoring was not enough, underestimation of necessary resources, underestimation of size and scope of project, inadequate project control mechanisms, changing in system specifications, and inadequate planning. Finally, Earl (1996) proclaimed that the leading causes for cost escalation are: lack of expertise in contract management, measurement problems, and supplier of the activities.

With respect to project cancellation, The Standish Group's (2009) Chaos Report shows that 24% of IT projects that failed are cancelled before completion or delivered and never been used. This rate is high, very costly and may cause huge wastes in budget and resources. Such phenomenon leads to high financial losses within organizations. One of the most important reasons for these cancellations are project team members who approved project objectives, still the differences and conflicts in ideas and opinions about the relative importance of time, scope, functionality and other factors can lead to canceling these projects (Wateridge, 1995).

Ahonen and Savolainen (2009) indicate that project cancellation is caused by some errors and mistakes occurring during project implementation as project outcomes are delivered to external customers who instigate project cancellation. The authors indicated several reasons for project cancellation, based on responses to a survey, such as: feasibility study factors, mostly finance and cost overrun, top management support, and miss-match between goals and objectives.

IT Project Managers

The project manager is often an IT executive or a member of the department who has the required experience in project management. The challenges that face the project manager in carrying out IT projects request both project management knowledge and practice. Some of the project manager responsibilities are: planning, organizing, coordinating and controlling tasks to ensure the successful completion of the project. In order to do these tasks, the project manager has to allocate the suitable human, financial, and informational resources to the project (Gottschalk & Karlsen, 2005).

Lack of leadership is the most cited reason for poor project performance; furthermore, lack of project management competencies during project can contribute to poor project performance. In addition, experience and ability to cope with stress are key factors that influence project success (Calisir & Gumussury, 2006; Crawford, 2005; Standing, et al., 2006; Smith, Passos, & Isaacs, 2010). IT project management requires a mix of skills including interpersonal abilities, technical competencies and emotional responses to be able to communicate with all parties involved in the project. This ensures a positive participation in the project and the ability to resolve conflicts between all parties (Pant & Baroudi, 2008).

Smith *et al.* (2010) suggested two strategies for IT project managers to cope with stress: adaptive (active coping, acceptance, project reframing, emotional support, instrumental support, and planning), and maladaptive (Self-distraction, venting, self-blame, behavioral disengagement, denial, and substance abuse). Project manager must have complete understanding of the methods and techniques being used for managing different parts of the project, this brings projects to successful completion (Rehman & Hussain, 2009).

The certification also is a very important factor to raise project manager's traits, skills, and high performance level. Muller and Turner (2007) assert that the best project managers are the certified ones, while non-certified project managers perform badly in some projects. They emphasize also that certification is not the absolute factor to guarantee success, other mentioned factors are important as well. Bedingfield and Thal (2008) explored the relationship between the "Big Five" personality traits on project success by surveying

United States Department of Defense project managers. They found that conscientiousness and openness were both good predictors of successful project managers. This result is important for the process of hiring and selecting project managers.

ICT Sector in Jordan

The ICT sector in Jordan is a booming industry. In 2003, there were 373 value adding IT companies, with activities covering wide ICT spectrum (ICT consulting, software development, communication and Internet services, hardware sales and technology provision, and licensing). The main areas of software development are accounting packages, Web-based applications, Arabization, banking, system integration, health insurance packages, and software conversion from 3rd to 4th generation. The ICT sector witnessed an annual growth of 50%, with foreign investments totaling $90 million in 2007. Amongst the foreign companies that have invested in Jordan's ICT sector are: Microsoft, Intel, Cisco Systems, France Telecom, IBM, Oracle, Dell, Compaq, HP, U.S. Robotics, and Apple.

The Ministry of information and Communication Technology (www.moict.gov.jo) conducted a survey to evaluate the Jordanian workforce in the ICT industry; results indicated the following: Employees in the ICT sector form 37% of the total of companies' employees, and 88% of them hold certificates in an ICT fields, female workforce account for 26% of total workforce, 82% of ICT employees hold a bachelor degree, and 36% of companies prefer hiring public universities' graduates, while 11% prefer private universities' graduates.

The survey also summarized main functional roles in the ICT sector in Jordan and demonstrated the following: 57% of companies indicated that the "Application Developer" is the most important functional role now and in the future, while the "Storage Service Specialist" had the least importance role. "System Programmer" was ranked first concerning the offering of big number of trained employees, followed by "Application Developer," and "Database Administrator" and "Project Manager." 80% of participants in the survey (highest percentage) indicated that the offered skills for fresh graduates for the role of "Project Manager" are considered "Advanced Skills." More than 40% of participants in the survey indicated that fresh graduates lack the needed skills for the role of "Business Analyst." The "Application Developer" post was ranked first regarding the availability of advanced-skills employees. And 15% of companies believe that their employees don't have the skills of "Content Manager Specialist." 20% of companies (highest percentage) indicated that the "Project Manager" certificate has been ranked as "low importance" now and in the future, while 21% indicated that this certificate has been ranked as "not important" now but it will be of "high importance" in the future. Finally, the most-demanded functional roles were listed as follows: Application Developer, Project Manager, Content Manager Specialist, Business Analyst, System Programmer, and Storage Manager Specialist.

THE CASE OF JORDAN: AN EMPIRICAL TEST

To explore the reality of Jordanian firms with respect to project success/failure and empirically test for the factors influencing the failure, success, escalation or cancelation process of IT projects, a questionnaire consisting of three parts was utilized. Part one gathered general information about respondents: gender, age, experience, job title, and organization's information. In addition, part 1 included one open question that measures the perceptions of respondents regarding the main criteria that the organization follows to decide when to cancel a project. Part two asked the respondents to rate (17) factors related to the causes of IT project failure utilizing a 7 point

Likert scale. Part three asked the respondents to rate (10) factors related to the reasons that causes IT project escalation (increase in time, cost and scope) utilizing the same 7 point Likert scale.

A sample of (108) questionnaires were distributed to IT specialists and employees working in diverse IT firms. 13 surveys were eliminated because they were severely incomplete. So (95) completed surveys were analyzed with a usable rate of (88%). Tables 1 and 2 show the demographics of the sample and some information about the firms.

Data Analysis and Results

The aim of this work revolves around distilling the main factors that influence the success or failure of IT projects. The previous literature review concluded to a list of reasons related to the failure/success of IT projects, and subsequently used this list to build an instrument that utilized a Likert scale (from 1-7) to measure the opinions of IT specialists and workers to know how they evaluate the influence of each factor.

The same was done to explore the reasons behind the escalation process in IT projects (increase in cost, time, and scope). The means of the items (factors) used were evaluated based on a scale shown in Table 3 with means higher than 5 to be labeled as high, and all means between 3 and 5 to be moderate. The Likert scale used (1-7) indicated 1 to mean "totally disagree" and 7 to mean "totally agree" for both the first and second list of factors.

Results indicated that all factors extracted from the literature were important with differential values. Table 4 indicates that five factors were evaluated highly as influencers of IT project success or failure. The factors 1, 2, 3, 5, and 11 are shown in Table 4 and were the highest among all. The highest perceived reason behind IT project failure was "poor planning." On the other hand, the lowest perceived factor influencing IT project success was "Conflicts in ideas and opinions between team members." Twelve factors included in the survey were perceived to be moderately influencing the success/failure of IT

Table 1. The demographics of the sample

Age			Gender		
Category	Frequency	%	Category	Frequency	%
< 25	5	5.3	Male	76	80
26-40 years	77	71.8	Female	19	20
41-60 years	13	13.7	Experience		
Job Categories			Category	Frequency	%
Category	Frequency	%	< 5	16	16.8
Programmer	14	14.7	6-10	52	54.7
Developer	26	27.4	11-20	25	26.3
Analyst	15	15.8	>20	2	2.1
System Engineer	11	11.6	Education		
Project Manager	22	23.2	Category	Frequency	%
CIO	5	5.3	High School	6	6.3
CEO	0	0	Bachelor	57	60
Other	2	2.1	Master	27	28.4
			PhD	5	5.3

Table 2. The firms' information

Sector			Number of Employees		
Category	Frequency	%	Category	Frequency	%
Public	26	27.4	<50	3	3.2
Private	69	72.6	50-100	35	36.8
			101-250	31	32.6
			251-500	8	8.4
			>500	16	18.9

Table 3. The categories of the scale used

#	Scale Range	Label of Evaluation
1	From 1-3	Low evaluation
2	From above 3 – below 5	Medium evaluation
3	From 5-7	High evaluation

Table 4. The means and standard deviations of the items (success/failure factors)

	Item (Factor)	Min	Max	Mean	Std Dev.
1	Poor planning*	2	7	5.27	1.60
2	Unclear goals and objectives	1	7	5.24	1.37
3	Changing objectives during the project	2	7	5.12	1.41
4	Unrealistic time and resource estimation	1	7	5.03	1.50
5	Lack of executive support	1	7	5.07	1.47
6	Lack of users involvement	1	7	4.94	1.34
7	Inappropriate skills and knowledge	2	7	4.98	1.39
8	Lack of control	2	7	4.71	1.40
9	Poor project management	1	7	4.55	1.56
10	Poor resource management	1	7	4.68	1.45
11	Poor cost management	2	7	5.03	1.39
12	Inability to meet project requirements and objectives	1	7	4.88	1.54
13	Lack of leadership	1	7	4.78	1.63
14	Lack of attention to human & organizational aspects of IT	1	7	4.55	1.58
15	Failure in communication between relevant parties	1	7	4.42	1.45
16	Conflicts in ideas and opinions between team members^	1	7	4.09	1.68
17	Errors and mistakes during project implementation	1	7	4.18	1.85

*Highest mean ^Lowest mean

projects (see Table 4). The range of perceptions of all items was 4.09-5.27, where none of the factors proposed were perceived to be low in influencing the success of IT projects.

Similarly, the factors that contribute to the escalation of IT projects are rated using the same scale and concluded to the fact that all factors were perceived to be moderately influencing the escalation of IT projects except the first factor "Inadequate planning," which was highly perceived by subjects (mean = 5.01). The lowest perceived factor to influence the escalation of IT projects was "Lack of experience of the supplier with the activity" (mean = 4.36). Table 5 shows the previously discussed results.

It is not a coincidence that the highest factors in both lists were related to planning, which emphasizes the importance of planning in project management process. Followed by "inadequate

planning," factors related to resource planning are perceived to be important when considering the escalation situation.

The open question related to canceling an IT project and the criteria used by firms in this regard was coded and summarized in Table 6. Results indicated that the highest criterion used to cancel or abort an IT project was cost overrun (or financial reasons). The second frequent reason was "top management support," then the "no match between goals and objectives," and finally, a "feasibility study." It is noticeable that 36 surveys did not report any answer for this question.

CONCLUSION AND FUTURE WORK

This chapter explored the literature related to the IT project management and tried to extract

Table 5. The mean and standard deviation of the items (success/failure factors)

	Item (Factor)	Min	Max	Mean	Std Dev.
1	Inadequate planning*	1	7	5.01	1.54
2	Underestimation of time for projects completion	1	7	4.84	1.39
3	Underestimation of necessary resources	1	7	4.74	1.44
4	Underestimation of size and scope of projects	2	7	4.83	1.35
5	Senior management monitoring did not enough	2	7	4.63	1.22
6	Inadequate project control mechanisms	2	7	4.62	1.26
7	Changing in system specifications	2	7	4.81	1.59
8	Lack of experience of the client with contract management	2	7	4.75	1.31
9	Measurement problems	2	7	4.61	1.19
10	Lack of experience of the supplier with the activity^	1	7	4.36	1.43

*Highest mean ^Lowest mean

Table 6. The criteria firms use to cancel a project (frequencies of open question responses)

Reason Behind Canceling a Project	Frequency	%
No response	36	37.9
Feasibility Study	8	8.4
Mostly financial reasons (cost overrun)	20	21.1
Top management support	18	18.9
No match between goals and objectives	12	13.7

the success and failure factors (or the critical success factors) that are commonly known to influence the IT project success. The literature review concluded to 17 important factors related to the success or failure of IT projects. The same was done to explore the escalation process and concluded to 10 important factors that cause IT projects to escalate (increase in time, cost, and scope).

The list was utilized in a descriptive empirical test in a Jordanian context. Ninety five IT specialist and workers in the IT field were probed to understand their perceptions towards the factors that causes IT project to fail or escalate. Results indicated that "poor planning was the highest perceived reason behind the failure of IT projects. On the other hand, and similarly, "inadequate planning" was perceived to be the highest factor behind IT project escalation.

It is important to acknowledge the factors behind IT project failure to guard against them and to plan well for avoiding them. All factors proposed in this review were perceived with a moderate level or above. None of the factors proposed were perceived to be low in influencing IT project failure, which indicates the congruence between the literature and the Jordanian experts' perceptions. Jordanian environment shows no difference against what is common in the literature and diverse countries of the world.

This result calls for a confirmation with a larger scale sample, where more IT professional are invited to participate in a comprehensive study. Also, the difference between IT project failure and IT project escalation needs to be emphasized. Finally, other research methods need to be conducted to reach similar conclusions like firms' records inspection or case studies and interviews with executives.

REFERENCES

Ahonen, J. J., & Savolainen, P. (2010). Software engineering projects may fail before they are started: Post-mortem analysis of five cancelled projects. *Journal of Systems and Software, 83,* 2175–2187. doi:10.1016/j.jss.2010.06.023

Al-Neimat, T. (2005). *Why IT projects fail.* Retrieved from www.projectperfect.com.au

Al-Tmeemy, S. M., Abdul-Rahman, H., & Harun, Z. (2011). Future criteria for success of building projects in Malaysia. *International Journal of Project Management, 29*(3), 241–356. doi:10.1016/j.ijproman.2010.03.003

Bakker, K. D., Boonstra, A., & Wortmann, H. (2010). Does risk management contribute to IT project success? A meta-analysis of empirical evidence. *International Journal of Project Management, 28,* 493–503. doi:10.1016/j.ijproman.2009.07.002

Bedingfield, J. D., & Thal, A. E. (2008). Project manager personality as a factor for success. In *Proceedings of Portland International Conference on Management of Engineering & Technology 2008 (PICMET 2008),* (pp. 1303-1314). Cape Town, South Africa: PICMET.

Cai, J., Ghali, S., Giannelia, M., Hughes, A., Johnson, A., & Khoo, T. (2003). *Identifying best practices in information technology project management.* Retrieved from http://www.pdfport.com/view/134033-identifying-best-practices-in-information-technology-project.html

Calisir, F., & Gumussoy, C. A. (2005). Determinants of budget overruns on IT projects. *Technovation, 25*(6), 631–636. doi:10.1016/j.technovation.2003.10.011

Cerpa, N., & Verner, J. M. (2009). Why did your project fail? *Communications of the ACM, 52*(12), 130–134. doi:10.1145/1610252.1610286

Chiong, J. (2008). *Predictors of project success: A Singapore study*. (Dissertation). University of Western Australia. Crawley, Australia.

Clark, R. B. (2010). *Information technology project management*. Helena, MT: State Government Publication.

Crawford, L. (2005). Senior management perceptions of project management competence. *International Journal of Project Management, 23*(1), 7–16. doi:10.1016/j.ijproman.2004.06.005

Ding, R., & Wang, Y. (2008). An empirical study on critical success factors based on governance for IT projects in China. In *Proceedings of the 4th International Conference on Wireless Communications, Networking and Mobile Computing, 2008*. WiCOM.

Earl, M. (1996, Spring). The risks of outsourcing IT. *Sloan Management Review*, 26–32.

Evans, M. W., Abela, A. M., & Beltz, T. (2002, April). Seven characteristics of dysfunctional software projects. *The Journal of Defense Software Engineering, CrossTalk*, 16-20.

Fan, D. (2010). Analysis of critical success factors in IT project management. In *Proceedings of the 2nd International Conference on Industrial and Information Systems*, (pp. 487-490). IEEE Press.

Gottschalk, P., & Karlsen, J. T. (2005). A comparison of leadership roles in internal IT projects versus outsourcing projects. *Industrial Management & Data Systems, 105*(9), 1137–1149. doi:10.1108/02635570510633220

Grant, K. A., & Qureshi, U. (2006). Knowledge management systems - Why so many failures? In *Proceedings of the Innovations in Information Technology Conference, 2006*. IEEE Press.

Hartman, F., & Ashrafi, R. (2004). Development of the SMART project planning framework. *International Journal of Project Management, 22*, 499–510. doi:10.1016/j.ijproman.2003.12.003

Hidding, G. J., & Nicholas, J. (2009). Reducing IT project management failures: A research proposal. In *Proceedings of the 42nd Hawaii International Conference on System Sciences – 2009*, (pp. 1-10). IEEE Press.

Huang, Z., Poli, M., & Mithiborwala, H. S. (2009). Project strategy: Success themes for strategic projects. In *Proceedings of Portland International Conference on Management of Engineering & Technology 2008 (PICMET 2009)*, (pp. 1282-1289). Portland, OR: PICMET.

Imamoglu, O., & Gozlu, S. (2008). The sources of success and failure of information technology projects: Project managers' perspective. In *Proceedings of Portland International Conference on Management of Engineering & Technology 2008 (PICMET 2008)*, (pp. 1430-1435). Cape Town, South Africa: PICMET.

Jugdev, K., & Muller, R. (2005). A retrospective look at our evolving understanding of project success. *Project Management Journal, 36*, 19–31.

Keil, M., & Mann, J. (1997). Understanding the nature and extent of IS project escalation: Results from a survey of IS audit and control professionals. In *Proceedings of the 30th Hawaii International Conference on System Sciences (HICSS)*, (Vol. 3, pp. 139-148). IEEE.

Keil, M., Rai, A., Mann, J., & Zhang, G. (2003). Why software projects escalate: The importance of project management constructs. *IEEE Transactions on Engineering Management, 50*(3), 251–261. doi:10.1109/TEM.2003.817312

KPGMS. (2003). *KPMG's international 2002-2003 programme management survey*. Retrieved from www.transformed.com.au/_.../Reports_-_Programm

Laplante, P. (2003, January-February). Remember the human element in IT project management. *IT Pro*, 46-50.

Latendresse, P., & Chen, J. (2003). *The information age and why it projects must not fail*. Retrieved from www.sbaer.uca.edu/research/swdsi/2003/.../045.pdf

Lubani, E., & Qirjo, M. (2002). *Developing skills for NGOS: Project management*. Szentendre, Hungary: The Regional Environmental Center For Central And Eastern Europe.

Meredith, J. R., & Mantel, S. L. (2003). *Project management-A managerial approach* (5th ed.). New York, NY: John Wiley & Sons.

Morgenshtern, O., Raz, T., & Dvir, D. (2007). Factors aVecting duration and eVort estimation errors in software development projects. *Information and Software Technology*, 49, 827–837. doi:10.1016/j.infsof.2006.09.006

Muller, R., & Turner, T. (2007). The influence of project managers on project success criteria and project success by type of project. *European Management Journal*, 25(4), 298–309. doi:10.1016/j.emj.2007.06.003

Munns, A. K., & Bjeirmi, B. F. (1996). The role of project management in achieving project success. *International Journal of Project Management*, 14(2), 81–87. doi:10.1016/0263-7863(95)00057-7

Nulden, U. (1996a). *Failing projects: Harder to abandon than to continue*. Bayonne, France: Projectics.

Nulden, U. (1996b). Escalation in IT projects: Can we afford to quit or do we have to continue. In *Proceedings of the Information Systems Conference of New Zealand*, (pp. 136-142). Palmerston North, New Zealand: IEEE Computer Society Press.

Pant, I., & Baroudi, B. (2008). Project management education: The human skills imperative. *International Journal of Project Management*, 26, 124–128. doi:10.1016/j.ijproman.2007.05.010

PMI. (2004). *A guide to the project management body of knowledge (PMBOK® guide)*. PMI.

Rehman, A., & Hussain, R. (2009). Software project management methodologies/frameworks dynamics: A comparative approach. In *Proceedings of the International Conference on Information and Emerging Technologies*. ICIET.

Reich, B. H. (2007). Managing knowledge and learning in IT projects: A conceptual framework and guideline for practices. *Project Management Institute*, 38(2), 5–17.

Repiso, R. L., Setchi, R., & Salmeron, J. L. (2007a). Modelling IT projects success with fuzzy cognitive maps. *Expert Systems with Applications*, 32, 543–559. doi:10.1016/j.eswa.2006.01.032

Repiso, R. L., Setchi, R., & Salmeron, J. L. (2007b). Modelling IT projects success: Emerging methodologies reviewed. *Technovation*, 27, 582–594. doi:10.1016/j.technovation.2006.12.006

Shou, Y., & Ying, Y. (2005). Critical failure factors of information system projects in Chinese enterprises. In *Proceedings of International Conference on Services Systems and Services Management*, (pp. 823-827). IEEE Press.

Smith, D., Passos, J., & Isaacs, R. (2010). How IT project managers cope with stress. In *Proceedings of the 48th Annual Conference on Computer Personnel Research*, (pp. 15-24). ACM Press.

Standing, C., Guilfoyle, A., Lin, C., & Love, P. E. (2006). The attribution of success and failure in IT projects. *Industrial Management & Data Systems, 106*(8), 1148–1165. doi:10.1108/02635570610710809

Standish Group International. (2009). *Website.* Retrieved from http://blog.standishgroup.com/

Storm, P., & Savelsbergh, C. (2005). *Lack of managerial learning as a potential cause of project failure.* Retrieved from http://www.ou.nl/Docs/Faculteiten/MW/Congres%20Papers/2005/18-20%20aug%202005.pdf

Sundari, R. T., Barwal, P. N., Prakash, R., Yadav, R., Garg, C., & Jain, D. K. (2009). An analysis of factors influencing success and failure of IT project. [Noida, India: CDAC.]. *Proceedings of ASCNT, 2009*, 152–158.

Taylor, M. A. (2002). *The 5 reasons why most projects fail and what steps you can take to prevent it.* Retrieved from http://www.idii.com/wp/Tse5reasonswhyprojectsfail.pdf

Thite, M. (1999). Leadership: A critical success factor in IT project management. In *Proceedings of Portland International Conference on Management of Engineering and Technology,* (pp. 298-303). PICMET.

Warne, L., & Hart, D. (1996). The impact of organizational politics on information systems project failure - A case study. In *Proceedings of the 29th Annual Hawaii International Conference on System Sciences,* (pp. 191-201). IEEE.

Wateridge, J. (1995). IT projects: A basis for success. *International Journal of Project Management, 13*(3), 169–172. doi:10.1016/0263-7863(95)00020-Q

Winklhofer, H. (2001). Organizational change as a contributing factor to IS failure. In *Proceedings of the 34th Hawaii International Conference on System Sciences,* (pp. 1-9). IEEE.

Wong, X., Jingchun, F., & Ming, L. (2009). Research on IT project life cycle. In *Proceedings of the IEEE Second International Conference on Intelligent Computation Technology and Automation,* (pp. 244-247). IEEE Press.

Yeo, K. T. (2002). Critical failure factors in information system projects. *International Journal of Project Management, 20*, 241–246. doi:10.1016/S0263-7863(01)00075-8

Yourdon, E. (1997). *Death march: The complete software developer's guide to surviving "mission impossible" projects.* Upper Saddle River, NJ: Prentice Hall Publications.

Chapter 12
Issues and Technologies of Effective Energy Management

Edward T. Chen
University of Massachusetts – Lowell, USA

ABSTRACT

The purpose of this chapter is to discuss critical issues and the role technology plays in today's energy sectors. Specific emphasis is placed on security, mobile dispatch solutions, and the so-called "Smart Grid." The industry continues to grow in both size and complexity, creating a multitude of challenges for companies as they struggle to keep the lights on. The utility business has traditionally lagged behind other sectors in the adoption and implementation of new technologies. However, mounting economic, environmental, social, and political pressures have thrust this once lumbering dinosaur out into the spotlight. Energy companies must look to innovative technology solutions to help them keep pace with our growing society. The chapter also touches upon how these issues create meaningful educational and employment opportunities.

INTRODUCTION

The utility industry has historically been slow to adapt to new technologies and business paradigms. However, mounting economic, environmental, social, and political pressures have thrust this lumbering dinosaur into the spotlight. The utility industry has witnessed a rapid increase in the rate of change in both technology and business prac-

tices. Energy companies need to find innovative technology solutions that provide better service to their customers while minimizing the impact on the environment and keeping costs under control (Chao, 2011; Kirschen & Strbac, 2004).

Organizations will need to invest in adequate infrastructure to support mobile workforce management and remote operation of the power grid. Furthermore, the fallout from the 9/11 terrorist attacks, Enron scandal, and major blackout events have created a sea of regulatory legislation that must be navigated if a company hopes to survive

DOI: 10.4018/978-1-4666-3658-3.ch012

in today's difficult business environment. Technology is poised to take a lead role in these efforts and competent. Qualified personnel will be needed to manage critical projects using smart technologies (Lui, Stirling, & Marcy, 2010).

This chapter examines existing issues, technologies, and business strategies as they relate to the energy industry and identify areas where potential problems exist. Examples will include both successful projects as well as those, which failed to live up to expectations. This chapter looks at several topics in an effort to provide the reader with an understanding of the state of the industry as it relates to technology and what the future holds.

It is a very dynamic time for the energy industry. Companies are scrambling to keep pace with changes in regulations and technology. Cost will become a major issue and additional funding will be needed (Chao, 2011; Kirschen & Strbac, 2004). Who ultimately pays for a smarter, more reliable power system is a hot debate. The increased interconnection of parties and availability of information requires solutions that are viable, consistent, and securely managed.

ENERGY INDUSTRY OVERVIEW

Electricity. It is amazing how something so seemingly simple can become so ingrained into our daily lives. Light bulbs, computers, smart phones, home appliances, and most recently consumer vehicles would all cease to function without the controlled flow of electrons. In order to fully appreciate the current state of this critical industry and predict its future direction, we must start at the beginning.

Society's first recordable encounters with electricity occurred about 2,600 years ago during the era of the ancient Greeks (Warkentin, 1998). In fact, interestingly enough the words electron and magnet are both subsequently of Greek origin. Given the primitive technology of the time, the Greeks were limited in their discoveries to the

observation of simple concepts like static electricity and the effects of magnetism. However, from these humble beginnings came a series of powerful experiments conducted by some of humanity's greatest scientific minds. Pioneers such as Maxwell, Faraday, and Volta helped to lay the groundwork for the industrial revolution and the eventual birth of the electric industry.

While there have been many contributors over the years, there are two in particular whose work cannot be ignored. Thomas Edison and George Westinghouse sought to bring electricity to the masses with their respective companies, eventually coming into conflict over what would be called "the war of currents." In 1882, Edison had established a direct current (DC) system to serve neighborhood incandescent lighting facilities. Centered in Manhattan, the Pearl Street Station became the first Investor Owned Utility (IOU) in the United States. Westinghouse on the other hand, had begun developing Alternating Current (AC) technology, boasting its superiority for transmitting power over longer distances to that of Edison's system. The two sides then engaged in a lengthy battle of propaganda and smear campaigns, attempting to win support from a largely uneducated populace. AC power eventually emerged as the dominant technology and has been used extensively in the construction and operation of the Bulk Electric System (BES).

The conclusion of the war of currents led to a blossoming of small utility companies primarily focused on serving local loads. After a period, it became clear that by interconnecting with their neighbors, these companies gained access to increased energy capacity and improved reliability. At present in the United States, there are five main types of power utilities participating in this widespread interconnection: Investor Owned (IOU's), Federally Owned, Publicly Owned, Co-operatively Owned, and the Independent Power Producers (IPPs). Each entity is unique in structure and function, yet they all must work together to keep the lights on and businesses running strong.

Representing the largest subsection, the IOU is usually either an individual corporation or a holding company serving as an 'umbrella' over a group of subsidiaries. These companies sell stock to fund capital projects and return profits to investors in the form of dividends. IOUs have been granted monopolistic control over specified geographic areas and are subject to state and federal regulations. Most regulated customers in an IOU's service territory do not have a great deal of control over their choice of provider and the rates are determined in periodic rate cases between the IOU and the local public service commission.

Federally owned utilities are operated on a non-profit basis by the United States government. There are currently only 10 in existence, under the jurisdiction of various internal departments, the most famous being the Tennessee Valley Authority (TVA). The majority of customers for this group are either large industrial entities or military installations. Less than 2% of the output of federal plants is sold to retail consumers (Warkentin, 1998). .

The remaining players in the market are from public companies, cooperatives, and IPPs. Such companies may be owned and operated by a community of residents, a municipal body, a corporation, or a group of venture capitalists (Denny & Dismukes, 2002). These bodies may be for profit or non-profit and they may or may not generate their own electricity. The concept of the IPP is starting to become more popular, especially with the advent of small distributed generating plants such as biomass, digesters, trash burners, etc. Operators of these facilities usually receive a tax incentive in exchange for their investment in "green energy" (El-Khattam & Salama, 2004).

Eventually as these entities interacted, the system grew to be so complex that it became difficult to manage and the threat of a large blackout loomed unnoticed on the horizon. When the lights finally went out on the East Coast in 1965, it became clear that change was needed in order to secure the reliability of the power grid. The National Electric Reliability Council (NERC) was formed to oversee each of the Regional Reliability Organizations (RROs) spread out over the three main interconnections--Eastern, Western, and ERCOT. The map in Figure 1 shows the lines of jurisdiction between each zone and it is important to note that NERC's authority is not limited to the United States but also includes portions of Canada and Mexico. The Federal Energy Regulatory Commission (FERC) was formed shortly thereafter in an effort to centralize authority over interstate electricity rates, hydroelectric facility licensing, and natural gas/oil pipeline systems. NERC and FERC work closely together to ensure policies are consistently enforced in all areas of the Bulk Electric System (BES).

As the primary regulating body, FERC introduced two very significant pieces of legislation in 1996, orders 888 and 889. Rule 888 dealt with open access rights and recovery of stranded costs, while 889 called for the establishment of the OASIS (open access same-time information system) for the electronic sharing of transmission information. These acts helped pave the way for the current energy market model we have in service today.

In conjunction with FERC's new policies, NERC began to recognize that changes would also be needed to the structure of the industry. The roles and functions of local control centers were changing as a result of the open access initiative, most notably:

1. Some utilities were separating their transmission from merchant functions and selling off generation. It is also called functional unbundling.

2. Some states and provinces were instituting "customer choice" options for selecting energy providers.

3. The developing power markets were requiring wide-area transmission reliability assessment and dispatch solutions far beyond the capabilities of an individual control center.

Figure 1. Map of North American electric reliability council and interconnections

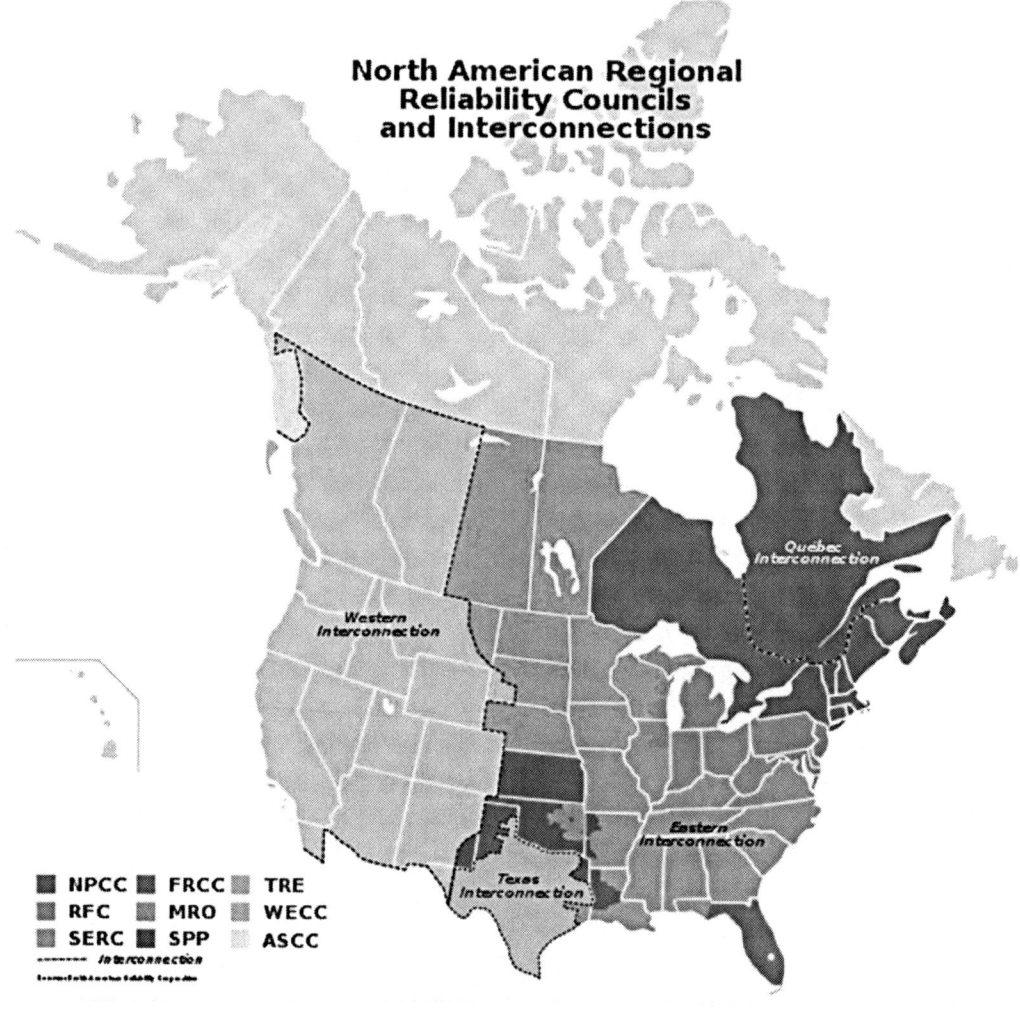

Source: NERC (2010)

NERC's changes resulted in a new division of tasks among members, with each being assigned one or more roles as shown in Figure 2.

The specific functions of each role are beyond the scope of this chapter, but it is important to understand that every player in the industry serves a unique purpose and it takes the combined effort of all parties to keep the system online.

Like the transportation and communication industries before it, the energy industry is in the process of a massive restructuring intent on creating open competition between players and lowering prices for consumers. This de-regulation, or perhaps more accurately, re-regulation, calls for specific guidelines on how companies and their systems are allowed to operate and interact (Zhong & Bhattacharya, 2002). As it stands, the operation of the Bulk Electric System can be divided into four main components: generation, transmission, distribution, and energy markets.

Figure 2. Functional model diagram

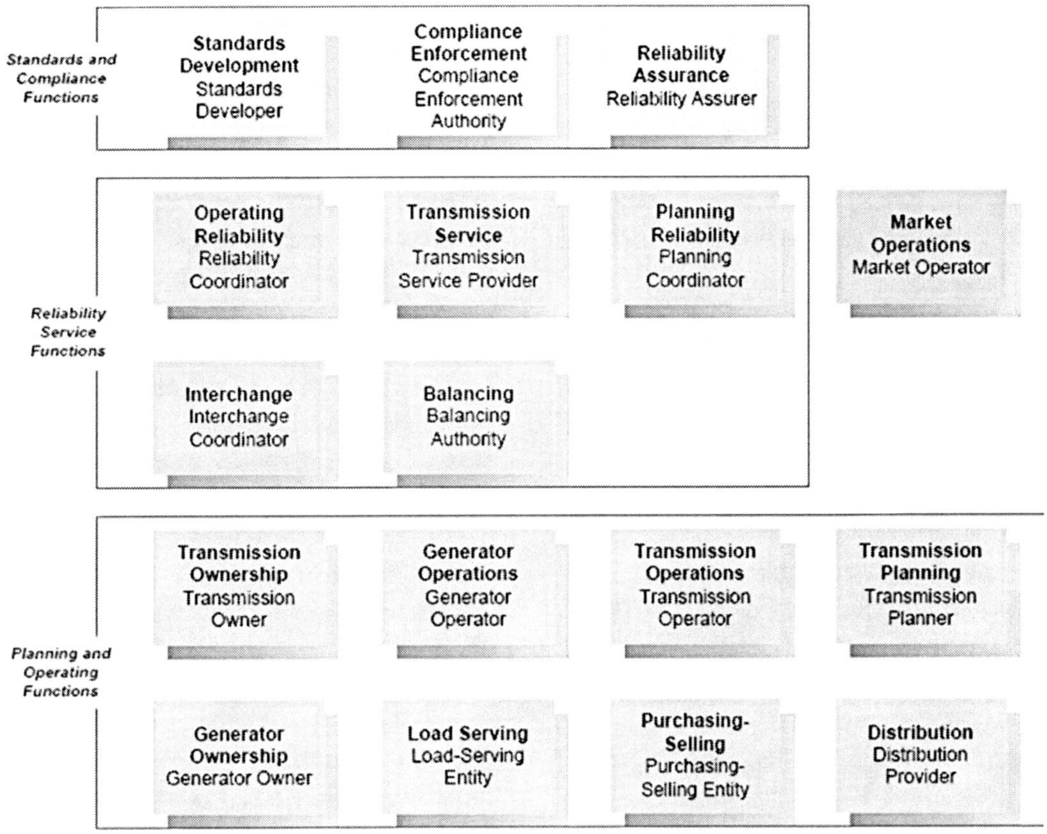

Source: NERC (2010)

Generation

Generators convert input fuels (fossil, nuclear, water, wind, solar, etc.) into electricity by way of combustion, mechanical, or chemical reactions. The output can range from a few kilowatts to thousands of megawatts depending on the facility. Recent environmental legislation has called for an increased investment is so-called "green energy" technologies, with some states requiring minimum standards for renewable resource portfolios (El-Khattam & Salama, 2004). This

creates challenges for utility companies who got by for years with relatively cheap outdated fossil units. The change will be gradual however and the majority of the nation's generators are still run on fossil or nuclear fuel.

Information technology becomes very important when it comes time to determine the best economical loadings for a utility company's generators. Optimum load dispatch points are established based on inputs such as fixed costs, variable costs, market prices, unit efficiencies, and incremental heat rates (Miller & Malinowski,

1994). Many of these factors change constantly and can be difficult to predict or control like weather, fuel costs, availability, etc. Thus, economic optimization of units is a continuous process (Ranatunga, Annakkage, & Kumble, 2003). Depending on the market, in some cases an organization may be better off simply buying power from a neighbor rather than producing it themselves (Lui, Stirling, & Marcy, 2010).

Computer programs are also used to adjust generation in response to changes in system load. These calculations are made through the use of an Area Control Error (ACE) computation that takes into account frequency and interchange deviations, frequency bias, and metering error. It is important to note that the ultimate goal of this process is to achieve a balance between generation and load, while maintaining a frequency of 60 Hz. These adjustments are made by an Automatic Generation Control (AGC) system, which can be operated in one of three possible modes: flat frequency, flat tie line, or tie line frequency bias (Miller & Malinowski, 1994). The AGC mode dictates which ACE parameters are to be used in controlling the unit. Flat frequency attempts to hold frequency constant while ignoring the interchange element, while flat tie line has the opposite effect. The tie line frequency bias mode includes all parameters and is considered the normal mode of operation for generator units.

Transmission

Transmission lines are responsible for transporting energy from the generating plants to distribution points where it is delivered to the customer. Voltage levels can range from 100kV upwards of 765kV or greater in some cases (Harris, 2006). The reason why the voltage is so high is to reduce the amount of current required to push the energy. Current levels contribute to the overall losses in the path. Power lost = I^2R where I is the current and R is the resistance. The majority of these lines are 3-phase AC but there are special cases where DC lines are more desirable.

One of the main issues with transmission systems is the concept of losses. Not all of the energy produced actually makes it to the end use customer. Some of it gets radiated away in the form of heat. Different factors can contribute to the amount of losses on line such as length, diameter, material, and weather. Hence, it is important that adequate systems are in place to calculate these effects. Each line has a loading limit based on thermal, voltage, or stability parameters and exceeding these limits causes tremendous strains on the system. When such a bottleneck occurs, market prices and loads are adjusted to reduce the flows on overloaded conductors. Periodic auctions are held wherein entities secure "rights" to these transmission facilities, thus determining whose deal will be more likely to survive a curtailment order.

It should be noted that in some cases a transmission line can also double as an information link between two substations. This technique is called the Power Line Carrier method and it functions by superimposing a stream of data onto the actual power signal. Coupling capacitors at the receiving ends (line traps) help to filter the data from the combination wave form. These circuits are specially tuned to block out interference and permit the flow of data (Miller & Malinowski, 1994).

Distribution

The distribution system picks up where transmission leaves off by transmitting the electricity to the end use customer. Distribution voltages are those below 100kV and the lines can be 3-phase or single phase depending on the situation. Unlike transmission grids, which are usually looped systems, distribution lines are usually radial in nature. The distribution sector represents the bottom of the energy food chain but ultimately has the highest degree of interaction with the customer (Harris, 2006). This interconnection occurs at the customer meter, which collects the load information used to generate the monthly billing statement.

Perhaps one of the most impressive technological changes on the distribution system is the development of automated meter reading technology. In years past, companies were forced to send employees to a physical location, read the customer's meter, and then transported the data back to the office where it was downloaded into a database once per month. Billing clerks and other office personnel would then look up the information as needed to generate statements and answer customer questions. This process was very labor intensive and caused headaches for utilities and customers alike. Safety, accuracy, and timeliness were all seriously jeopardized under the old system. Automated meter reading relies on Radio Frequency (RF) signals to transmit data to remote collection centers, eliminating the need for a personal site visit (Khan, Aditi, Sreeram, & Iu, 2010).

With the arrival of Automated Meter Reading (AMR), Advanced Metering Infrastructure (AMI), and Meter Data Management Systems (MDMS), employees and customers now have constant access to real time load data. In addition to billing matters, having this kind of data stream also helps utilities manage power outages. Meters can be "pinged" to test for connectivity, allowing a dispatcher or analyst to confirm the presence of the outage. Distribution Automation (DA) and Distribution Management Systems (DMS) are also made possible through the use of AMI (Hatfield, 2011). The data gathered from meters can be used to identify and isolate faults on the system, helping the network "heal itself" in times of trouble. This is a critical component in the modern Smart Grid systems currently under development (Vijayapriya & Kothari, 2011). The main drawbacks to this technology are the high up-front costs and potentially long transition period (Seymour & Kadrmas, 2011).

Energy Markets

Energy markets represent the final component of the BES and it is in this dynamic environment that the power marketer reigns. In the early onset of the industry, energy was consumed relatively close to where it was produced. Communities had their own power plants and life was good. As the system began to expand, interconnections formed, and it became possible to transmit power between regions. This concept was eventually termed "wheeling" and has become a very important part of the power grid's operation. Power marketers facilitate wholesale and retail transactions between buyers and sellers in the market. These transactions are typically done via "tags" which identify the source, destination, amount of power, start time, stop time, and the path the energy will ideally take.

Power markets function very similarly to other financial markets, the goal is to buy low and sell high through the use of a variety of hedging tools, options, futures contracts, and swaps (Denny & Dismukes, 2002). The inner workings of these instruments will not be covered here, but marketers have the ability to deal in the day-ahead (forward) or real time (spot) markets depending on the timeframe in question. Traders usually employ complex forecasting programs to predict prices and estimate future demand. Each of the major energy markets is administered by an Independent System Operator (ISO), which ensures fair business practice and electric reliability are maintained at all time. Prices at predetermined locations (i.e., busses or nodes) are calculated using a method called Locational Marginal Price (LMP). LMP takes into account load centers, transmission constraints, and generator outputs to determine the relative cost of energy at any point on the system, which becomes the market price for the next unit of energy at that location.

ISSUES AND PROBLEMS

Companies ultimately need to find innovative technological solutions that result in better service to customers while keeping costs under control. This is easier said than done and it presents special challenges to the installation of high tech security, mobile dispatch, and smart grid systems.

Security

As technology advances so does the need to protect it from hazards such as terrorist acts, vandalism, or other malicious activities. Security is divided into two main categories: physical and electronic or cyber. NERC has formal requirements for securing so-called "critical assets" or those, which are instrumental in maintaining the integrity of the Bulk Electric System. Fines for violations of these standards can be steep, upwards of a million dollars per violation per day, but the monetary punishment pales in comparison to the potential fallout after a large-scale attack (Bahgat, 2010).

When designing a security program, McDonald (2007) suggests a systemic approach including the following steps:

1. Threat assessment
2. System analysis
   a. Criticality assessment
   b. Vulnerability assessment
   c. Risk assessment
3. Risk management
4. Implementation

The purpose of the threat assessment is to identify elements that pose a significant danger to the system. This requires an examination of the presence, capability, and intent of the parties in question. Groups under consideration should include the general public, thieves, vandals, disgruntled employees, and terrorist cells. It is advisable that companies form partnerships with both local and federal law enforcement when developing security programs. Having these contacts and communication lines in place will save precious time in the case of an emergency. Additionally, all utility companies should secure access to the Electric Sector Information Sharing and Analysis Center (ESISAC), an organization that serves as a go-between for companies, government agencies, and law enforcement. The ESISAC is a portal through which utilities can report real or suspected threats on critical infrastructure and view similar postings from neighboring companies. A solid physical security system should include fences or walls, locked gates, adequate lighting, proper signage, access controls, door alarms, and video cameras.

Physical security is only one-half of the equation. Electronic or cyber security is also critical in preserving the integrity of our electric system. Complex computer systems are fundamental in monitoring and controlling the power grid, making them tempting targets for the would-be saboteur. Worms, viruses, and hackers can attack without notice and considering the degree in which these systems are interconnected, the damage can spread rapidly.

One of the reasons cyber terrorism can be so difficult to spot is that the culprit does not need to be in close proximity to the target when launching the attack. The target could be an isolated substation, generating plant, or a control center's Supervisory Control and Data Acquisition (SCADA) system. The SCADA system is the very heart of a utility's operations. This software combines the communications, alarm management, system status, and control functions of the power grid; when lost, the organization is essentially running blind. The loss of the SCADA system is one of the most debilitating injuries an area can suffer, but any attack has the potential to lead to widespread outages, equipment failures, or even fatalities. Some of the more common vulnerabilities to automated systems are highlighted in Table 1.

Table 1. Substation automation system vulnerabilities

Slow processors with stringent real-time constraints	Older technologies are unable to handle resource-intensive 'block encryption' and 'authenticated' data transmissions. (Lack sufficient speed, bandwidth, memory, etc.)
Real-time operating systems that preclude security	Inferior software embedded in system components already in service.
Insecure communications media	Data traffic to the outside world may pass through open networks.
Open protocols	Protocols used in the industry (IED, Modbus, DNP3, etc.) are well known and commercially available to the public.
Lack of authentication	Systems exchanging data lack the ability to confirm the identity of the user.
Low priority cyber security	Management does not provide enough support for cyber security initiatives.
Lack of centralized system administration	Failure to properly oversee and control access to the system.
Large number of remote devices	High cost associated with upgrading multiple units impedes scheduled maintenance.
Substation diagnostic systems	Some programs may become a "back door" into the system.

Source: McDonald (2007)

In addition to addressing the vulnerabilities, personnel responsible for the corporate security plan should also consider the following items when developing critical policies and procedures (McDonald, 2007):

1. Remove all default user IDs and passwords on installed systems.
2. Ensure all accounts have strong password.
3. Close unused ports and unnecessary services.
4. Install security patches in a timely manner.
5. Remove all sample scripts in browsers.
6. Implement robust firewalls.
7. Implement intrusion detecting systems, logging devices, and thoroughly investigate suspicious activity.

Mobile Dispatch

One area that has been receiving a great deal of attention in utility circles lately is mobile dispatch and workforce management. Wireless technology is improving at a rapid pace and companies stand to benefit through improved response times and added flexibility. The economy is placing a serious strain on an organization's bottom line. "Do more with less" has become the normal mode of operation. One way to achieve this goal is through the efficiency gained in the logistics of mobile workforce operations. Technology makes this all possible by way of smarter metering, GPS installations, wireless cards, diagnostic setups, and advanced outage management software (Khan, Aditi, Sreeram, & Iu, 2010). Mobile units can now freely exchange critical documents and information with field office personnel.

The increased range of wireless connectivity between the field and the home office has led to the creation of a "mobile branch office" environment, wherein utility vehicles can be transformed into mini work management centers. When operating in this mode, a line worker's devices (laptops, tablets, PDAs, and smart phones.) can all be securely connected to a corporate network. This gives workers access to the same applications and information as their office counterparts, facilitating better communication, coordination, and customer service. An example of this technology is the In Motion Technology onBoard Mobile Gateway system, which has been shown to increase efficiencies by reducing operating costs, improving scheduling, managing customer contacts, and maximizing asset utilization (Hughes-White, 2011).

The final component in the mobilization of workforce management is a powerful software solution with the ability to organize and prioritize customer outage calls and work requests. Dubbed Outage Management Systems (OMS), these tools keep track of where problems are on the system and can even predict the probable cause location. The system interfaces with a company's customer information and geographic mapping databases to create a clear picture of the company facilities. Companies with GPS capable trucks will see dispatched unit locations update in real time, facilitating better logistics for assigning work orders. While every system is unique, they each are similar in their fundamental design.

The software can also be configured to take advantage of Automated Meter Reading technology and "ping" customer meters to verify the presence of power. Outages can be rolled into the next up or down stream device as needed once additional information becomes available. In some cases, OMS can incorporate SCADA controls or load flow simulation modules to create an all-inclusive distribution management system. Information regarding a customer's outage is passed back to the customer information system in real time, providing outage duration and estimated restoration times. The data can also be exported to a Web-based reporting application, which displays information for the entire service territory. This is especially helpful for call center employees or management personnel who may be contacted by curious customers or the media.

Smart Grid

Smart Grid is perhaps the most interesting way in which technology is changing the face of the energy industry. The exact form this smart grid will ultimately take is still evolving and unknown. However, it is clear that several different technologies will need to work together to help automate the system. This automation will entail demand response, consumer load management, fault analysis, and remote switching (Davis, 2011; Vijayapriya & Kothari, 2011).

The long-term plan for the implementation of smart grid is to provide a solution for rising energy demands, aging infrastructures, renewable energy sources, and electric vehicles. This transition from Jurassic grid to smart grid will require additional communication systems, real time energy monitoring capabilities, and the bi-directional flow of both data and energy. As this technology changes, the traditional relationship between utilities, and their customers must also change. They will in effect become partners in managing the nation's power grid, helping to jointly maximize reliability and cost effectiveness (Seymour & Kadrmas, 2011; Vijayapriya & Kothari, 2011).

At the heart of many of these smart grid designs lies the microprocessor-based relay control unit. This technology incorporates protection devices, programmable functionality, and digital communication capability into one compact unit, freeing up large amounts of real estate inside substations that was previously occupied by a jungle of wires, boxes, electromechanical devices, and switches.

The ability to program the unit adds a great deal of functionality to the relay package design and data capturing ability. The primary programming tools offered in these units are: Boolean operators, control equation elements, binary elements, analog quantities, and math operators (Blackburn & Domin, 2007). Boolean algebra is a system of 1s and 0s used to denote the "on" or "off" position of control states or switches. These values are then operated upon by inputting them into logic gates as AND, OR, NOR, NOT, etc. The system then interprets the result and takes a prescribed course of action. This is in essence the basis for the "smart" in smart grid as it gives the system a degree of thought (e.g., decision making) when presented with a specific situation (e.g., set of inputs). In addition to Boolean logic, these units also offer the user with customizable features including data storage, operations counters, time-delay curves, fault locating, and event reporting (Vijayapriya & Kothari, 2011).

The advantages and disadvantages of this microprocessor-based relay control technology are summarized in Table 2. Microprocessor relays have led to more advanced protection schemes for distribution and transmission systems as well as an overall increase in substation automation. Examples of some of these systems include:

1. Centralized load-shedding schemes to address under-frequency or overvoltage conditions.
2. Automatic changing of relay settings for various loading or system conditions.
3. Wide area controls that prevent angular instability.

Communication with the unit can be achieved through a variety of protocols, including: ASCII, MODBUS, MODBUS PLUS, and UCA/MMS. The communications for the entire smart grid however, are much more involved as no single network can provide a complete solution (Blackburn & Domin, 2007).

The various components and power management systems making up the smart grid have different criteria when it comes to communications performance. Primary performance measures for smart grid communications solutions are based

Table 2. Advantages and disadvantages of microprocessor control units

Advantages	Disadvantages
1. More protection for less cost 2. Wiring simplification 3. Greater flexibility 4. Lower maintenance 5. Reduced space requirements 6. Recording capability 7. Fault locating 8. Data collection/metering 9. Logic control/automation 10. Self-checking capability 11. Communications schemes 12. Remote control operation 13. Dynamic setting changes	1. A single failure can disrupt many protective functions. 2. Complicated instruction manuals. 3. Excessive data input requirements. 4. Frequent firmware upgrades. 5. Software compatibility problems.

on their approximate bandwidth and latency requirements (Cornish, 2011). There are several alternative solutions by increasing the bandwidth or the latency, or both (Tsang, 2009; Tung, et al., 2011). Each of these solutions requires a different level of performance. Attempting to select a single communications network would mean either sacrificing performance or over-paying for the system. Cornish (2011) recommend that a physical approach be setup with a four-tier network to improve the performance as shown in Table 3.

The example above represents a typical solution but each situation is unique and companies should perform their own internal analysis to determine the best fit for their needs. In this case,

Table 3. Physical approaches in a four-tier smart grid network design

1. Backbone	Corporate fiber	Corporate information flow
	Microwave	System protection
2. Backhaul	WiMAX	Substation automation
		Video security
		Smart meter backhaul
		Distribution automation
3. LAN/NAN	Single-purpose RF	System monitoring
	Smart meter networks	Smart meters
4. HAN	Zigbee	In-premise devices

Tier 1 would handle high-bandwidth, low-latency applications, including the backbone of utility communications infrastructure. The requirements become less stringent as the level increases, until finally Tier 4 is reached, which is best suited for interconnections with the customer.

ORGANIZATIONAL IMPACTS

Organizations will need to invest in adequate infrastructure to support mobile work force management and remote operation of the power grid. Furthermore, the fallout from the 9/11 terrorist attacks, Enron scandal, and major blackout events have created a sea of regulatory red tape that must be navigated if a company hopes to survive in today's difficult business environment. Technology is poised to take a lead role in these efforts and competent, qualified personnel will be needed to manage critical projects.

Unfortunately finding the right people for the job will be increasingly difficult in the coming years. According to research done by Reyes (2011), an estimated 30 – 40% of industry workers will be eligible for retirement by 2013. This creates serious workforce management issues, including:

1. Massive retirement across all ranks and skill sets.
2. Recruitment challenges stemming from less interest in power fields.
3. Declining numbers and quality of middle management.
4. Wide division between worker ages and expectations.
5. Inadequate educational structure.
6. Lack of emphasis on workforce planning.

In order to combat these issues, employers will need to get creative when managing and recruiting new and existing talent. This may entail changes to the culture or structure of the organization. The focus should be on creating a learner atmosphere where skill development and succession planning are held in high regard.

SUGGESTIONS

Most utility companies rely on some form of rate relief to support regular business operations and fund major capital projects. Given the state of the economy, customers will not react favorably to any perceived unnecessary price increases. Organizations will be under immense pressure to spend every dollar wisely and failed projects will not be taken lightly. Developing solid metrics to gauge success is critical in these situations. When mismanaged, IT projects can rapidly deplete an organization's resources. One company that has figured this out is FirstEnergy (Pearlson & Saunders, 2010). The workers at this utility juggernaut have found that by using an MIS scorecard, they can provide better service to their internal customers while keeping a close eye on the bottom line.

Besides winning the favor of clients, the First Energy team identified three primary drivers for adding value to their business: reliability, finance, and culture. They then translated these drivers into observable metrics as follows:

1. Percentage of projects completed on time and under budget.
2. Percentage of projects released by agreed-upon delivery date.
3. Customer satisfaction surveys at the conclusion of the project.

This is a great example of how attention to detail can help generate a culture where timeliness, financial responsibility, and customer satisfaction are elevated as top priorities.

In addition to improving project management, finding ways to improve everyday operations can also help companies get the most out of limited resources. Hughes-White (2011) has identified several strategies for accelerating the return on a wireless workforce investment as listed in Table 4. By incorporating these points into the operational business strategy, companies are able to get the most out of their investments.

The maximization of asset investment is certainly an important goal for every company, but the need to protect these investments cannot be understated. Having a comprehensive security program in place ensures that advanced technologies are not damaged or manipulated by undesirable parties. McDonald (2007) points out that the development of such security program should minimally include the following items:

1. Asset identification
2. Threat identification
3. Threat detection
4. Incident response
5. Training and documentation
6. Administration
7. Software management

Special arrangements should be made to ensure that in the event of a large scale disaster, a backup control center can be established for continuing operations. This facility must be isolated from the primary facility and have provisions for proper security as well as redundant voice and data capabilities. Auditable drills should be conducted annually to verify the viability of such a plan and the results used for compliance documentation and process improvement.

The need for vigilance is not limited to the threat of terrorist attack or sabotage but must also include monitoring of peers within the industry. When proper oversight is not maintained, disaster can strike. Such was the case in California in 2000 and 2001 when traders like Enron began capitalizing on loopholes in the market structure of the California ISO (Denny & Dismukes, 2002). The level of corruption involved in the scandal was all inclusive: traders, plant operators, executives, politicians, and banks all had their hand in the cookie jar. Through clever schemes and faulty accounting practices, Enron was able to manipulate the market and drive wholesale prices through the roof. California's regulated utility companies were unable to pass these costs onto their customers due to retail price caps and a lack of sufficient internal generation forced them to buy from the market. After a series of bankruptcies and rolling blackouts order was finally restored but it was a hard lesson

Table 4. Maximizing returns on wireless workforce

Create a more productive shift	Save on downtime and paper costs by distributing work orders electronically.
Save on airtime	Manage data traffic appropriately to avoid charges for minute overages.
Remove carrier barriers	Use carrier provided network cards in a protective housing.
Stay on the road	Monitor a vehicles status remotely to determine upcoming repairs/maintenance.
Choose easy-to-use networking technologies	Free up field crews to focus on their jobs, not learning how to use the technology.
Keep data secure	Choose the proper architecture to better protect critical information.
Have network flexibility	Give crews the ability to roam across different networks depending on location and signal strength.
Mobile branch office must be rugged	Make sure the equipment can survive temperature extremes and is not easily damaged or stolen.
Manage the mobile workforce	Provide crews with a Web based dashboard of real time data to keep them running at peak efficiency.
Have the right tool for the job	Mobile branch offices can provide superior functionality to traditional laptop units.
Communication platform must be cost effective	Mobile branch offices usually cost less than traditional laptop units.

for the industry and the residents of California to learn. That is why it is so important that ISO's have cutting edge technology in place to monitor system parameters and keep companies honest.

FUTURE IMPLICATIONS

Energy companies face a great deal of uncertainty in the coming years. Will the economy ever completely recover? How will fluctuations in fuel costs and environmental legislation impact a utility's ability to remain competitive in the market? Do the benefits of smart grid outweigh the costs? Are more frequent mergers and acquisitions becoming a necessity for survival? These are just some of the issues organizations must evaluate when making important business decisions. Regardless of how the market changes, it is clear that successful parties will need to invest in two things: people and technology (Zhong & Bhattacharya, 2002).

The foundations for such investment begin at the educational level. Utilities need to create a steady stream of talent and ideas to replace an aging workforce looking to retire in the next 5 – 10 years. Improvements in technology will require more complicated skill sets than in the past and by taking an active role in educational institutions, companies can ensure that incoming employees are better able to "hit the ground running." Career fairs, internships, and co-ops will be even more important in the days ahead. This is an exciting time for the industry and raising awareness and interest among young people will help to ensure its survival over the long term.

CONCLUSION

There are three primary drivers towards expanding the competition of the energy industry in the future:

1. Evolving philosophies on government and success on deregulation of other utilities.
2. New technologies, better generating efficiencies, and an increase in independent power producers.
3. Widely varying differences in unit revenues from electric power sales.

Of these, technology becomes the most important factor as it ultimately permits or restricts the future possibilities in market reformation and smart grid flexibility. Designing, installing, and controlling this equipment will likely require a more advanced skill set than in years past. Ideally, this will generate interest in the power technology field and create abundant educational and employment opportunities for young professionals.

Companies must find a way to reduce costs without sacrificing their most precious asset: people. A great deal of experience is exiting the industry at an alarming rate and new bodies will be needed to fill critical positions. In addition to investing in the education of incoming workers, utilities must strive to keep the public educated on the state of the industry and the progress on futuristic goals such as smart grid. Cost is certainly a concern, but it is important that safety and reliability continue to be top priorities in the energy industry.

Armed with new technologies and powerful tools, it will take a combined effort from governments, utilities, customers, scientists, and environmental groups to create the energy future. This future is brightest when electricity remains clean, reliable, and affordable for everyone.

REFERENCES

Bahgat, G. (2010). United States' energy security: challenges and opportunities. *The Journal of Social, Political, and Economic Studies, 35*(4), 409–425.

Blackburn, J. L., & Domin, T. J. (2007). *Protective relaying principles and applications*. Boca Raton, FL: Taylor & Francis Group.

Chao, H. P. (2011). Demand response in wholesale electricity markets: The choice of customer baseline. *Journal of Regulatory Economics, 39*(1), 68–88. doi:10.1007/s11149-010-9135-y

Cornish, K. (2011). Communications technology for the smart grid. *Electric Light and Power, 89*(3), 62–63.

Davis, P. (2011). Smart grid, evolution of DR and the impact of FERC 745. *Electric Light and Power, 89*(3), 46–47.

Denny, F. I., & Dismukes, D. E. (2002). *Power system operations and electricity markets*. Boca Raton, FL: CRC Press.

El-khattam, W., & Salama, M. M. A. (2004). Distributed generation technologies, definitions and benefits. *Electric Power Systems Research, 71*(2), 119–128. doi:10.1016/j.epsr.2004.01.006

Harris, C. (2006). *Electricity markets pricing, structures, and economics*. Hoboken, NJ: John Wiley & Sons.

Hatfield, M. (2011). Meter data management systems and the paradigms of time. *Utility Products, 15*(7), 48–50.

Hughes-White, S. (2011). Fleet efficiency in a wireless world. *Utility Products, 15*(6), 10–14.

Khan, R. H., Aditi, T. F., Sreeram, V., & Iu, H. H. C. (2010). A prepaid smart metering scheme based on WiMAX prepaid accounting model. *Smart Grid and Renewable Energy, 1*(2), 63-69. Doi:10.4236/sgre/2010/12010doi:10.4236/sgre.2010.12010

Kirschen, D., & Strbac, G. (2004). *Fundamentals of power system economics*. Chichester, UK: Wiley. doi:10.1002/0470020598

Lui, T. J., Stirling, W., & Marcy, H. O. (2010). Get smart. *Power and Energy Magazine, 8*(3), 66–78. doi:10.1109/MPE.2010.936353

McDonald, J. D. (2007). *Electric power substations engineering*. Boca Raton, FL: Taylor & Francis Group. doi:10.1201/9781420007312

Miller, R. H., & Malinowski, J. H. (1994). *Power system operation*. Boston, MA: McGraw-Hill.

North American Electric Reliability Corporation. (2010). *Reliability functional model*. Retrieved March 10, 2012 from http://www.nerc.com/files/Functional_Model_V5_Final_2009Dec1.pdf

Pearlson, K. E., & Saunders, C. S. (2010). *Managing and using information systems*. Hoboken, NJ: John Wiley & Sons.

Ranatunga, R. A. S. K., Annakkage, U. D., & Kumble, C. S. (2003). Algorithms for incorporating reactive power into market dispatch. *Electric Power Systems Research, 65*(3), 179–186. doi:10.1016/S0378-7796(02)00217-1

Reyes, V. (2011). Human resources transformation for power and utilities Companies. *Electric Light and Power, 89*(3), 20–23.

Seymour, T., & Kadrmas, W. (2011). Smart grid U.S. transmission grid: Issues and opportunities. *The Review of Business Information Systems, 15*(3), 1–7.

Tsang, K. F., Tung, H. Y., & Lam, K. L. (2009). *ZigBee: From basics to designs and applications*. Upper Saddle River, NJ: Prentice Hall.

Tung, H. Y., Tsang, K. F., Lam, K. L., Tung, H. C., Zheng, R. J., Ko, K. T., & Lai, L. L. (2011). A WiMAX-ZigBee energy management system for green education. *Smart Grid and Renewable Energy, 2*(4), 338–348. doi:10.4236/sgre.2011.24039

Vijayapriya, T., & Kothari, D. P. (2011). Smart grid: An overview. *Smart Grid and Renewable Energy, 2*(4), 305–311. doi:10.4236/sgre.2011.24035

Warkentin, D. (1998). *Electric power industry in nontechnical language.* Tulsa, OK: PennWell.

Zhong, J., & Bhattacharya, K. (2002). Towards competitive market for reactive power. *IEEE Transactions on Power Systems, 17*(4), 1206–1215. doi:10.1109/TPWRS.2002.805025

KEY TERMS AND DEFINITIONS

Functional Model Diagram: A management technique to achieve customer service excellence by identifying the standards and compliance functions, reliability service functions, and planning and operating functions.

Green Energy: Green energy is used to describe that sources of power are known to be non-polluting energy sources. It is also an environmentally friendly means of locating and finding power that will remedy the effects of pollutants on our environment, as well as prevent future global warming.

Risk Management: A systematic approach of identification, assessment, and prioritization of risks to which a company might be subject to. It is then followed by coordinated and economical application of resources to minimize, monitor, and control the probability or impact of the risk.

Smart Grid: The electric delivery network, from suppliers to consumers, integrated with sensors, software, and two-way communications technologies to improve grid reliability, security, efficiency, and reduce cost.

Supervisory Control and Data Acquisition (SCADA) System: The SCADA system combines the communications, alarm management, system status, and control functions of the power grid for process control. SCADA system collects data in real time from remote locations in order to control equipment and conditions. SCADA system is the very heart of a utility's operations.

War of Currents: The battle between Thomas Edison, who had established a Direct Current (DC) system and George Westinghouse, who had begun developing Alternating Current (AC) technology since 1882. AC power eventually emerged as the dominant technology.

Workforce Management: A function of human resource management. The purpose of workforce management is to optimally maintain a productive manpower in an organization.

Chapter 13
An Examination of the Decision Making Styles of Egyptian Managers

Hisham M. Abdelsalam
Cairo University, Egypt

Reem H. Dawoud
Financial Consultant, Egypt

Hatem A. ElKadi
Cairo University, Egypt

ABSTRACT

Many factors play roles in the success of managers. However, the manager's decision-making style is one factor that highly contributes to that success and, therefore, to the success of their organization. In this chapter, a survey that includes a sample of 138 Egyptian managers in different organizational levels (junior, middle, and senior) is conducted to explore their decision-making styles. The research, then, investigates the relation between the variety of managers' decision styles and seven variables: gender, age, ethnicity, educational level, educational major, administrative experience, and current position. Based on the findings, this research is able to provide baseline information to improve on the implications of decision-making styles on the selection and design of decision-support systems in Egypt.

1. INTRODUCTION

One important key in a company's performance and success, or failure, is its management and the chief role of management; that is of planning and decision making (Yuki, 1994). In order to be effective and productive in an organization, all managers at different managerial levels must possess the ability to make valued decisions. Decision-making is, thus, a crucial element of a manager's responsibility in day-to-day operations as much as in long-term oriented strategic planning.

DOI: 10.4018/978-1-4666-3658-3.ch013

1.1. Decision Making Styles

While management is eternal and all embracing, debates continue to rage as to what does and should actually constitute management. The host of definitions now available cannot cloud the central fact that management is about decision-making (Fitzgerald, 2002). Decision-making is a fundamental function of management; it can be described as synonymous with managing (Tam et al., 1994), the essence of the manager's job (Robbins, 1999), and the essence of management (Dearlove, 2001).

A decision has been defined as "an answer to some question or a choice between two or more alternatives" (Rowe, Boulgarides, & McGrath, 1984). At a very fundamental level, the ability to make a decision relates to making choices within a pool of alternatives (Hammond, 1999). Various classifications of decisions are cited in the literature: routine, creative, and negotiated (Rowe & Boulgarides, 1992); programmed and non-programmed decisions (Gibson, Ivancevich, & Donnelly, 1994); operational decisions and strategic decisions (Dearlove, 1998); and day-to-day decisions, tactical decisions and strategic decisions (Dearlove, 1998).

The majority of decision-making research ascribes to the belief that decision-making is a process. In order to reach an outcome of a proper decision attention need to be awarded to the process of decision-making. Thus, various decision-making process models have been developed in literature: general model for decision-making (Gore, 1964); ideal decision-making model (Hill, 1979); the five point plan (Adair, 1985); traditional analytical model (Geernberg & Baron, 1993); synoptic model (Dearlove, 1998); six step rational decision-making model (Robbins, 1998); and the contingency model (Allwood & Selart, 2001). All of these processes protract similarities with regard to the basic approach to decision-making; rational decision-making.

Decision-making is a human act; within the decision making process information is structurally managed by an organization through the human manipulation of information (Streufert & Streufert, 1978). The human actors are decision makers who ultimately determine the choice among the alternatives in the decision making process. Significant research has demonstrated variations among individual decision processing; individuals within the decision making process can act very differently (Jill, 2006). The personalization of such information processing has been defined under the term cognitive style. Throughout the years, literature has used the term cognitive style to mean many different conceptualizations (Leonard, Scholl, & Kowalski, 1999). However, due to the complexity and variation in use, Rowe and Mason (Rowe & Mason, 1987) proposed the term decision style to mean the way a person uses information to formulate a decision.

Decision style models classify an individual's cognitive process by integrating his/her ability to understand, organize, think, process, and formulate information (Jill, 2006). Decision style is a cognitive process, which represents the way an individual approaches a problem (Rowe & Mason, 1987). It reflects the way a person perceives, thinks, and interprets situations. Related research has revealed two key factors in how individuals vary in making decisions; information use and focus (Jill, 2006). During the past 40 years many frameworks for investigating decision making styles have emerged (Vroom & Yetton, 1973; McKenney & Keen, 1974; Arroba, 1977; Harren, 1979; Merrill & Reid, 1981; Rowe & Mason, 1987; Nutt, 1990; Driver, Brousseau, & Hunsaker, 1993; Hersey, Blanchard, & Johnson, 1996). In reviewing the various frameworks applied to the decision making process, there remains a need to investigate the implications of decision-making styles among various decision-making contexts.

Moreover, a managerial decision typically commits organizational resources to a course of action in order to accomplish something that the organization (and/or the manager) desires and values (Fitzgerald, 2002). As international interactions increase in frequency and importance, there is a growing need to know how managers make decisions in different parts of the world, and how IT applications may support their decision-making activities (Maris & Robert, 2007).

The decision-making styles of managers was a focus point in the literature. For example, Fox and Spence (1999) surveyed a group of over 200 project managers from across the United States attempting to measure their decision-making styles. The results of their study indicated that project managers, on an individual basis, have very clearly defined differences in their preferred style of decision-making. However, taken as a group, project managers do tend to support the suggested need for a 'whole-brain' approach to project management.

Steinberg (2003) investigated decision-making styles of members in three managerial levels within the South African Military Health Service. Research findings indicated that throughout the three different managerial levels, the behavioral decision-making style was dominating. Alqarani (2003) explored the managerial decision styles of the managers of Florida's state university libraries and examined the relation between the variety of managers' decision styles and seven demographic variables. As in the previous case, the behavioral decision style was the predominant style for the majority of managers, followed by the conceptual decision style. It was also found that there was no relationship between managers' decision style and their gender, age, or highest academic degree. On the other hand, years of administrative experience, ethnicity, position, and educational major of these managers were indeed related to the decision style or styles used by these managers.

Jacoby (2006) investigated to what extent a principal's decision style influences his/her acceptance and use of technology. The findings indicated that a principal's decision style has no bearing on his/her acceptance and use of technol-

ogy. Finally, Maris and Robert (2007) attempted to analyze the distinctively American, Japanese, and Chinese leaders' styles of strategic decision making and the reflection of these differences on information systems used.

1.2. Systems that Support Decision Making

The term decision support represents all the means (models, methods, tools, concepts) that are available to the decision-maker in order to make easier the decision-making. To improve the cognitive process of the decision-makers, it is necessary to have decision support aids through an adapted Information System (IS). Every decision support aid is equipped with its own realization and implementation methods. In this research will focus on five information systems: Management Information Systems (MIS); Decision Support Systems (DSS); Executive Information Systems (EIS); Group Support Systems (GSS); and Organizational Decision Support Systems (ODSS).

Decision Support Systems (DSS) were first developed in the 1970s, and have been used widely since the PC revolution in the 1980s (Maris & Robert, 2007). DSS can be described as "computer-based systems that help decision makers confront ill-structured problems through direct interaction with data and analysis models" (McNurlin & Sprague, 2004). DSS aim to enhance the decision making process via providing tools that facilitate the processing and analysis of large amounts of data. DSS were originally developed as tools for managers, but they are now also used by many non-management employees (Maris & Robert, 2007).

Executive Information Systems (EIS) are intended specifically for executives. They have been used to monitor and communicate company performance data and to scan the business environment (McNurlin & Sprague, 2004; Ba, Lang, & Whinston, 1997; Elam & Leidner, 1995). An EIS can be described as a DSS that "(1) provides access to (mostly) summary performance data, (2) uses graphics to display and visualize the data in an easy-to-use fashion, and (3) has a minimum of analysis or modeling beyond the capability to 'drill down' in summary data to examine components (McNurlin & Sprague, 2004)."

Group Support Systems (GSS), formerly known as Group DSS, are networked systems that facilitate discussion by groups of proximate or distributed individuals synchronously or asynchronously (Maris & Robert, 2007). A GSS includes software tools designed to focus and structure group deliberation, reducing the cognitive costs of communication as group members work collectively towards a goal (Maris & Robert, 2007). GSS are designed to support decision making of a group of people (a team) engaged in a decision-related task. They are supposed to reduce communication barriers, stimulate or hasten exchange of messages, reduce uncertainty or noise in group's decision process, and drive or regulate the group's decision process (Desanctis & Gallupe, 1987; Schmidt, 1991).

An Organizational Decision Support System (ODSS) supports and organizes the division of labor for decision-making inside a firm. It focuses on an organizational process which cuts across organizational functions and hierarchical layers (Davenport, 1993). It supports interrelated but autonomous local decisions, but its main help is to coordinate these multiple local decisions with the objective of optimizing organizational decision. An ODSS shares some characteristics with other management information systems such as DSS, GDSS, and EIS, but it has distinctly different objectives and a broader scope (Holsapple & Whinston, 1996; Kroenke & Hatch, 1994). It has a strong organizational component not present in a DSS or a GDSS and a coordination component not present in an EIS.

This chapter aims to explore the decision-making styles among Egyptian managers in different managerial levels. The results will provide information needed on the expected needs on

different information systems used for decision support. Following the introduction section, the rest of this chapter is organized as follows. Section 2 provides the methodology used in this research followed by results in section 3. Discussions are provided in section 4, and finally, conclusions are given in section 5.

2. METHODOLOGY

A survey made of 600 managers in Egypt was conducted to gather information related to various decision-making styles of Egyptian managers in different organizational levels. Respondents were asked to complete a two parts questionnaire: demographics and the Decision Style Inventory (DSI) developed by (Rowe & Mason, 1987).

Rowe and Mason (1987) took a management perspective when attempting to understand, assess, and improve decision-making, and defined four decision making styles based on two dimensions of thinking: cognitive complexity and values orientation as shown in Figure 1. Cognitive complexity refers to a person's tolerance for ambiguity as opposed to need for structure while values orientation refers to a person's task as opposed to relational concerns. "The DSI, with fewer and more managerially oriented questions, also measures style on the basis of its own theory, and it also correlates highly and consistently with Jung's concepts as measured by the Myers Briggs Type Indicator (Rowe & Mason, 1987)."

The DSI was developed to measure the relative propensity to make use of four decision-making styles: directive, analytical, conceptual, and Behavioral. The instrument consists of 20 sentence-beginnings and four possible sentence-endings. For each sentence beginning, the subject is asked to rank each of the sentence endings by 8, 4, 2, or 1, identifying which one is most like them, next most like them, etc. A ranking of 8 indicates the response that you most prefer, a 4 indicates a response that you consider often, a 2 indicates a

Figure 1. Decision style model (adapted from Rowe & Boulgarides, 1994)

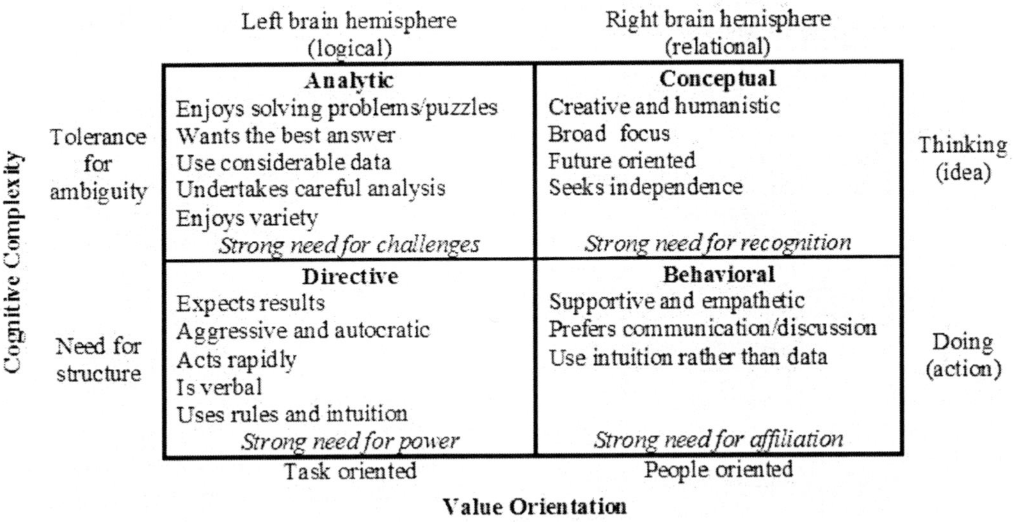

	Left brain hemisphere (logical)	Right brain hemisphere (relational)	
Tolerance for ambiguity	**Analytic** Enjoys solving problems/puzzles Wants the best answer Use considerable data Undertakes careful analysis Enjoys variety *Strong need for challenges*	**Conceptual** Creative and humanistic Broad focus Future oriented Seeks independence *Strong need for recognition*	**Thinking (idea)**
Need for structure	**Directive** Expects results Aggressive and autocratic Acts rapidly Is verbal Uses rules and intuition *Strong need for power*	**Behavioral** Supportive and empathetic Prefers communication/discussion Use intuition rather than data *Strong need for affiliation*	**Doing (action)**
	Task oriented	People oriented	

Cognitive Complexity (left vertical axis label)

Value Orientation

response that you consider on occasion, and a 1 indicates the response that you least prefer. From these rankings, a score is summed up for each decision making style to determine the propensity of respondent towards each decision style. The instrument, further, identifies: (1) an individual's propensity towards each style as either 'very dominant,' 'dominant,' 'backup,' or 'least'; (2) an individual's orientation towards ideas versus actions; and (3) an individual's orientation towards executive, middle-management, or staff level decision making.

The DSI has been tested extensively for validity and reliability (Leonard, Scholl, & Kowalski, 1999; Robey & Taggart, 1981). It has "a very high face validity and reliability. Respondents have almost invariably agreed with their decision styles as shown on the test instrument" (Rowe & Boulgarides, 1994).

2.1. Theoretical Framework

The decision style adapted by a manager may be influenced by a number of variables. This research aimed to study the effect of seven variables divided

into two categories as shown in Figure 2 on the decision making style of Egyptian managers. The research will, thus, examine the following seven hypotheses:

$H_0 1$: There is no relation between the decision making style and the age (Mech, 1993).

$H_0 2$: There is no relation between the decision making style and the gender (Mech, 1993).

$H_0 3$: There is no relation between the decision making style and the educational level (Benson, 1986; Yousef, 1998; Goodyear, 1987).

$H_0 4$: There is no relation between the decision making style and the total years of experience (Mech, 1993; Benson, 1986; Goodyear, 1987).

$H_0 5$: There is no relation between the decision making style and the level of management (Benson, 1986).

$H_0 6$: There is no relation between the decision making style and the business type (Ali, 1989).

$H_0 7$: There is no relation between the decision making style and the total number of employees.

Figure 2. Theoretical framework

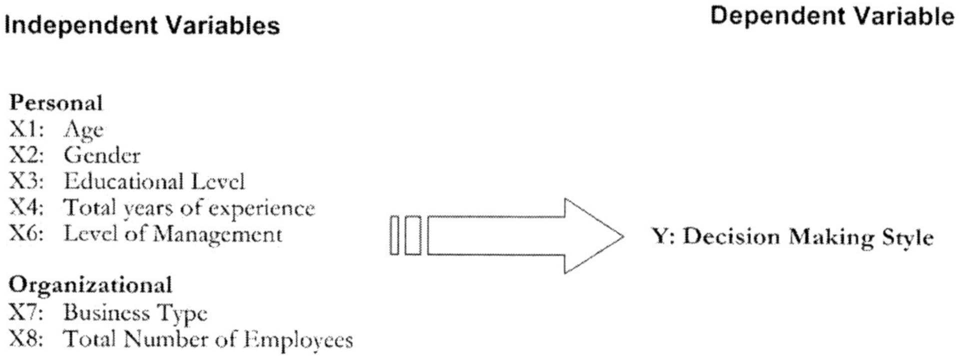

Independent Variables   **Dependent Variable**

Personal
X1:   Age
X2:   Gender
X3:   Educational Level
X4:   Total years of experience
X6:   Level of Management

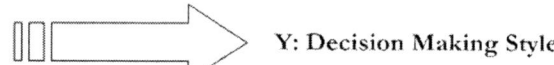

Y: Decision Making Style

Organizational
X7:   Business Type
X8:   Total Number of Employees

3. RESULTS

Data collected was analyzed through both descriptive and inferential analysis. Descriptive analysis was used to describe the segmentation of demographics of the survey and to: (1) identify the decision-making styles of the Egyptian managers; (2) determine their level of inclination towards being idea or action oriented; and (3) determine their level of orientation towards a certain level of decision-making.

Inferential analysis was used to analyze the relationship between the demographic information (seven variables) of the individuals under study and their decision making style. Chi-Square and Contingency Coefficient test were used to test the seven hypotheses mentioned above.

3.1. Sample Characteristics

The sampling frame of this study was based on a database of randomly selected Egyptian managers working in different companies and diverse sectors. Data was collected from the Egyptian managers working in different fields such as financial, construction and real estate development, education and research, and other sectors. The mail survey method was used to collect the data from respondents, in addition to hard copies that were distributed to managers in different organizations to increase the number of respondents.

The questionnaire was distributed to approximately 600 Egyptian managers—Junior, Middle, Senior—in different sectors of the industry. A total of 156 responses were received yielding a response rate of approximately 26%, however from the 156 responses, only 138 were useable, the remaining 18 were rejected as they were missing variables. Table 1 presents the demographic characteristics of the sample.

As the table shows, approximately 50% of the respondents are between the age of 30 and 35, and a total of 75% are between the ages of 30 and 40. On the other hand, a total of 88% of the sample are middle and senior managers. This finding can be also associated with the fact that almost 80% of sample had less than 20 years of experience. This leads to the conclusion that the age bracket of senior and middle managers has decreased among Egyptian managers, which is different from how it was in the previous years, when one of the main drivers to move up the organization ladder was the age and number of years of experience. The promotion of managers, now, can be attributed to other factors such as education. This can be easily concluded from the table as 55% of the managers hold a university degree and almost 40% hold a masters degree.

Table 1. Demographics characteristics of the respondents

	n=138			n=138
Age (years)			**Managerial level**	
30 - 35	44.9%		Junior	12.3%
35-40	23.9%		Middle	47.1%
40-50	23.2%		Senior	40.6%
50-60	6.5%		**Years of Experience (years)**	
above 60	1.4%		0-9	21.0%
Gender			10-19	58.7%
Male	58%		20-29	15.2%
Female	42%		30 or more	5.1%
Education			**Number of Employees**	
University Graduate	55.1%		less than 100	26.8%
Masters	38.4%		101 to 200	5.1%
PhD	6.5%		201 to 400	10.1%
Business			401 to 600	18.8%
Banking & Finance	30.4%		above 601	39.1%
Construction & Real Est.	18.8%			
IT & Telecommunication	16.7%			
Education & Research	13.8%			
Other	20.3%			

3.2. Decision Style

The mean scores and standard deviations for the sample on the four decision styles are presented in Table 2. The Table, also, presents corresponding scores obtained in Rowe and Mason (1987) revealing almost identical nature of sample which, in turn, allows us to use further analysis illustrated by Rowe and Mason (1987). It must be noted that these values do not represent dominance or least favorable decision-making styles as it only reflects mean scores. In order to interpret the scores attained through the instrument in terms of dominance, the scores must be mapped against the different decision style intensity levels. According to (Rowe and Mason, 1987), the typical scores for each style are: directive – 75; analytical – 90; conceptual – 80; and behavioral – 55. Differences in scores from the typical scores can therefore be interpreted from intensity or style dominance according to the values reported in Table 3.

The propensity of managers to each of the four decision styles is presented in Table 4, which shows a matrix comparing different decision styles with measured propensity. Most managers expressed a propensity towards the behavioral and directive styles much more frequently than those expressing a propensity towards either the directive or analytical styles. These results also reveal the highest percentage of the 'least preferred' propensity towards the conceptual followed by the analytical styles.

However, it is interesting to note that the percentage of decision styles at the 'backup' level is much closer across decision styles. In other words, while managers distinctly preferring either a behavioral or a directive approach to decision making, they can comfortably adapt to another style as a backup.

Table 5 presents the data from the perspective of propensity levels for each decision style. Looking at the conceptual style, the 'least preferred' and the 'back up' levels of propensity were reported most often for this style. On the other hand, the directive and the behavioral styles were reported most often as either in the 'back up' or the 'very

Table 2. Mean scores on the decision style inventory

Decision style	This research		(Rowe and Mason, 1987)	
	Mean	Standard deviation	Mean	Standard deviation
Analytical	86.8	14.5	90.5	16.5
Behavioral	60.7	16.3	56.6	15.5
Conceptual	73.9	12.8	74.8	15.2
Directive	81.4	13.9	78.0	15.3

Table 3. Decision-style intensity levels (adopted from Rowe & Boulgarides, 1992)

Decision style	Intensity			
	Least preferred	Back-up	Dominant	Very dominant
Analytical	Below 83	83 to 97	98 to 104	Over 104
Behavioral	Below 48	48 to 62	63 to 70	Over 70
Conceptual	Below 73	73 to 87	88 to 94	Over 94
Directive	Below 68	68 to 82	83 to 90	Over 90

Table 4. Propensity toward decision style

Decision style level	Decision style				Total
	Analytical	Behavioral	Conceptual	Directive	
Very dominant	14	37	8	32	91
	15.4%	40.7%	8.8%	35.2%	100%
Dominant	21	20	10	29	80
	26.5%	25%	12.5%	36.25%	100%
Backup	53	49	53	55	210
	25.2%	23.3%	25.2%	26.2%	100%
Least Preferred	50	32	67	22	171
	29.2%	18.7%	39.2%	12.9%	100
Total	138	138	138	138	

dominant' level. Finally, the analytical style, the highest percentage of propensity is at the 'back up' level. Table 6 provides propensity levels for each decision style per different managerial levels. Surprisingly, the behavioral and directive styles were the dominating styles on all managerial levels.

3.3. Brain Dominance

The next analysis of the DSI is concerned with the determination of brain dominance, or the tendency for one side of the brain to be more dominant than the other. The right hemisphere is the more creative and perceives things as a whole. An individual with 'right-brain' dominant would, thus, tend to have a strong concern for individuals

Table 5. Decision style percentage by propensity

Decision style level	Decision style			
	Analytical (%)	Behavioral (%)	Conceptual (%)	Directive (%)
Very dominant	10.1	26.8	5.8	23.2
Dominant	15.2	14.5	7.2	21.0
Backup	38.4	35.5	38.4	39.9
Least Preferred	36.2	23.2	48.6	15.9
Total	100	100	100	100

Table 6. Matrix of managers by decision style propensity

Management level	Decision style level	Decision style				Total
		Analytical	Behavioral	Conceptual	Directive	
Junior	Very dominant	2	4	2	0	8
	Dominant	1	1	2	6	10
	Backup	10	9	4	6	29
	Least Preferred	4	3	9	5	21
Middle	Very dominant	8	14	1	20	43
	Dominant	11	11	4	12	38
	Backup	27	23	25	27	102
	Least Preferred	19	17	35	6	77
Senior	Very dominant	4	19	5	12	40
	Dominant	9	8	4	11	32
	Backup	16	17	24	22	79
	Least Preferred	27	12	23	11	73

and prefer broad thinking and creative approaches. The people who think using this side of brain have a comprehensive sense of timing and they can encompass many thoughts at the same time using parallel processing of information. They are also more artistic, appreciate space, imagery, fantasy, and music (Alqarni, 2003).

On the other hand, the left hemisphere controls logical and analytic thought and processes information consecutively. It handles speech, pointing and smiling as well as the abstract logic needed for mathematics and verbal thinking (Alqarni, 2003). An individual with 'left-brain' dominant would tend to have a strong technical focus and be inclined towards logical thinking. In general, right brain

thinkers exhibit intuition, while left-brain thinkers are more rational (Rowe & Boulgarides, 1992).

By adding the analytical and directive scores from the DSI, individuals' 'left brain' score can be derived, and by adding the conceptual and behavioral scores, the 'right-brain' score is determined. A respondent is either: 'left-brain' dominant if the corresponding score is more than 165; 'right-brain' dominant' if the corresponding score is more than 135, or 'mixed' dominant. Of the 138 respondents, 82 were found to be left-brain dominant, 54 were found to be right-brain dominant, and 2 were found to have 'mixed' dominance. Table 7 presents the distribution of respondent's brain-dominance with respect to different managerial levels. While the ratio of left-brain personnel to right-brain personnel is the double in the middle management, this ratio is almost 1 to 1 in junior and senior managerial levels. The table, further, shows an inclination towards left-brain dominance (60% of the respondents). While senior and junior managers are almost evenly split between left- and right-brain dominant individuals, middle managers showed a great incline towards left-rain dominance. In the three levels, very few (1.5%) were found to be mixed-dominant.

3.3. Idea vs. Action Orientation

An individual's inclination towards either an idea or an action orientation can also be extracted form the DS; combining an individual's analytical and conceptual scores provides an indication of their inclination toward an idea orientation, while combining their directive and behavioral scores indicates their tendency towards an action orientation (Fox & Spence, 1999). As shown in Table 9, 101 individuals possessed an idea orientation while only 37 individuals possessed an action orientation. Also, there was a clear tendency among managers—in the three managerial levels—with regard to idea/action orientation as the majority of them showed an idea-orientation; 73% versus 27%, respectively.

3.4. Level of Decision Making Orientation

The final DSI analysis identifies an individual's propensity toward executive, middle management, or staff level decision-making orientation. The measure of the inclination towards executive-level decision making is determined by adding an individual's conceptual and directive style scores. An individual's inclination towards middle-management decision making is established by combining the directive, analytical, and conceptual decision styles. The measure of an individual's

Table 7. Brain dominance in different managerial levels

Management level	Brain dominance		
	Left	Right	None
Junior	9	7	1
Middle	44	21	0
Senior	29	26	1
Total	82	54	2

inclination towards staff-level decision making is designated by the combination of analytical and behavioral decision-style scores (Fox & Spence, 1999). The respondents of this study exhibited the distribution shown in Table 8.

As presented in the table, in the senior management level, only 32% of managers showed an executive decision making orientation while about 40% showed an orientation towards middle management decision making and 28% showed an orientation towards staff level decision making. The same applies to the middle managers where around 22% showed an orientation towards staff level decision making.

3.5. Hypotheses Testing

Two statistical tests were used by to test the hypotheses outlined earlier: (1) Chi-Square Test was used for the purpose of testing the independency level between two variables (Null Hypothesis of the test: the two variables are independent); and (2) Contingency Coefficient Test was used for the purpose of determining the correlation between variables regardless of the trend or power of this correlation (Null Hypothesis of the test: there is no relation between the two variables).

With the results of both tests—reported in Table 9—and the descriptive analysis illustrating the distribution of decision making styles per different independent variables—Figures 3-9, this research was able to reject 4 null hypotheses.

Table 8. Orientation of managers in different levels

Management level	Action/Idea Orientation		Level of Decision Making Orientation		
	Action	Idea	Executive	Middle management	Staff
Junior	2	15	3	10	4
Middle	18	47	14	37	14
Senior	17	39	18	22	16
Total	37	101	35	69	34

Table 9. Hypotheses testing

Hypothesis		Statistics			Critical Chi			
No.	Variable	Chi	p	df	0.05	Null hypothesis	0.10	Null hypothesis
1	Age	18.03	0.058	12	21.03	cannot reject	18.55	cannot reject
2	Gender	8.21	0.021	3	6.25	reject		
3	Educational Level	9.12	0.083	6	12.59	cannot reject	10.65	cannot reject
4	Total years of experience	17.87	0.018	9	16.92	reject		
5	Level of Management	10	0.063	6	12.59	cannot reject	10.65	cannot reject
6	Business Type	19.18	0.042	12	21.03	cannot reject	18.55	reject
7	Total Number of Employees	23.7	0.011	12	21.03	reject		

Thus, it can be concluded that the decision making style of Egyptian manager is affected by two personal characteristics (gender and total years of experience) and two organizational characteristics (business type and total number of employees in the organization).

One of the surprising findings of this study was the no significant relation between the individuals' educational level and their decision making style. Literature implies; that the manager who has a lower degree is more directive than the one who has a Ph.D." (Goodyear, 1987). The research findings revealed that most Ph.D. holders have an analytical decision making style followed by the behavioral. The conceptual and behavioral decision styles were dominated by the junior and middle managers.

4. DISCUSSIONS AND IMPLICATIONS FOR DECISION SUPPORT SYSTEMS

Marakas (2003) pointed out the importance of knowing managers' decision style on the design of a decision support system to provide appropriate support for a particular style. Martinsons and Davison (2007) argued that DSS, GSS, and EIS are each more compatible with some decision styles than others and illustrated how each of these systems may lend itself more readily to different decision makers as shown in Figure 10. This research adopts their model and extends the discussions to include the other two tools: MIS and ODSS.

The results of this research confirmed the diversity of decision-making styles among Egyptian managers leading to the conclusion that there is no generic DSS; no DSS fits all. Respondents, however, appeared capable of shifting to other

Figure 3. Decision making styles per age

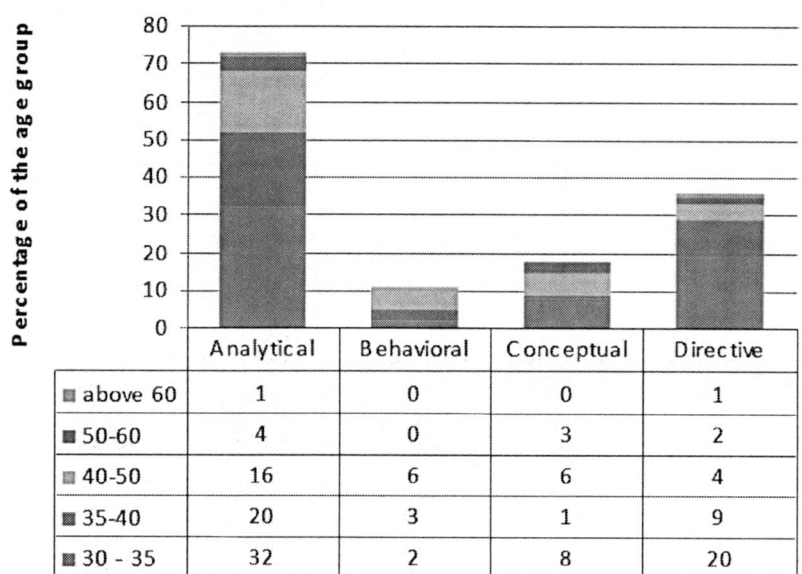

	Analytical	Behavioral	Conceptual	Directive
above 60	1	0	0	1
50-60	4	0	3	2
40-50	16	6	6	4
35-40	20	3	1	9
30 - 35	32	2	8	20

Decision making style

Figure 4. Decision making styles per gender

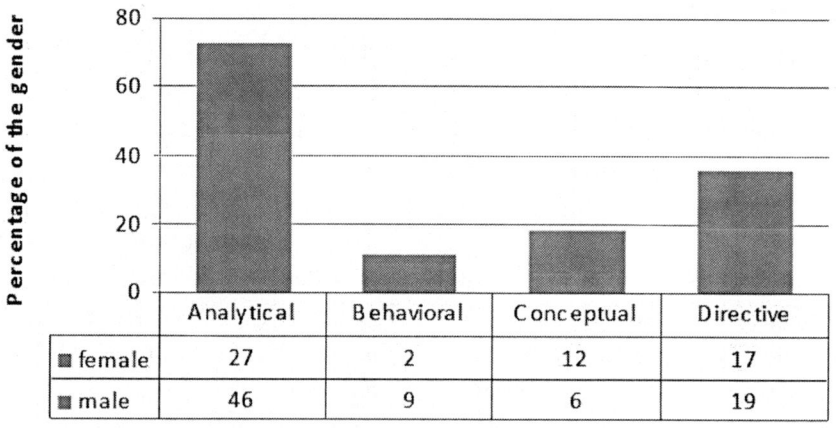

	Analytical	Behavioral	Conceptual	Directive
female	27	2	12	17
male	46	9	6	19

Decision making style

decision styles as a backup, when the situation warrants, regardless of their dominant style. This result confirms findings by Fox and Spence (1999) and Driver et al. (1996) that managers change their preferred decision style over time, and when faced with more complex problems, they tended to migrate toward an 'integrated' decision style.

This result is, also, aligned with the Situational Leadership Theory (Hersey, Blanchard, & Johnson, 2001) arguing that managers must use different leadership styles depending on the situation.

In the research conducted by (Steinberg, 2003; Alqarni, 2003) it was found that the behavioral decision making style was dominating among senior

Figure 5. Decision making styles per education level

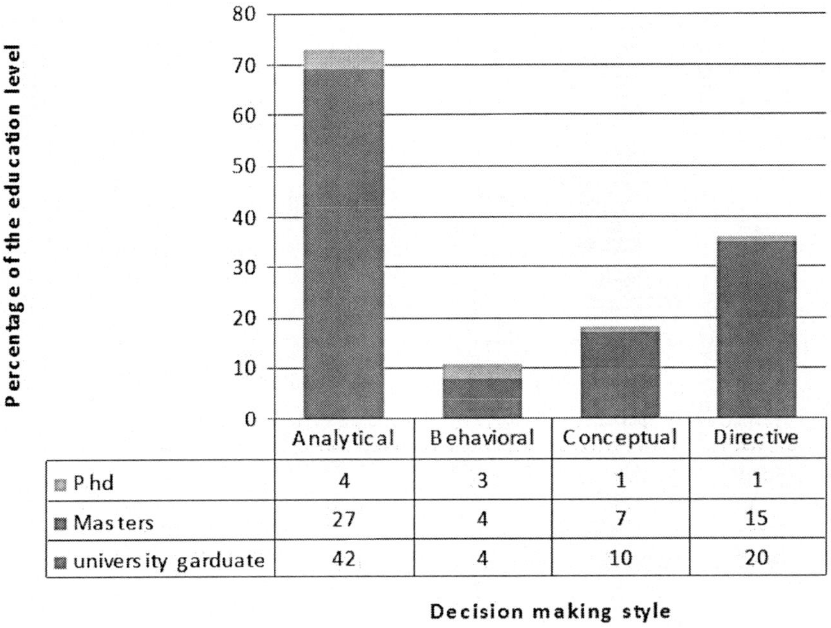

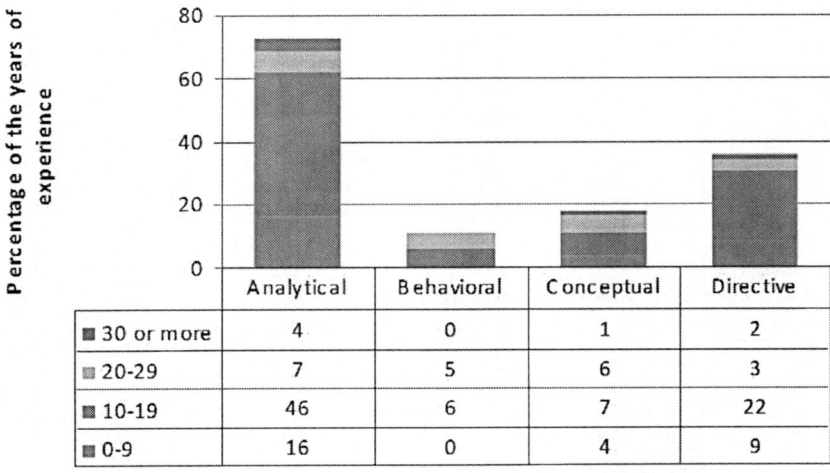

	Analytical	Behavioral	Conceptual	Directive
▣ Phd	4	3	1	1
▣ Masters	27	4	7	15
▣ university garduate	42	4	10	20

Decision making style

Figure 6. Decision making styles per years of experience

	Analytical	Behavioral	Conceptual	Directive
▣ 30 or more	4	0	1	2
▣ 20-29	7	5	6	3
▣ 10-19	46	6	7	22
▣ 0-9	16	0	4	9

Decision making style

managers. The same case applies in the findings of this research. This research concluded that the behavioral decision making style is the dominating style among Egyptian managers followed by the directive style and surprisingly, the analytical style came as the least preferred. Generally, the results showed diversity in decision-making styles of Egyptian managers.

Egyptian managers, thus, are expected to have limited interest in data processing and build their decision mostly on intuition and relationships. Their propensity to adopt GSS is, thus, much

Figure 7. Decision making styles per level of management

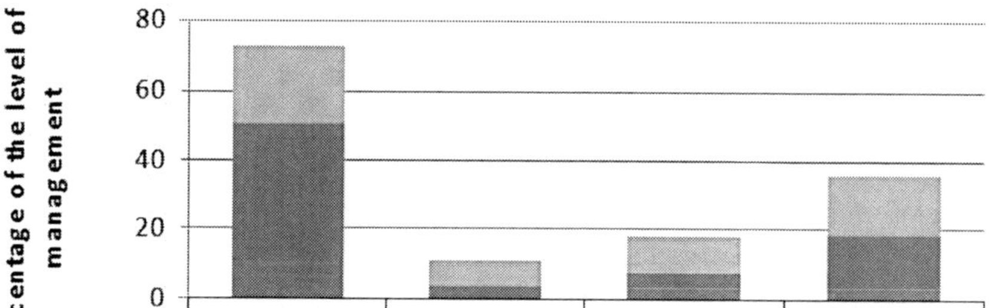

	Analytical	Behavioral	Conceptual	Directive
Senior	22	7	10	17
Middle	40	4	5	16
Junior	11	0	3	3

Decision making style

Figure 8. Decision making styles per business type

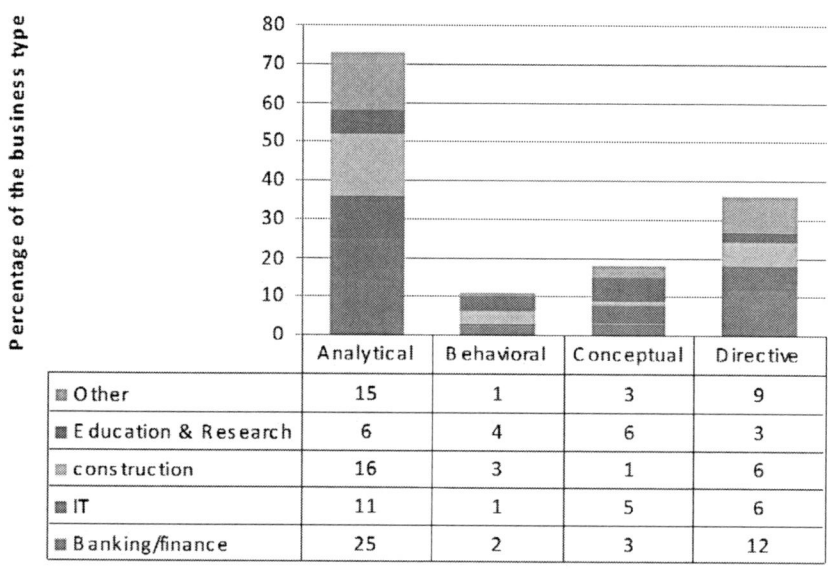

	Analytical	Behavioral	Conceptual	Directive
Other	15	1	3	9
Education & Research	6	4	6	3
construction	16	3	1	6
IT	11	1	5	6
Banking/finance	25	2	3	12

Decision making style

higher than to adopt DSS; a GSS will provide tools that aid multi-participant decision makers in identifying and addressing different issues. On the other hand, an ODSS with advanced technologies to facilitate communication will provide an enhanced support that accommodates for different levels in the organization and cuts the boundaries among different organizational levels and functional units. On the other hand, MIS is expected to fit with the findings as it produces summary scheduled operational reports that can be used to provide information, advice, and explanations to support specific decisions.

Figure 9. Decision making styles per number of employees

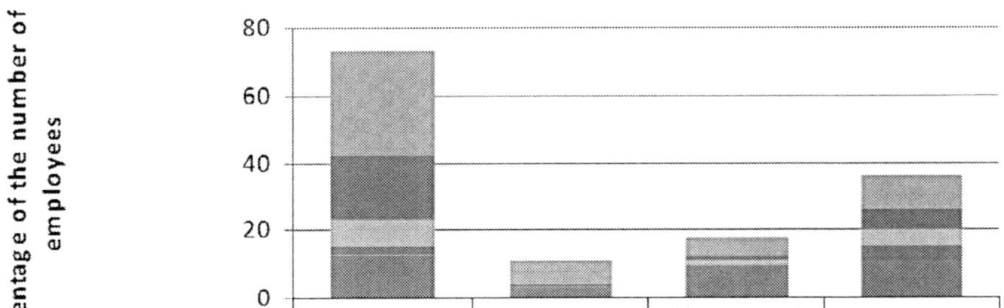

	Analytical	Behavioral	Conceptual	Directive
above 600	31	7	6	10
400 to 600	19	0	1	6
200 to 400	8	0	1	5
100 to 200	2	1	0	4
less than 100	13	3	10	11

Decision making style

Figure 10. Information technology applications and decision styles (adopted from Maris & Robert, 2007)

Propensity to Adopt GSS

	Low	High
High **Propensity to Adopt DSS**	**Analytic** Reliance on careful analysis of large volumes of codified data Interest in both the details of situation and the larger, aggregated profile.	**Conceptual** Data processing is useful to consider long-term what-if scenarios. Some IT applications may be useful for interpersonal relationships.
Low	**Directive** Limited interest in large-scale data analysis or human relationships. Interest in the aggregated profile of the situation to support a rapid decision making process	**Behavioral** Limited interest in data processing since information and rules are less important than intuition and relationships. Tools that facilitate communication and consensus-building will be used.
	High	**Low**

Propensity to Adopt EIS

5. CONCLUSION

With the increasing business potential in the Egyptian market and as organizations become more decentralized and sound decision making is pushed down the organizational ladder, it becomes imperative to understand the variety of decision making styles and the factors shaping those styles. From an organizational perspective, sound decision-making is a key success, if not survival factor. This research tried to explore the different decision making styles of the Egyptian managers, and the elements affecting those decision styles. The results would provide baseline information to improve our understanding of Egyptian managers and management.

REFERENCES

Adair, J. (1985). *Management decision-making.* Aldershot, UK: Gower Publishing.

Ali, A. (1989). Decision styles and work satisfaction of Arab executives: A cross-national study. *International Studies of Management and Organization, 19*(2), 22–37.

Allwood, C. M., & Selart, M. (2001). *Decision-making: Social and creative dimensions.* Dordrecht, The Netherlands: Kluwer.

Alqarni, A. (2003). *The managerial decision styles of Florida's State University libraries' mangers.* (Unpublished Dissertation). Florida State University. Tallahassee, FL.

Arroba, T. (1977). Styles of decision-making and their use: an empirical study. *British Journal of Guidance & Counselling, 5,* 149–158. doi:10.1080/03069887708258110

Ba, S., Lang, K. R., & Whinston, A. B. (1997). Enterprise decision support using intranet technology. *Decision Support Systems, 20*(2), 99–134. doi:10.1016/S0167-9236(96)00068-1

Benson, B. E. (1986). *Self-reported decision styles for chief nurses and assistant chief nurses in veterans administration field hospitals.* (Unpublished Doctoral Dissertation). Kansas State University. Lawrence, KS.

Davenport, T. H. (1993). *Process innovation: Reengineering work through information technology.* Boston, MA: Harvard Business School.

Dearlove, D. (1998). *Key management decisions: Tools and techniques of the executive decision-maker.* London, UK: Pitman Publishing.

Dearlove, D. (2001). *The ultimate book of business thinking: Harnessing the power of the world's greatest business ideas.* Oxford, UK: Capstone Publishing.

Desanctis, G., & Gallupe, R. B. (1987). A foundation for the study of group decision support systems. *Management Science, 33*(5). doi:10.1287/mnsc.33.5.589

Driver, M. J., Brousseau, K. E., & Hunsaker, P. L. (1993). *The dynamic decision maker.* San Francisco, CA: Jossey-Bass Publishers.

Driver, M. J., Svensson, K., Amato, R. P., & Pate, L. E. (1996). A human information- Processing approach to strategic change. *International Studies of Management and Organization, 26*(1), 41–58.

Elam, J. J., & Leidner, D. G. (1995). EIS adoption, use and impact: The executive perspective. *Decision Support Systems, 14*(2), 89–103. doi:10.1016/0167-9236(94)00004-C

Fitzgerald, S. P. (2002). *Decision making.* Oxford, UK: Capstone Publishing.

Fox, T. L., & Spence, J. W. (1999). An examination of the decision styles of project managers: Evidence of significant diversity. *Information & Management, 36,* 313–320. doi:10.1016/S0378-7206(99)00025-7

Gibson, J. L., Ivancevich, J. M., & Donnelly, J. H. Jr. (1994). *Organizations* (8th ed.). Burr Ridge, IL: Richard D. Irwin.

Goodyear, R. (1987). *A descriptive correlational study of the decision-making patterns of nurse practitioners in primary care.* (Unpublished Doctoral Dissertation). University of San Diego. San Diego, CA.

Gore, W. J. (1964). *Administrative decision-making: A heuristic model.* New York, NY: John Wiley and Sons.

Greenberg, J., & Baron, A. (1993). *Behavior in organizations* (4th ed.). Boston, MA: Allyn and Bacon.

Hammond, J. S. (1999). *Smart choices.* Boston, MA: Harvard Business School Press.

Harren, V. A. (1979). A model of career decision-making for college students. *Journal of Vocational Behavior, 14,* 119–133. doi:10.1016/0001-8791(79)90065-4

Hersey, P., Blanchard, K., & Johnson, D. (2001). *Management of organizational behavior: Leading human resources* (8th ed.). Upper Saddle River, NJ: Prentice Hall.

Hersey, P., Blanchard, K. H., & Johnson, D. E. (1996). *Management of organizational behavior: Utilizing human resources* (7th ed.). Upper Saddle River, NJ: Prentice Hall.

Hill, P. H. (1979). *Making decisions: A multidisciplinary introduction.* London, UK: Addison-Wesley Pub. Co.

Holsapple, C. W., & Whinston, A. B. (1996). *Decision support systems: A knowledge based approach.* New York, NY: West Publishing Company.

Jill, M. J. (2006). *Relationship between principals' decision making styles and technology acceptance & use.* (Dissertation). University of Pittsburgh. Pittsburgh, PA.

Kroenke, D., & Hatch, R. (1994). *Management information systems.* New York, NY: McGraw-Hill.

Leonard, N. H., Scholl, R. W., & Kowalski, K. B. (1999). Information processing style and decision making. *Journal of Organizational Behavior, 20,* 407–420. doi:10.1002/(SICI)1099-1379(199905)20:3<407::AID-JOB891>3.0.CO;2-3

Marakas, G. M. (2003). *Decision support systems in the 21st century* (2nd ed.). Upper Saddle River, NJ: Prentice Hall.

Maris, G. M., & Robert, M. D. (2007). Strategic decision making and support systems: Comparing American, Japanese and Chinese management. *Decision Support Systems, 43,* 284–300. doi:10.1016/j.dss.2006.10.005

McKenney, J. L., & Keen, P. G. W. (1974). How manager's minds work. *Harvard Business Review, 52*(3), 79.

McNurlin, B. C., & Sprague, R. H. (2004). *Information systems management in practice* (6th ed.). Englewood Cliffs, NJ: Prentice Hall.

Mech, T. F. (1993). The managerial decision styles of academic library director. *College & Research Libraries, 54*(5), 375–386.

Merrill, D. W., & Reid, R. H. (1981). *Personal styles and effective performance: Making your style work for you.* Radnor, PA: Chilton Book Co.

Nutt, P. C. (1990). Strategic decision made by top executive and middle managers with data and process dominant styles. *Journal of Management Studies, 27*(2), 172–194. doi:10.1111/j.1467-6486.1990.tb00759.x

Robbins, S. P. (1998). *Organizational behavior* (8th ed.). Upper Saddle River, NJ: Prentice Hall.

Robbins, S. P. (1999). *Management* (6th ed.). Englewood Cliffs, NJ: Prentice Hall.

Robey, D., & Taggart, W. (1981). Measuring managers' minds: The assessment of style in human information processing. *Academy of Management Review*, 6, 375–383.

Rowe, A. J., & Boulgarides, J. D. (1992). *Managerial decision making: A guide to successful business decisions*. New York, NY: McMillan.

Rowe, A. J., & Boulgarides, J. D. (1994). *Managerial decision making*. Englewood Cliffs, NJ: Prentice-Hall.

Rowe, A. J., Boulgarides, J. D., & McGrath, M. R. (1984). *Managerial decision making*. Chicago, IL: Science Research Associates.

Rowe, A. J., & Mason, R. O. (1987). *Managing with style: A guide to understanding, assessing, and improving decision making*. San Francisco, CA: Jossey Bass.

Schmidt, K. (1991). Cooperative work: A conceptual framework. In Rasmussen, J., Brehmer, B., & Leplat, J. (Eds.), *Distributed Decision Making: Cognitive Models for Cooperative Work*. New York, NY: John Wiley & Sons Ltd.

Steinberg, P. W. (2003). *Decision making styles within different hierarchical levels in the South African military health service*. (Thesis). Technikon Pretoria. Pretoria, South Africa.

Streufert, S., & Streufert, S. (1978). *Behavior in the complex environment*. Washington, DC: Winston-Wiley.

Tam, M. M. C., Chung, W. W. C., Yung, K. L., David, A. K., & Saxena, K. B. C. (1994). Managing organizational DSS development in small manufacturing enterprises. *Information & Management*, 26(1), 33–47. doi:10.1016/0378-7206(94)90005-1

Vroom, V. H., & Yetton, P. W. (1973). *Leadership and decision making*. Pittsburgh, PA: University of Pittsburgh Press.

Yousef, D. A. (1998). Predictors of decision-making styles in a non-western country. *Leadership and Organization Development Journal*, 19(7), 366–373. doi:10.1108/01437739810242522

Yuki, G. (1994). *Leadership in organization* (3rd ed.). Englewood Cliffs, NJ: Prentice-Hall.

KEY TERMS AND DEFINITIONS

Decision-Making: A choice within a pool (two or more) of alternatives.

Decision Style: Is a cognitive process which represents the way an individual approaches a problem. It reflects the way a person perceives, thinks, and interprets situations. Related research has revealed two key factors in how individuals vary in making decisions: information use and focus.

Decision Style Model: A model used to classify an individual's cognitive process by integrating his/her ability to understand, organize, think, process, and formulate information.

Decision Support Systems: Computer-based systems that help decision makers confront ill-structured problems through direct interaction with data and analysis models.

Executive Information Systems (EIS): A class of decision support systems intended specifically for executives.

Group Support Systems (GSS): A class of decision support systems that facilitate discussion by groups of proximate or distributed individuals synchronously or asynchronously.

Chapter 14

Identification of Major FOD Contributors in Aviation Industry

Hammad Ahmed Rafiq
Centre for Advance Studies in Engineering (CASE), Pakistan

Irfan Anjum Manarvi
HITECH University Taxila, Pakistan

Assad Iqbal
Bahria University Islamabad, Pakistan

ABSTRACT

Aviation safety is considered of paramount importance, and the Foreign Object Debris and the resulting Foreign Object Damage (FOD) is one of the major causes that put aviation safety at risk. FOD Prevention is thus a continual challenge for all aircraft operators and maintenance crew. It costs the aviation industry millions of dollars every year. This financial effect is a result of direct costs, such as harm to aircraft structures or damage of aircraft engines, as well as the indirect costs, which include flight schedule delays, cancellations, disruptions, and additional effort for the employees. In addition, on occasion, more critical than the financial impact, is the safety impact and potential loss of human life associated with occurrences caused by FOD. It is therefore ranked as the most likely potential ground-based cause that can lead to a catastrophic aviation event. The present chapter is based on statistical analysis of aircraft occurrences attributed to various types of FOD during the last ten years of operations in an aviation organization. Eight major cause factors contributing towards these cases have been identified. A broad FOD prevention and control plan is thus proposed to address the foremost cause factors and improve organizational response to FOD. The objective of the research is to promote ground and flight safety and the preservation of assets by reducing FOD.

DOI: 10.4018/978-1-4666-3658-3.ch014

INTRODUCTION

Aviation maintenance is a complex organization in which individuals perform varied tasks in an environment with time pressures, minimal feedback, and sometimes difficult ambient conditions. Aircraft, as well as inspection and maintenance equipment are thus becoming more complex. As the commercial aviation fleet ages, and work force of maintenance personnel diminishes, maintenance workload is increasing (Latorella & Prabhu, 2000). Air transport policies have aimed at increasing system capacity on the one hand and reducing acceptable risk and safety thresholds on the other (Netjasov & Janic, 2008). The term FOD is an aviation term. Typically, it is used to describe any small item, particle or debris that does not belong on an airport pavement surface and has the capability to cause harm or damage to an aircraft that passes by. It includes a wide range of materials such as loose hardware, pavement fragments, catering supplies, building material, rocks, sand, pieces of luggage, and even wildlife. Besides airport runway, FOD can also be found at aircraft hangers, terminal gates, cargo aprons, taxiways and flight decks. The sources of FOD include many kinds. The characteristic and feature of FOD may vary obviously depending on the attribute and size of foreign object. According to the nature of foreign object and potential damnification to engine, foreign object can be as diverse as sand particles, metal materials and soft materials, such as birds or verts (Chen, Lu, Li, Fu, Wang, Chai, & Xu, 2009). Recently, Foreign Object Debris (FOD) detection on airport runways has become of increasing interest. A basic motivation for this is the fatal accident with a Concorde aircraft a few years ago due to a metal part lost by an aircraft on the runway some time before (Feil1, Menzel1, Nguyen, Pichot, & Migliaccio, 2008).

Each year commercial, private and military aircraft jet engines are damaged by the ingestion of foreign objects. Annual engine repair costs for ingestion damage is in the tens of millions of dollars (Greneker, 1999). There is no simple way of ensuring that FOD incidents and accidents do not occur. However, an important step towards industry wide and maintenance crew awareness is the recognition that maintenance quality errors, omission, and lapses may be indicators of wider organizational problems (Mason, Kraus, Johnson, & Watson, 2001). FOD has become of increasing international interest, with many countries initiating their own research to FOD studying (Xu, Ning, & Chen, 2009).

This chapter is focused on establishing the classification, error analysis and finalization of FOD related occurrences of an aviation organization along with identifying the areas requiring attention of the management to limit and address further losses incurred through FOD. A prevention plan consisting of appropriate proactive and protective measures is also discussed, which will also be beneficial in improving organizational responses against FOD.

METHODOLOGY

The FOD related occurrences data for last ten years of an aviation organization was collected. This data was used to identify the month wise classification of FOD cases along with relationship of FOD cases with the hours flown by the organization. Moreover classification of FOD cases with respect to either air or ground was established. Similarly the error analysis of the FOD occurrences was made to identify if the occurrences were unavoidable or could be evaded through improved practices or workmanship. The finalization analysis of data was made to categorize the FOD cases of last ten years into following major groups:

1. Engine FOD Ingestion Cases
2. Miscellaneous Cases
3. Material Factor Cases
4. IOD Cases
5. Human Factor (Ground Crew) Cases

6. Engine Bird Ingestion Cases
7. Human Factor (Air Crew) Cases
8. Tyre Damage Cases

A number of statistical tools were employed to analyze the data. Based upon the analysis, an elaborative FOD Prevention Plan was discussed with major focus on those FOD contributory factors which were seemed more significant through subject study.

IDENTIFICATION OF MAJOR FOD CONTRIBUTORS

FOD Occurences in the Last Ten Years

The FOD occurrence of an organization for last ten years of flying operations was gathered. A total of 346 cases were observed in ten years. Total occurrences were plotted for each year as shown in Figure 1.

The following is inferred from Figure 1:

1. In Year 2002 least FOD occurrences were encountered i.e. only 16. The financial loss thus incurred on the organization through FOD was minimum in this year.

2. Year 2009 saw the maximum number of FOD occurrences in any single year. The total of 54 occurrences was equal to the total occurrences of year 2001, 2002, and 2003. Thus the financial losses through FOD for year 2009 alone were equal to the combine loss of three years i.e. 2001, 2002, and 2003.

3. A distinctive trend of high number of FOD cases could be noticed from year 2005 onward.

4. Since the data before year 2001 was not available, it could not be exactly established that if there is a sinusoidal behavior of FOD trends on a larger scale.

5. The data helped in establishing that the FOD prevention efforts are not proving fruitful in the organization, as the number of FOD cases kept increasing without any noticeable downward trend.

MONTH WISE CLASSIFICATION

To explore any trend related to either seasonal or climatic changes, the data of FOD occurrences for last ten years was then computed and plotted month wise, as shown in Figure 2.

Figure 1. Year wise comparison of FOD occurrences

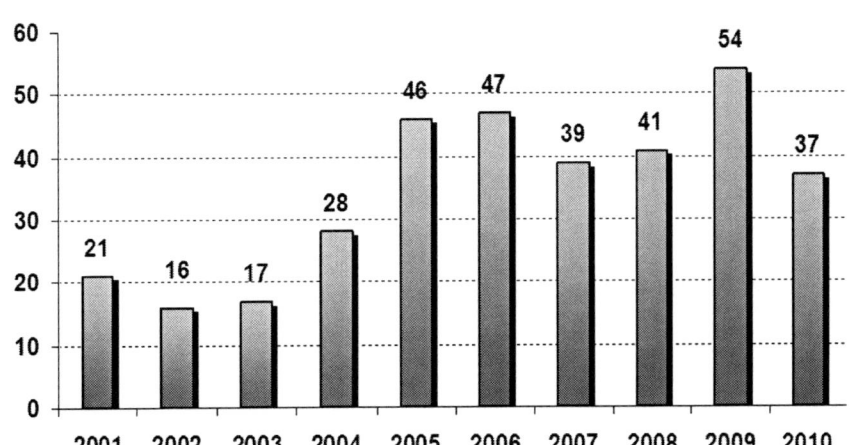

The figures lead to the following inferences:

1. FOD occurrences were found almost equally distributed to each month of year. The month of January, March and August though had highest number of occurrences, but was not substantial enough to lead to any conclusion.
2. No specific effect of seasonal change could be appreciated on FOD occurrences. Weather changes which could have its consequences on wild growth, icing conditions, work practices or environmental changes were not seen to have any specific influence on number of FOD cases.

FLYING HOUR RELATIONSHIP WITH FOD

The relationship of FOD occurrences with flying hours flown in last ten years was then analyzed. The rate was calculated for per 10000 flying hours basis. The FOD rate calculated for duration between 2001 to 2010 is shown in Figure 3.

The figure shows the following results:

1. The FOD rate was observed to be precisely corresponding with number of FOD occurrences for last ten years. More the number of occurrences high was the rate calculated.

Figure 2. Month wise comparison of FOD occurrences (2001-2010)

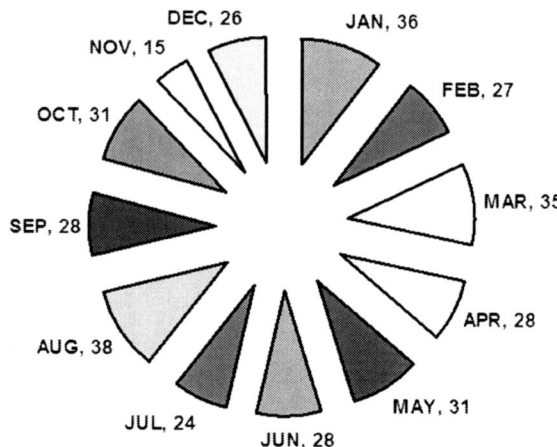

2. In no year, with high number of FOD occurrences, the rate dropped. This clearly indicated that the effectiveness of FOD Prevention Plan at the organization was not able to arrest the rise in FOD rates.
3. With highest number of FOD cases in year 2009, the FOD rate was also highest as 6.04 FOD occurrences per 10000 flying hours.
4. The organization had the lowest FOD occurrences rate in year 2003, i.e. 2.24 per 10000 flying hours; which is one third of the rate of year 2009.

Figure 3. Rate per 10000 Flg Hrs of FOD occurrences

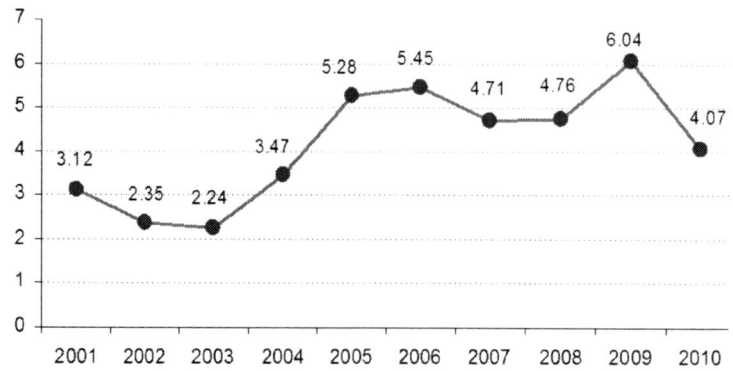

CLASSIFICATION AND ERROR ANALYSIS

The FOD occurrences were plotted for their classification of whether noticed in air and on ground. The in air occurrences were those cases where an abnormal situation is faced by the pilot in the air; which forced him to discontinue his mission and come back. The on ground occurrences on the other hand were those cases which were encountered on ground during routine inspection, while tackling various aircraft defects or during aircraft ground operations. These results are depicted below in Figure 4a and 4b, individually, and in 4c showing a comparison.

The figures lead to the following inferences:

1. The maximum number of in air FOD occurrences were encountered in year 2001. In year 2009 where maximum number of total FOD occurrences was encountered, the number of in air occurrences came out to be only 4.
2. The on ground occurrences have an approximate equal distribution for year 2005 to 2010. For year 2001 to 2004 there were a relative low number of on ground occurrences.
3. The on ground occurrences are significantly higher than the in air occurrences. The reason being that most of the occurrences could not be appreciated during flight and were noticed during aircraft maintenance on ground. Moreover there had been occurrences which were encountered during aircraft operations at ramp or runway whereby aborting mission. Similarly there had also been occurrences which were faced during engine run up operations.

Data pertaining of FOD occurrences were plotted with respect to their finalized error analysis i.e. whether the occurrences were avoidable or unavoidable. The avoidable occurrences took place due to bad maintenance, poor workmanship and lack of required knowledge or not following the right procedure. All such cases which could any way be evaded were all categorized as avoidable. The unavoidable occurrences were the one which were beyond the human control. Even tightly controlled procedures and quality maintenance could not stop these occurrences from taking place. Figure 4d describes the error analysis of FOD occurrences.

The figure leads to the following inferences:

1. Most of the occurrences were avoidable, since they could be either stopped or avoided. The highest number of avoidable occurrences were encountered in year 2009 with unavoidable occurrences as only 02 in same year.
2. The unavoidable occurrences were very small in number as compared to avoidable ones. Year 2010 though saw the maximum 06 unavoidable occurrences against 31 avoidable ones.
3. There had been years like 2003, 2004 and 2007 when there were no unavoidable FOD cases. i.e. all cases were avoidable. Thus employment of a better effort and implementation of a good FOD Prevention Plan could have evaded these occurrences.

FINALIZATION ANALYSIS

The finalization of FOD occurrences are the closing cause factor allocated to the cases upon completion of investigation process. There are total of eight finalization cause factors identified, which have been mostly allocated upon conclusion of an investigation process of an FOD related occurrence.

The first contributory cause factor discussed is the Engine FOD ingestion cases. The Engine FOD Cases are those cases where an ingested FOD has resulted into damage to compressor or turbine blades. In a number of cases the exact nature of FOD could not be identified during

Figure 4. (a-b) Classification of in Air FOD occurrences; (c) analysis for classification of FOD occurrences (2001-2010); (d) error analysis of FOD occurrence

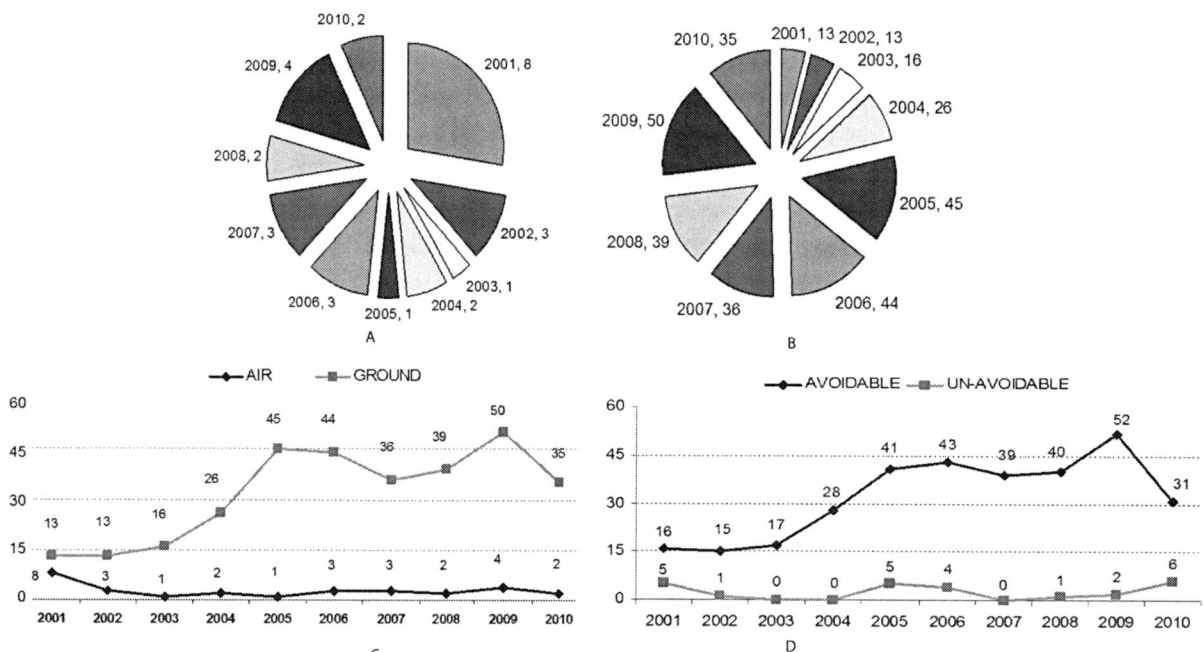

investigations. The result of FOD has been damage, nick marks, bending, abrasions, fracturing, tearing apart of blades. At most of the occasions the damages are noted during engine inspections after or before flights and during second level maintenance through Boroscope or disassembly process. Although there had been cases where the damage is so substantial that it resulted into evident power loss, compressor stalls, engine flame out or rise in EGT or TIT; where the mission had to be aborted. No aircraft crash during same period could be finalized as an FOD case due to lack of concrete evidence.

The repair action following the Engine FOD cases ranges from simple blade rework or blending to replacement of selected or complete blades or even major overhaul of entire engine; depending upon the nature and severity of occurrence. The period of FOD exposure in most of these cases is identified as Engine run up for start up or mainte-

nance ground runs, aircraft taxi over ramps, taxi ways and most importantly aircraft take rolls at high power settings on the runways.

The source of FOD are metal or wire clippings, solder balls and debris lying in the vicinity of aircraft, electrical terminals, connectors, small items, tools, unaccounted hardware, debris lying on runways, taxi ways and aprons, protective covers, caps, personal items such as badges, hats, pins, pens, pencils, cell phones, pocket lights, knives, aircraft and engine fasteners, nuts, bolts, washers and safety wire and paper and plastic debris from freight pallets, luggage parts, and other ramp equipment and transport. Figure 5a depicts the yearly analyses of cases finalized as Engine FOD.

The figures provide the following observations:

1. There had been a constant increase in Engine FOD rates from year 2002 to 2006. This can be attributed to deterioration of airport

Figure 5. (a) FOD occurrences finalized as engine FOD; (b) FOD occurrences finalized as miscellaneous;(c)FODoccurrencesfinalizedasmaterialfactor;(d)FODoccurrencesfinalizedasIOD;(e)FOD occurrences finalized as human factor (ground crew) year (2001-2010); (f) FOD occurrences finalized as bird hit; (g) FOD occurrences finalized as human factor (air crew) year (2001-2010); (h) FOD occurrences finalized tyre damage; (i) comparative analysis for finalization factors of FOD cases (2001-2010)

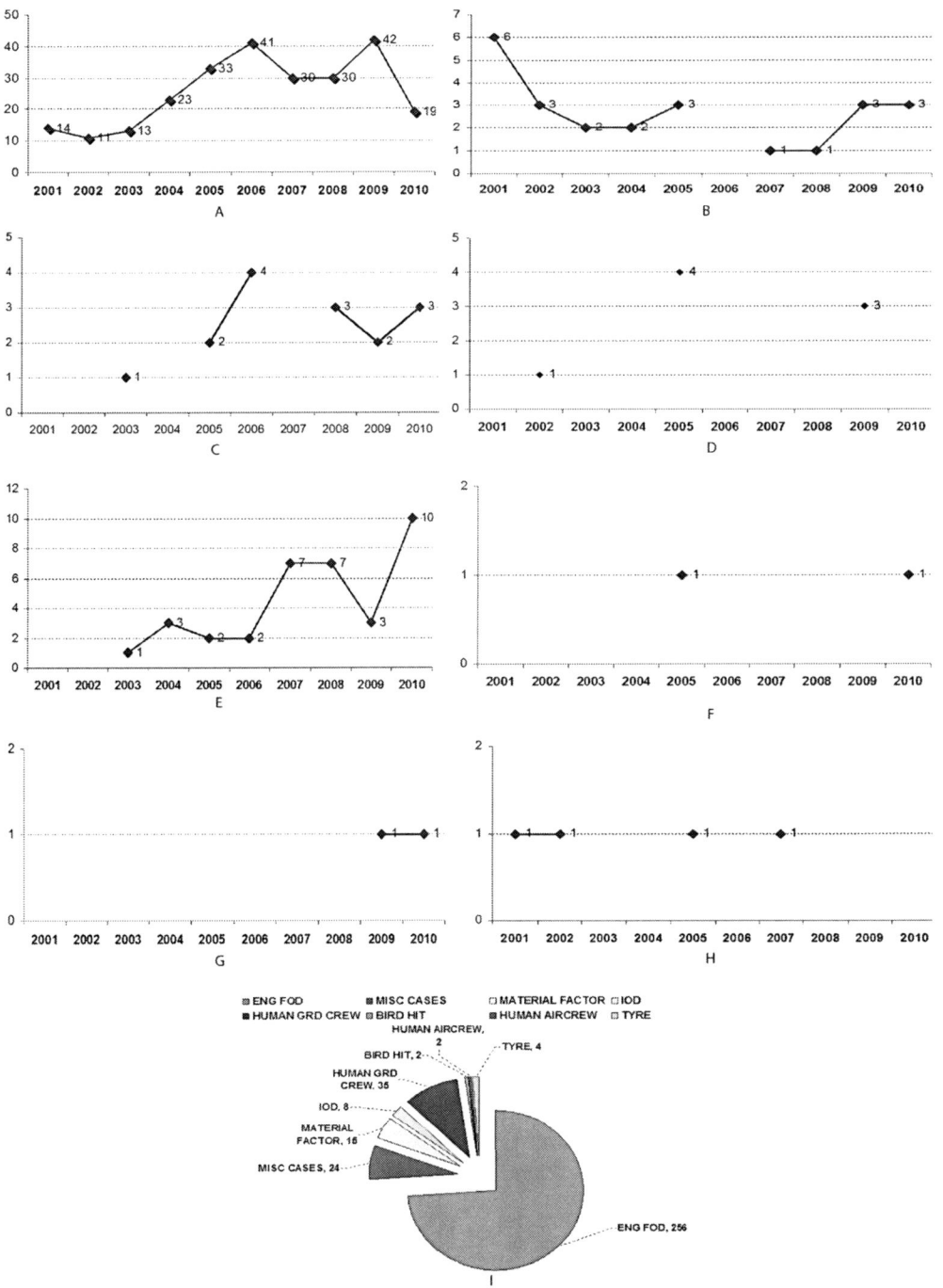

facilities, lack of required attention towards clearing off FOD or degeneration of work ethics and practices of maintenance work force.

2. Year 2006 depicted the highest number of Engine FOD cases, but after which two years with substantially reduced FOD cases were seen. This could be due to increased attention of management towards FOD trends amid high rates of year 2006.

3. Year 2009 again depicted a very high Engine FOD cases but after which it gain substantially dropped in year 2010. This yet again indicates towards management focus to address concerns on FOD.

4. The sinusoidal pattern of Engine FOD cases depicts lack of consistent effort to address the FOD problem at the organization.

The second cause discussed is the Miscellaneous Factor. It includes those occurrences where the damage is not inflicted to Engine, but to aircraft structure, flight controls, landing gears or pitot etc. This can be due to hitting of any object from ground or through tearing apart or disengaging from aircraft itself. The high speed aircraft roll or taxi makes it vulnerable to damage by any foreign object from either ground or from aircraft. This can inflict a significant damage to the aircraft. Along with cleaning of runway surfaces, it is also vital to ensure a quality maintenance and inspection of the aircraft before releasing it for flight whereas to preclude every probability of any aircraft part or hardware disengaging undesirably from its place. Figure 5b depicts the yearly analyses of FOD cases finalized as Miscellaneous Factor.

The figures lead to the following inferences:

1. Year 2001 had the highest miscellaneous factor cases, whereas year 2006 did not encounter any such occurrence.

2. There is no specific trend of miscellaneous factor FOD cases, but the occasional occur-

rences do describe that the organization is not able to address the problem completely. The sporadic cases indicate that the individuals are yet missing the inspections items or the runway inspection procedure has inherent flaws, where it is not ensured for meticulous cleaning before take offs.

The next factor is the Material FOD cases. These are occurrences attributed to material failure of aircraft skin, hardware, component or any other; resulting into damage to either aircraft structure or even engine and categorized as unavoidable unless it could be noticed upon its initiation by naked eye during inspection.

The analysis of Material Factor FOD occurrences for last ten years is shown in Figure 5c. The figure leads to the following inferences:

1. The Material Factor FOD did not hold a specific pattern.

2. Year 2006 had the maximum of 04 occurrences, whereas during a number of years, not even a single occurrence was reported.

3. In the last three years of operations, frequent cases of material failure were encountered. This led to an indication that the aging toll is being faced by the aircraft and underlines importance of improved maintenance practices and enhanced inspection items based on aging of fleet.

The IOD, Internal Object Damage are the cases where harm is being inflicted on the engine due to failure of its own hardware, part or accessory. This can be a rivet shearing off from engine intake, a propeller portion shaving off and ingesting into engine, a compressor blade giving away and causing damage to the engine or an engine probe or even accessory ingesting inside. The analysis of IOD occurrences for last ten years is shown in Figure 5d.

The figure leads to the following inferences:

1. There is no specific pattern of IOD cases. Year 2005 had a maximum of 05 IOD cases, whereas year 2009 had 03 cases.
2. The high number of IOD cases at any point warrants meticulous inspection and correct follow up of all work items.

The Human Factor (Ground Crew) cases are those occurrences which are finalized as attributed to maintenance technicians. These occur when an obvious negligence is made by the individual working on the aircraft. These mistakes ranges from an unintentional placing of an undesired object in the engine intake or its close vicinity to not following the procedures and thus resulting into ingestion of a personal item or tool inside the engine or not removing the intake plugs before engine startup and causing damage upon engine start up. Also failing to detect an obvious defect which caused an FOD or failing to perform a vital inspection step which led to an FOD related occurrences are also covered in Human Factor (Ground Crew) FOD occurrences. The analysis of Human Factor (Ground Crew) FOD occurrences for last ten years are shown in Figure 5e.

The figure provides the following observations:

1. There is a currently an increasing trend of occurrences attributed to Human Factor (Ground Crew).
2. The trend though seems sinusoidal, but year 2010 had depicted an alarmingly high number of cases attributed to ground crew.

The high number of occurrences warrants training and preparation of human resource and inculcation of a sense of awareness against hazards of FOD.

The Bird Hit cause factor covered here are those occurrences which upon detailed investigation at a later stage, concluded as bird hit cases causing damage to the engine. Those obviously noted cases which are observed during aircraft post fight inspections are straight away categorized as Bird Hit Occurrences and not rounded off as FOD cases.

Since the domain and subsequent follow up action of Bird Hit cases is a vast field; they are covered as a separate entity. Figure 5f depicts the yearly analyses of Bird Hit FOD cases.

Human Factor (Air Crew) cases are those occurrences which are finalized as attributed to the pilot of the aircraft. These occur when an obvious intentional negligence is made by the pilot which ultimately led to an Engine FOD occurrence. These mistakes include not securing their personal gear and causing it to ingest inside engine or inappropriate practice or adopted procedure which caused an FOD to get sucked in and inflict damage to the engine. The analysis of Human Factor (Air Crew) FOD occurrences for last ten years are shown in Figure 5g.

The figure indicates that there were two isolated cases in the last ten years where a mistake by the pilot resulted into an FOD, though never the less, the aircrew may be repeatedly briefed to ensure strict compliance of procedures to evade any such undesired situation.

The last factor investigated was the FOD occurrences are those which result into an aircraft tyre damage. These are no complex phenomenon. Any undesired object on the runway or ramp which causes a damage to the tyre in form of tyre burst, layer separation, FOD embedding or tread failure are all categorized as tyre related FOD occurrences.

These occurrences are caused not only due to negligence of maintenance or other staff having access to operational areas, but also to poor maintenance whereas the worker failed to remove a separable item from aircraft and the eventual failure to detect presence of such items on ground. Figure 5h depicts the yearly analyses of FOD case finalized as Tyre damage.

It was seen that tyre damage cases are though less in number but are consistently faced after few years, thus indicating failure of organization to implement an efficient FOD Prevention Plan.

A comparative analysis of all the finalized factors for last ten years of flying operations of the organizations was made to find out the major contributory finalization factor towards FOD occurrences in the organization. The comparative analysis of Finalization Factors of FOD occurrences for last ten years is shown in Figure 5i.

The figure provides the following observations:

1. Engine FOD ingestion cases were the highest contributor towards FOD occurrences of last ten years. 77% of all the occurrences had been Engine FOD cases.
2. Human Factor (ground Crew) had the second largest share towards FOD occurrences in the last ten years.
3. Material Factor and Miscellaneous Factor were other prominent areas contributing to FOD related occurrences in the organization.
4. The remaining four categories were not substantial contributors to the total of FOD occurrences for the last ten years.

The high number of Engine FOD occurrences and the Human Factor (Ground Crew) occurrences are indicator of a weak FOD prevention effort by the organization. Lack of awareness of individuals toward FOD eradication and the hazards associated with FOD, non adherence to work procedures, poor work ethics, lack of safety awareness and failure of nets to identify and clean FOD are the basic reasons for such high number of occurrences and huge financial implications associated to FOD cases.

FOD PREVENTION MEASURES

Eradication of FOD may be an uphill and overwhelming task, but it is not unequivocally an unachievable ambition. A sincere resolve and an honest effort are the only two prerequisites to completely diminish this rival. An earnest effort can literally bring down their FOD rate to either zero or very close to zero.

The most vital factor to initiate any drive is the commitment of management. Management support is the key to success of any FOD programme that is developed. Management commitment to the cause should be more than lip service. On the contrary, it should include the creation of an FOD prevention/elimination programme with adequate funding, appointment of a responsible point of contact with authority to carry out the FOD programme and full support and encouragement of an FOD prevention culture that crosses all boundaries within the organization.

There are two major but simple actions to tackle the FOD problems. The first step is the proactive approach and the second one is the protective measures. The same are discussed in detail in subsequent paragraphs.

PROACTIVE APPROACH

The Proactive Approach is one where efforts are made in the direction that by no means, an FOD is generated. This can be pursued by creating awareness among all tiers of personnel and accomplishment of their work ethics. This is a Proactive Approach, where it is endeavored that the problem does not even develop. The major areas to focus in the Proactive Approach is the training, work practices, housekeeping and material handling of the maintenance technicians.

All maintenance personnel should receive training in the identification and elimination of FOD, including the potential consequences of ignoring it. As part of the initial orientation, all newly inducted technical work force should receive FOD prevention indoctrination. It may be enforced that no maintenance personnel may be allowed to work on an aircraft or associated areas, till they receive FOD training. Records should be maintained on all personnel receiving FOD training. Moreover, recurrent training be undertaken at regular intervals through refresher lectures or briefs. The effective training should stress safety of personnel, the hazards to equipment, the direct costs associated with FOD damage and the indirect costs associated with mission delays and aborts.

A clean working environment is fundamental to FOD prevention. In the aviation maintenance environment, individuals must be emphasized to follow the concept of "clean as you go." When finished or when work cannot continue, they must be asked to clean the immediate work area and work stands. The debris that have the potential to migrate into an out-of-sight or inaccessible location must be picked. The area before inspections and after work completed is cleaned. The cleaning equipment, hoses, drop lights and power cords are to be returned to their proper storage area. The fundamental process to prevent foreign object damage is to perform all maintenance tasks "by the book." This includes all procedures from removing excess grease from a component to blanking aircraft ports and open connections with approved covers.

The responsibility of proper housekeeping resides, not only with the management but also with the individual as well. If debris are noticed, don't walk over it, pick it up and dispose it off properly. As part of the FOD prevention and elimination programme, FOD containers and bins should be placed in easily accessible locations throughout the maintenance environment. The containers should be painted or marked for easy recognition. In order to prevent FOD migration from these bins, they should be of appropriate size and enclosed to prevent overflow. If the container is located outside the hangar area, then they should be watertight to prevent leaching. Finally, they should be emptied on a regular basis as well as when requested. There should be regular scheduled FOD walks of hangars, aircraft tarmacs and parking bays. Consideration should be given to using specialized arrangements to clean areas.

Material handling, include consumables such as issued equipment like glue, paint, sealant, rags, sandpaper, brushes, applications and stock items (i.e., rivets, washers, fasteners, etc.). A well established plan for material handling and parts protection can eliminate many potential FOD hazards. Similarly for tools handling an efficient tool accounting and management procedure must be in vogue. The main objective of a positive tool accountability programme is to eliminate accidents/incidents and loss of life or equipment due to tool FOD. There are number of processes to facilitate accountability, which includes use of shadow boards, bar coding, tool counters, chit system tool tags or consolidated tool kits. The maintenance technician should avoid relying solely on visual inspection of the work areas to account for tools used during a maintenance task. Properly designed procedures for tool control and accountability are designed to preclude FOD.

PROTECTIVE APPROACH

The second approach is the protective measures, whereas efforts are made towards timely detection and clearance of FOD, before it inflicts any damage to the equipment or system. This method will evolve institution of a range of measures with participation of all personnel to effectively remove any FOD in the area. This includes FOD control techniques, performance measurement procedures, a feed back process, FOD investigations and a formal FOD prevention plan.

The FOD control elements would include measures to ensure removal of any FOD present in areas which can subsequently cause an occurrence. The activities at aircraft operating areas primarily involve maintenance/inspections of aircraft, care and upkeep of tarmacs and movements of specialized vehicles on ramps/runways/taxiways. Similarly movement of vehicles must be ensured on paved surfaces only according to a defined vehicular traffic patterns.

Various equipment and systems can be utilized consistently or as per predefined frequency for clearing the FOD from different operational and maintenance areas. Mechanical Sweeper is one of the most effective equipment for removing FOD from aircraft movement areas. The sweeper removes debris from cracks and pavement joints, and is used in all areas except for those that can be reached only with a hand broom. Magnetic Bars can be suspended beneath Mechanical Sweepers, GPUs and trucks as well to pick up metallic material fallen on the surface. Rumble Strips of varied lengths can be used near aircraft movement areas, as they are beneficial in dislodges FOD from vehicle undercarriages. Friction Mat Sweepers are rectangular assemblies which can be towed behind a vehicle that employs a series of bristle brushes and friction to sweep FOD into sets of capture scoops, which are covered by a retaining mesh to hold collected debris. Similarly in the modern equipment Tarsier Radar system uses high-resolution millimeter wave radar, to detect small, potentially hazardous objects on a runway, to within an accuracy of three meters at a range of up to two kilometers. The system is also able to detect a range of different materials, including metal, plastic, glass, wood, fiber-glass and animal remains. The FOD Finder System consists of millimeter band radar mounted on a vehicle. It includes a radar sensor with video capture capabilities, and on-board data processing controlled by a tablet PC, which serves as the interface with the user. The radar incorporates a 78-81 GHz sensor mounted on a reciprocating platform that allows scanning a field of approximately 80 deg in front of the vehicle.

Various performance measuring tools such as statistical graphics derived from audit or incident data, trend analysis, and performance review of worker to ensure conformance to standards or expectations can be utilized. More over, where the number of FOD occurrences is an indicator of performance, total cost value lost due to FOD is another count for performance review. It can be worked out and disseminated for consumption and awareness.

A feedback process is vital for workers to pass them the specific information about what is wrong, before they can be expected to improve processes. Feedback is vital for the management to continually play critical role on their account.

All incidents of FOD must be reported and investigated. When an FOD incident occurs, an investigation must be initiated to determine the cause. Cause and corrective action should be attained in a timely manner to preclude similar occurrences from happening in the future. Cause may be determined by visual observation, forensic analysis, or by location of the object.

A standardized document covering all aspects of FOD counter and preventive measures, must be prepared by each organization which should cover all the processes and details required to effectively counter and eliminate FOD. The document must include detail pertaining to the responsibilities of each individual involved in the process with optimum involvement of the top management.

FINDINGS

This research was based on the analysis of FOD occurrences of an aviation organization for a period of ten years. The large sample size helped in bringing a realistic picture of classification, error analysis and major contributory factors of FOD occurrences in the organization. The analysis carried out helped the researchers to arrive at the following findings:

1. Year 2009 saw the maximum number of FOD occurrences in any single year. The total of 54 occurrences was equal to the total occurrences of year 2001, 2002 and 2003. Thus

the huge financial losses through FOD for year 2009 alone were equal to the combine loss of three years i.e. 2001, 2002 and 2003. With highest number of FOD cases in year 2009, the FOD rate was also highest as 6.04 per 10000 flying hours.

2. The on ground occurrences are significantly higher than the in air occurrences. The reason being that most of the occurrences could not be appreciated during flight and were noticed during aircraft maintenance on ground. Moreover occurrences which were encountered during aircraft operations at ramp or runway and resulting into abort of mission, are also reason for high number of on ground occurrences. This thus indicates FOD to be a silent enemy.

3. Most of the occurrences were avoidable, as they could be either stopped or avoided. The highest number of avoidable occurrences was encountered in year 2009 against number of unavoidable occurrences as only 02 in same year.

4. Engine FOD ingestion cases were the highest contributor towards FOD occurrences of last ten years. 77% of all the occurrences had been Engine FOD cases. Also a high number of Human factor (Ground Crew) occurrences were encountered during last ten years.

5. The high number of Engine FOD occurrences and the Human Factor (Ground Crew) occurrences are indicator of a weak FOD prevention effort by the organization. Lack of awareness of individual toward eradication and hazards associated with FOD, non adherence to work procedures, poor work ethics, lack of safety awareness and failure of nets to identify and clean FOD are the basic reasons for such high number of occurrences

and huge financial implications associated to FOD cases. More so if an aging toll is being faced by the fleet, it underlines importance of improved maintenance practices and enhanced inspection items based on fleet aging.

Eradication of FOD may be an uphill and overwhelming task, but it is not unequivocally an unachievable ambition. The most vital factor to initiate any drive is the commitment of management. Management support is the key to success of any FOD programme that is developed.

CONCLUSION

The research data was adequate to infer significant information about the major FOD contributors in an aviation organization. It is though felt that further detailed analysis of Engine FOD ingestion occurrences to find out details of damages incurred on engine blades vis-à-vis the cost factor can be more beneficial to refer the cost factor associated with FOD. This can also be helpful in identifying the amount lost through FOD related occurrences and its comparison with spending on FOD counter measures, may be in shape of improving human resource or inducting most modern FOD countering equipment. Researchers in this area could use similar and more advanced statistical tools to establish the correlations of FOD occurrences and their contributory factors. The endeavors towards FOD prevention should be clear and comprehensible. The effort should address such items as an understanding of the importance of FOD elimination and how does FOD prevention affect safety, quality, costs, and operational readiness.

REFERENCES

Chen, R., Lu, F., Li, Q., Fu, G., Wang, H., Chai, Z., & Xu, Z. (2009). Failure analysis on foreign object damage of aero-engine compressor blade. In *Proceedings of 8th International Conference on Reliability, Maintainability and Safety,* (pp.1085-1088). IEEE.

Feil, P., Menzel, W., Nguyen, T. P., Pichot, C., & Migliaccio, C. (2008). Foreign objects debris detection (FOD) on airport runways using a broadband 78 GHz sensor. In *Proceedings of 38th European Microwave Conference,* (pp. 1608-1611). IEEE.

Greneker, G. (1999). *Radar to detect foreign object ingestion by a jet engine.* Retrieved from http://digitalcommons.unl.edu/birdstrike1999/17

Latorella, K. A., & Prabhu, P. V. (2000). A review of human error in aviation maintenance and inspection. *International Journal of Industrial Ergonomics, 1*(2), 133–161. doi:10.1016/S0169-8141(99)00063-3

Mason, F. A., Kraus, D. C., Johnson, W. B., & Watson, J. (2001). *Reducing foreign object damage through improved human performance: Best practices.* Washington, DC: Galaxy Scientific Corporation.

Netjasov, F., & Janic, M. (2008). A review of research on risk and safety modelling in civil aviation. *Journal of Air Transport Management, 1*(4), 213–220. doi:10.1016/j.jairtraman.2008.04.008

Remawi, H., Bates, P., & Dix, I. (2011). The relationship between the implementation of a safety management system and the attitudes of employees towards unsafe acts in aviation. *Journal of Safety Science, 1*(5), 625–632. doi:10.1016/j.ssci.2010.09.014

Xu, Q., Ning, H., & Chen, W. (2009). Video-based foreign object debris detection. In *Proceedings of IEEE International Workshop on Imaging Systems and Techniques (IST 2009),* (pp.119-122). IEEE Press.

Section 3
Technology Management

Chapter 15
Media Management in Disaster Events:
A Case Study of a Japanese Earthquake

Eleonora Benecchi
University of Lugano, Switzerland

Vincenzo De Masi
University of Zurich, Switzerland

ABSTRACT

According to a survey by Goo Research (April 2011), the average Japanese person appears to have relied primarily on television news for gathering information in times of disaster, and as unlike a lot of overseas media, the public broadcaster NHK's news broadcasts were defined as very calm and measured. This chapter focuses on the NHK coverage of the earthquake and nuclear crisis in March 2011 compared with private channels' and specific websites' coverage with regard to specific events. The aim is to enlighten the ways and the tools through which Japanese Public Television played a double role: on one side it became a primary source of information for hard news and played a "service" role for the population in need; on the other side and with special regard to the coverage of the nuclear crisis, the duty to inform was balanced by the duty to reassure the public and promote harmony so that NHK privileged government and corporate statements about the Fukushima situation. The authors corroborate their study through an analysis of NHK's programming and private channels' changing schedules and advertising during the recent disaster. This chapter provides a concrete example of the potential television role in disaster mitigation, taking into account both the positive and critical aspects.

DOI: 10.4018/978-1-4666-3658-3.ch015

INTRODUCTION

The object of this study is the use of Japanese Public Television to manage and mitigate the earthquake consequences and the Fukushima situation. A preliminary analysis of the tools and strategies used by NHK to describe the earthquake and the tsunami occurred on March 11th and to inform people about their consequences, will demonstrate that the Japanese Pubcaster entered a "natural disaster mode" during the first week from the natural disaster. A follow up analysis of the official statements and documents released by NHK Media Department during the first week from the earthquake will help enlighten the main characteristics and functions of the so-called "natural disaster mode" and will demonstrate how this kind of coverage is innate to the NHK.

A sample analysis of the programming and contents of the week from March 11th to March 19th will demonstrate how in the NHK coverage of the disaster the duty to inform was balanced by the duty to reassure and promote harmony. With special regard to this last point we will perform a comparative analysis of specific events connected to the Fukushima incident as reported both by NHK and independent journalists on the Net.

This study aim at documenting a type of TV coverage we have defined as the "NHK natural disaster mode," in the belief that this model could represent a good practice to follow when covering a natural disaster and its immediate consequences. Despite its good features, though, this model has also shown some critical aspects when confronted with controversial situations, such as the Fukushima incident. As Japan's nuclear energy crisis continues to unfold at the Fukushima Daiichi power station, the news media have struggled to sort through confusing, and often conflicting, information about damage to the crippled plant and its threat to public safety.

BACKGROUND

Media Coverage of Japan's Disaster

The scholarship shows that the media can play a critical role before, during, and after crisis and/or disaster. The media are essential, for example, for warnings to be effective and may be the single most important source of public information in the wake of a disaster (Quarantelli, 1991). The scholarship, though, also shows that media reports that distort what happens in a disaster and lead to misunderstandings (Parker, 1980). Failure by officials to issue a warning, for example, may be a result of myths created by the media. Death, economic loss, human suffering, and social disruption are the standard themes in the media's portrayal of disaster (Smith, 1998). For the audience, the apparent image is one of total destruction (Wenger, James, & Faupel, 1980). Generally, researchers argue that media content tends to overemphasize the chaotic, non-social, irrational, and non-traditional aspects of the event (Adams, 1986).

Moreover, a review of Journalism textbooks suggests that the authors who deal with disaster coverage often state as fact what disaster scholars have shown to be inaccurate (Scanlon, 1998). For instance, researches have demonstrated that people find it easier to cope with the truth, with clear factual accounts of what is known about what is happening. It is lack of clarity and confusion not accuracy that makes persons uneasy (Quarantelli & Dynes, 1972). And yet, scholars researching the field found reporters felt it was their duty to shape their stories to avoid panic (Kueneman & Wright, 1976) and may downplay negative stories, especially in their own communities or be sympathetic to officials with the same goal (Rogers & Sood, 1981).

Another aspect to be taken into consideration is that most media did not have disaster plans for their own organizations—no plans as to how they would continue to operate in such conditions, no plans as to how they would deal with the demands of disaster coverage (Wenger & Quarantelli, 1989).

Despite all of this, scholars argue that depending on how they cover the disaster, the media can either help or hinder relief efforts (Scanlon & Alldred, 1982). Immediately after the disaster occurs, media could be invaluable during the initial assessments, search, and rescue. Timely, accurate, and sensitive communications in the face of natural hazards are demonstrated, cost-effective means of saving lives, reducing property damage, and increasing public understanding. Such communications can educate, warn, inform, and empower people to take practical steps to protect themselves from natural hazards. After the first assessments, it often informs the international community and citizens about the conditions in the country of disaster. The positive impact of media is twofold: firstly, media raises people's awareness on the disaster and, secondly, it helps to acquire the funding (Scanlon, Osborne, & McClellan, 1996).

The Tohoku earthquake and tsunami on March 11th together with the Fukushima incident which is still on-going enlightened both how television can play an important role in disaster mitigation and how it can reinforce false myths or it can fail to correctly report on the current situation.

Foreign media for instance has copped criticism for some of their coverage of the Japanese disaster from within Japan. The examples of perceived poor journalism, including sensationalist headlines, inappropriate references to World war Two and caption errors, are featured on a "Wall of Shame" on the online forum JPQuake, where Japanese residents have collated examples of erroneous reporting on the Japanese disaster.

There are of course exceptions to this oversensationalism. For instance an Italian journalist, Pio D'Emilia, correspondent for the Italian TV of Sky TG24, decided to go and visit the areas struck by the calamity soon after the disaster in order to give the news of the earthquake and the tsunami to the Italian television as soon as possible. In his book he reports about his first-eyes experience in the "danger-zone," but he also write about the real meaning of reporting such an event "The tsunami was a good test to teach us what 'reporting' really means: checking facts, being there, distinguishing between opinions and gossip" (D'Emilia, 2011).

The case of Pio D'Emilia is also relevant to this work because the Italian journalist denounced the "tsunami of information" that swallowed both the foreign and Japanese media and exposed a widespread lack of journalistic integrity (Knight, 2011). With special regard to the television's coverage of the events occurred on March 2011 he underlined how Japanese Public Television was slow and not proactive enough in seeking the truth. According to him, the fact that mainstream media "toed the government's official line or that of Tokyo Electric Power Co., operator of the devastated Fukushima No. 1 nuclear power plant" created a void which was filled by "who witnessed the compulsive lying and omissions of the so-called mainstream media" (Knight, 2011). The *Times of London's* Leo Lewis, a former Tokyo correspondent back in Japan to cover the disaster, confirmed that the Japanese public has "not always been well-served by its domestic media" (Harper, 2011).

Bulding from this framework, we decided to perform an independent analysis of the NHK coverage during the first week since the March 2011 earthquake. We used a methodology created and used to analyze RAI (Italian Public Television) coverage of natural disasters and emergency events (Lasagni & Richeri, 1999). This methodology is based on a content analysis of the TV programming, performed using a semi-structured form. We analyzed in detail the programming of March 11th 2011 and a casual sample of the following days until March 18th. We also performed an additional in depth analysis of programs specifically dedicated to the disaster such as news special and news magazines.

The objective of this first step of analysis was to describe NHK's coverage of the earthquake and nuclear crisis with concrete examples. The following step was to compare what NHK showed to its public with the official statements and documents of NHK' members talking about NHK's coverage of the events.

Some critical pints were also enlightened through a direct comparison with alternative sources of information, specifically the Web, with special regard to the Fukushima situation.

MAIN FOCUS

Media Background and the Earthquake

On 11th March 2011 at 14:46:18: an earthquake measuring 9 Richter strikes the coasts of the Tōhoku Region, in Japan, provoking a tsunami with 10 metres waves high which hit the coast for a lot of kilometres. The tsunami also strikes the power station of Fukushima, run by the Tokyo Electric Power Company (TEPCO), provoking, after some hours, a first blow up of one of the six reactors. Japan experiences its first nuclear crisis and, in the span of a year, the Japanese Government is to close all the nuclear reactors in the country, at least for the moment. After the nuclear disaster, mainstream media in Japan were accused of not giving fair information about it, but to be highly influenced by the commercial market and TEPCO which is not only the operator of the devastated Fukushima No. 1 nuclear power plant but also one of the greatest sponsors of this market. After some weeks the Japanese people asked more detailed information without any kind of "filter," but only alternative programs such as *Gachibatoru 2011* by Beat Takeshi (ビートたけ しのガチバトル*2011* at 21:00 on TBS), as far as its author Takeshi Kitano, was a special polemic "battle talk" program about Nuclear crisis at the end of the year (28 December 2011). It was in fact a new and "independent journalism" which found its main channel in the Web which made people more aware of the possible problems caused by

the nuclear radiations and contributed to reinforce the Japanese antinuclear group that was in the minority before the disaster.

This research tries to analyze the way the Japanese TVs have reacted to the earthquake, the tsunami and then to the nuclear crisis, taking into consideration the 6 main TVs of the Kanto Region, aired all over Japan. The choice to study the TVs of Kanto is strategic for two main reasons: firstly, the Kanto Region borders north with the region of Tōhoku; secondly, in the Kanto Region and exactly in Tokyo there are the studios of the 6 main TVs linked in real time to the National weather forecast, and Kanto is also the most populated region in Japan (population of about 43 million).

First Step of Analysis

The first step of our analysis consists in comparing the programming of the six channels of the Kanto Region chosen for this study soon after the earthquake struck Japan. NHK is the first to show an earthquake warning (a flashing message at the bottom of the screen) after 28 seconds from the beginning of the tremors (14:46:18): "This is an Earthquake Early Warning. Please prepare for powerful tremors." (「緊急地震速 報です。強い揺れに警戒して下さい。」 *Kinkyū Jishin Sokuhō desu. Tsuyoi yure ni keikai shite kudasai*.) This timing can be explained by the fact that NHK is directly connected to the Earthquake Early Warning System. In the specific case of the earthquake on the 11th of March, at 2.46 pm instruments detected the earthquake 15 seconds after the tremor in the epicenter off the coasts of the Tōhoku Region, and a signal was instantly transmitted to the electric power substations, which instantly sent the alert signal to hospitals, public loudspeakers, factories, TVs, mobile phones etc.., in order to prevent accidents and stop all the activities. For instance the Shinkansen begun slowing down after 3 seconds, then it took 73 seconds to reach 100km and other 30 seconds to reach a complete stop. Going back to NHK, after 45 seconds (14:47:31) a notice of the earthquake is superimposed on top of the screen. In this second case the message is not a warning but a detection of the occurred earthquake. After 2 minutes and 4 seconds from the first tremor (14:46:22), the NHK has already cut to the studio to give live news and information about the earthquake. The other TV stations are not able to give a running message superimposed on top of the screen because they are not directly connected to the Earthquake Early Warning System: the first warning on Nihon TV and TBS come after 11 seconds from the start of live broadcasting on NHK. Because of the disaster, commercial TV stations are under the pressure to change the tone of their broadcasts. Among commercial televisions, NTV is the first to give news of the earthquake cutting from news about Tokyo Mayor Ishihara Shintaro to a live studio shot. After this brief message, though, it returns to pre-booked commercials, specifically a model demonstrating how to use mascara remover (Boirè) and a spot about an insurance company (American Home Direct). Almost in contemporary TBS gives news of the earthquake running a text superimposed on top of the screen, clocking in just few instants later than NTV. No interruption of regular programming (*San nen B-gumi Kinpachi-sensei, 3年B組金八先生*). Terebi ASAHI broadcasts a running warning message about the earthquake superimposed on the screen after 39 seconds from the live cut of NHK (at 14:49:01). On Terebi Tokyo, 3 minutes after Nihon Terebi and TBS, a timely earthquake alert flashes on the screen. There is no interruption of regular programming (14:49:17). Up until this point, Fuji TV has not given any warning yet. At 14:49:37 sec. Nihon TV cuts to a live image of the studio which is shaking; 4 seconds after that, also Fuji TV tries a live update, but it shows only an empty news desk with chairs rolling around and immediately after a graph of the earthquake, switching finally to a Korean drama. After 3 minutes and 54 sec since the first tremor (14:50:12), NHK makes us see a graph with the areas that will be struck by

the tsunami. After 4 minutes and a half, TBS goes live (14:50:45), followed after 2 seconds by Nihon TV, and after 10 seconds by Fuji TV. TV Asahi continues broadcasting its regular programming but after 4 minutes and 57 seconds, it shows a map of the tsunami alert superimposed on the screen and after some seconds (14:51:34) goes live too. While the other channels broadcast the tsunami alert, Fuji TV shows, quite late, the areas struck by the earthquake and later on the tsunami alert (14:53:15). Also TV Tokyo continues its regular programming but with the map of the tsunami alert superimposed (14:52:17). After about 8 minutes and 5 seconds, TV Tokyo is indeed the last commercial TV to cut regular programming (fiction about three TV personalities who decide to try their hand at fishing) switching to full-time earthquake coverage.

NHK Natural Disaster Mode

The role of television news in disasters is varied. In local settings or in the immediate area within which disaster has struck or is striking, television news is one of the primary means of disseminating information often vital to the physical and emotional health and safety of community residents. In the process, the medium sometimes serves a vital function, informing and instructing viewers in matters pertaining to safety and recovery.

When on 11th March the earthquake struck Japan, the NHK, thanks to its experience in dealing with the disasters and thanks to the live link-up with the EEW, was immediately able to call the 14 helicopters that were on permanent standby, 88 mobile broadcasting units, whose 18 of them dedicated to the satellite programs. Besides NHK has got 5 local stations in the areas struck by the disaster, 46 additional local stations and about 600 members of the staff were on the spot ready to commentate and inform about the crisis situation, and last but not least, it has also got the chance to control over 460 weather cameras in real time. Therefore, thanks to that NHK natural disaster mode, the NHK was very fast in broadcasting the news. In fact, in 90 seconds after the warning the cut from regular programming (coverage of Diet deliberations) to announcer Ito in the still-shaking Shibuya studio was swiftly made and minutes before every competing channels.

30 minutes before the Tsunami struck, NHK's domestic channels started beaming in live footage from its robot cameras at ports in Miyagi and Iwate Prefectures. While the Tsunami struck, the helicopter NHK crew started beaming live images of the tsunami in Miyagi, those same images were used from TV all around the world. Therefore, the NHK wins praise for its purpose and understated coverage of the earthquake and nuclear crisis.

In this context, then, television news does not merely convey information about disasters. It has the power to *define* disaster. As already stated, Television coverage of natural disasters is often framed in such a way as to convey hopelessness, presenting them as battles between powerless humans and powerful nature. The analysis of NHK coverage of the earthquake and tsunami on March 11th is a demonstration of an alternative way to cover a disaster in television.

As in cases registered during RAI research (Lasagni & Richeri, 1999-2004) the programming is completely revolutionated to give space to emergency and vital information.

Even when the real Time footage of Tsunami beamed from a NHK chopper was broadcasted, the horror of the images was under-sensationalized through the anchor's words as shown in Table 1.

From the analysis performed on the programming of the first hours after the disaster we can say that there were no reference to people as victims, no images of dead people broadcasted, no expression of strong emotions from the reporters, no trace of Hype.

Another interesting aspect is the fact that during disaster coverage, NHK rotated its anchors off camera every hour to ensure their namelessness. In fact, the Cardboard model and Diorama of Nuclear Reactor are more recognizable than the on-air personality. They are in some way tiny and familiar while what is at stake is huge and unknown.

In contrast with a typical mainstream media approach though, NHK programming and coverage show distinguish characteristics as shown in Table 2.

Alternative Sources of Information

After the disaster, all the TVs have followed the NHK's way of working: for example the Fuji TV aired for 61 hours only news and documentaries about the earthquake, without any commercial breaks, because any companies wanted to link their commercial breaks to the disaster. Tokyo TV is the first in broadcasting some commercial

Table 1. Anchor's words and style of coverage

Anchorman words	Style of coverage
"We are showing you the current situation"	Factuality+Actuality
"We can see how houses are being pushed away by the tsunami"	No Panic-Inducers
"We can see buildings and cars. Black waves gulping down buildings and farms"	Actuality+No Panic
"This is the current situation at the mouth of the Natori River in Senday City	Factuality+Actuality

Table 2. NHK programming and coverage's distinguishing characteristics

COVERAGE	MEANS/TOOLS
Actual	Events in real time thanks to training and direct connection to EEW System
Balanced	Images both from most affected areas and of recovery efforts in massive areas
Factual	No panic inducers ("massive," "devastating" "severe") Absence of hype and emotion
Respectful	No close-up images of deceased people Disaster etiquette for reporters ("What do you need" instead of "What do you feel")

breaks after 33 hours of public service. TV Asahi was the TV with more programming of public service; in fact, the first commercial break was aired after 74 hours. The other commercial TVs were on line with about 60 hours of programming such as the Tbs after 62 hours and NTV after 61 hours. In the next days and exactly on 14th March, the first commercial breaks start being aired on the TVs, commercials is replaced by private non-profit organization the Advertising Council Japan (Ac Japan). AC Japan is the distributor of breaks of governmental agencies, Public Service Announcement (PSA) and other non-profit organizations. Normally, on the Japanese TV, there are 4000 commercial breaks for 500 different companies and 1000 products but on March 12th there were only 54 commercial breaks and from 12th to 19th March Ac Japan broadcasted about 20000 commercial breaks. After the disaster, the commercial breaks of AC Japan strongly entered the houses of the Japanese people who judged most of them quite inappropriate considering the period of crisis they were experiencing. In fact the aired commercial breaks used to show smiling animated characters with happy background and music. Besides the final jingles were similar for all the commercial breaks so the audience used to link them to the disaster. The audience, tired of these unsuitable commercial breaks of the AC, started to complain with the TV stations that, after a few days, cut out the music of the jingles. Then the AC Japan had to write an official letter where it gave its excuses and it stated that it would have changed and revised all the commercial breaks. A very interesting point of this event is that, after a few days, the Japanese young people started to make fun of the characters of this commercial break on the Web, causing a real race of parody videos such as transformations of the characters into transformer robots, and then musical parodies and so on.

In general, Japanese media coverage seems to have led readers and viewers to be extremely skeptical of the degree of reliability of reported information. This is because Japan's media almost solely depend on the prime minister's office and Tepco for information, because it is likely Tepco has not revealed everything they know, and because the Japanese media has been playing down the gravity of the situation because they do not want to fan people's fears.

With special regard to the Fukushima incident it appears that NHK privileged government and corporate statements and that in this case specifically the duty to inform was balanced by the duty to reassure and promote harmony.

While commercial channels basically retransmitted NHK images and coverage of the events, Internet became a real alternative source of information on the Fukushima situation.

On 11th March 2011, for instance, Tepco, announced on NHK that no radiation had been released into the atmosphere. Hajime Shiraishi (Our Planet TV) broadcast via Web a live interview with 5 Japanese reporters in Futaba City told the experts "held up Geiger counters showing the level of radiation was almost beyond calculation, they had never seen anything like it." Since this show, OPTV increased viewers from 1,000 up to 100,000.

Nuclear physicist Ryugo Hayano tweeted "whether TV news or the government, people are now criticizing authority in fundamental ways they didn't before." It is a fact that is twitter account providing information on the Fukushima situation and nuclear emergency went up from 3,000 to 150,000 followers after March 11th.

Independent journalists and blogger launched news about supposed "cover-ups" of Fukushima danger by NHK through the Web. For instance on 12th March News, 11:40 am director supposedly stops an announcer reading the news about exposed fuel rod assemblies and the file audio with evidence of the "cover-up" is uploaded on many youtube channels and blogs such as "Fukushima Diary," by Moshizuki Iori.

Solutions and Recommendations

This research tries to highlight how the "alternative" communication sources, in primis the Internet, have played an important role in giving the true and real information to the Japanese people. However, these "alternative" communication sources did not properly exploit the Internet, so there was the birth of many websites dealing with the disaster topic causing in this way confusion for the Internet users (Auf der Heide, 1989). Even the big Web companies launched their way to give information and helped people in finding lost persons and things, using for example the Google Finder. They were very useful services but also unorganized because they were too many and not coordinated by only one institution and so they didn't give their best. The evolution of these services will surely be more efficient if they are managed by only one institution able to involve all the other governmental institutes, using only one platform in order to have a clear idea of the crisis situation without any kind of "filter."

FUTURE RESEARCH DIRECTIONS

The earthquake and tsunami in Japan (11/3/2011) is very useful in order to understand the future development of the media and above all the failures and successes of both the Japanese government and media.

During the tsunami in Japan, the rescuers in cooperation with the weather bureau used even the Google Earth in order to find the lost people. But other remarkable examples of this kind can also be given by the cooperations occurred soon after the earthquake in Haiti and in China where the volunteers created sort of platforms on the Web showing maps of the areas of crisis soon after the disaster, so in real time. They even used the *tag* to gather information through the data sent by the users (for example texts, pictures,

videos, etc.) living in the internal areas struck by the earthquake. They sent all the information by text messages, Web, email, radio, phone, Twitter, Facebook, television, list-serve and live-streams: it was a real common platform of 'Person Finder,' a space where everybody could insert useful information for the researches and the people to find or the people already found. However, it was not enough, in fact during moments of crisis all the mass media should be united and organized in a sort of global platform in order to facilitate the rescue actions and give more detailed information to the people. A negative example during the Japanese earthquake occurred to lot of foreign workers, who used to live in the areas struck by the disaster, because they didn't speak Japanese or even English and so they did not have any ways to communicate with the others, in fact all the communication means were out of order. They were rescued but they did not know what happened and they were obviously frightened and confused. All those problems could have been avoided if only the governmental institutions had used the communication in a proper way and given information in different languages and had asked for some help from the foreign embassies and consulates.

CONCLUSION

We observed how NHK covered the disaster of March 11th in an alternative way with respect to international televisions.

For instance, NHK disallowed the online transmission of earthquake footage by other news media outlets. NHK also allowed newspapers to publish images of their footage as long as they were credited. This means that people could watch the disaster almost solely through NHK channels or NHK branded footage/images.

Despite the fact that most of modern televisions have now the EEW system incorporated, before only NHK was directly connected to the system and could therefore broadcast earthquake warnings immediately. However, we are convinced that direct connection with EEW should not be an exclusive of NHK.

In general NHK tendency to privilege corporate and government' statements produced a shift in the trust younger Japanese place in government and media so that Internet became in fact an alternative sources of information.

REFERENCES

Adams, W. C. (1986, Spring). Whose lives count? TV coverage of natural disasters. *The Journal of Communication*. doi:10.1111/j.1460-2466.1986.tb01429.x

Auf der Heide, E. (1989). *Disaster response: Principles of preparation and coordination*. St. Louis, MO: CV Mosby.

D'Emilia, P. (2011). *Tsunami nucleare, I 30 giorni che sconvolsero il giappone*. Rome, Italy: Il Manifesto Libri.

Harper, P. (2011, March 21). *Media slammed for Japan crisis coverage*. Retrieved from http://www.nzherald.co.nz/japan-tsunami/news/article.cfm?c_id=1503051&objectid=10713939

Knight, S. (2011). Italian journalist denounces media's disaster coverage. *Asahi.com*. Retrieved from http://www.asahi.com/english/TKY201107050261.html

Kueneman, R. M., & Wright, J. E. (1975). New policies of broadcast stations for civil disturbances and disasters. *The Journalism Quarterly*, *52*(4). doi:10.1177/107769907505200409

Lasagni, M., & Richeri, G. (2004). *Un anno di temi sociali nella programmazione RAI*. Retrieved from http://www.segretariatosociale.rai.it/palinsesto/indice_dati_program_soc.html

Lombardi, M. (1993). *Tsunami: 'Crisis management' della comunicazione*. Milan, Italy: Vita & Pensiero.

Lombardi, M. (2005a). *Comunicare nell'emergenza*. Milan, Italy: Vita & Pensiero.

Lombardi, M. (2005b). *La comunicazione dei rischi internazionali: Un confronto internazionale*. Milan, Italy: Vita & Pensiero.

Parker, E. C. (1980). What is right and wrong with media coverage of disaster? In *Disasters and the Mass Media*. Washington, DC: The National Research Council.

Quarantelli, E. L. (1991). *Lessons from research: Findings on mass communications system behavior in the pre, trans and postimpact periods*. Newark, NJ: Disaster Research Center.

Quarantelli, E. L., & Dynes, R. (1972). When disaster strikes: It isn't much like what you've heard and read about. *Psychology Today*, *5*(9).

Rogers, E. M., & Sood, R. (1981). *Mass media operations in a quick-onset natural disaster: Hurricane David in Dominica*. Boulder, CO: Natural Hazards Research and Applications Information Center.

Scanlon, J. (1998). The search for non-existent facts in the reporting of disaster. *Journalism and Mass Communication Educator*, *53*(2). doi:10.1177/107769589805300205

Scanlon, T. J. (1992). *Disaster preparedness some myths and misconceptions*. Easingwold, UK: The Emergency Planning College.

Scanlon, T. J., & Alldred, S. (1982). Media coverage of disasters: The same old story. In Jones, B. G., & Tomazevic, M. (Eds.), *Social and Economic Aspects of Earthquakes*. Ithaca, NY: Cornell University.

Scanlon, T. J., Dixon, K., & McClellan, S. (1982). *The Miramichi earthquakes: The media respond to an invisible emergency*. Ottawa, Canada: Emergency Communications Research Unit.

Smith, C. (1998). Visual evidence in environmental catastrophe TV stories. *Journal of Mass Media Ethics*, *13*(4). doi:10.1207/s15327728jmme1304_4

Sood, R., Stockdale, G., & Rogers, E. M. (1987). How the news media operate in natural disasters. *The Journal of Communication*, 37.

Turner, R. (1980). The mass media and preparations for natural disaster. In *Disasters and the Mass Media*. Washington, DC: The National Research Council.

Wenger, D., James, T., & Faupel, C. (1980). *Disaster beliefs and emergency planning*. Newark, NJ: Disaster Research Center.

Wenger, D., & Quarantelli, E. L. (1989). *Local mass media operations, problems and products in disasters report series # 19*. Newark, NJ: Disaster Research Center.

Wolf, M. (1993). *Teoria delle comunicazioni di massa*. Milan, Italy: Bompiani.

Chapter 16

The Impact of Virtual Community (Web 2.0) in the Economic, Social, and Political Environment of Traditional Society

Irene Samanta

Technological Educational Institute of Piraues, Greece

ABSTRACT

The chapter enhances the scientific research in the area of the new digital era with a focus on diversity created in real society from the influence of social media. Specifically, it reveals the effects of social media on economic, political, and real society affairs. The latest riots in Middle East countries demonstrate that virtual social communities wield an influence on the citizens, and the changes they implemented show these countries will never be the same again. The effects of social media in real society are examined in highly developed countries such as the EU and North America (USA and Canada).

INTRODUCTION

The widespread use of Web technologies, especially in economy, society, and governance, creates an unprecedented volume of data and information, which are exchanged and traded daily on the Internet.

DOI: 10.4018/978-1-4666-3658-3.ch016

The recent elections for the U.S. presidency in 2008, as well the latest riots in Middle East countries with the widespread use of social networking tools and technologies (YouTube, MySpace, Facebook, etc.) have highlighted the powerful penetration of the Internet in the political process.

This raises questions as to the degree to which the real community affected by what happens in a virtual community.

The transition to Web 2.0 marks the transition to the age of Politics 2.0. The traditional one-way relationship between society and politicians gives way to a new form of "interactive democracy" dominated by freedom of expression. The ability to access (and use) a variety of information sources and Web tools allow the establishment of truly open processes and the export of meaningful dialogue. The potential political communities take the form of an inclusive public sphere, and new forms of political speech and action, influence, and political culture are created. By extension, new political-social-commercial-technological identities and relationships are formed, whose content and format is shaped and transformed dynamically in the realm of the social fabric and is mediated by both the limits and scope of the tools and by the willingness of users-citizens itself.

The first to record the phenomenon of globalization were Swanson and Mancini, who focused on political communication and noted that politicians and candidates in various democratic countries of the world follow common or similar election practices (Mancini & Swanson, 1996).

These practices originate from the U.S., and thus this phenomenon was dubbed the "Americanization" of political communication. As pointed out by Kaid et al. (1995) today globalization means "Americanization." This phenomenon is also known under the rather specious terms "professionalism," "upgrade," "streamlining," and "modernization" of political communication.

The main characteristics of this form of political communication is the dominant role of all modern media (Internet, email, mobile telephony, fax, etc.) in the development of a communication policy (Swanson & Mancini, 1996; Kaid, et al., 1995; Negrine, 1996). Nevertheless, each country simultaneously presents significant differences, as each country has its own distinct political and communication system and its own culture and social tradition.

Living in this new digital environment means that, in many countries, the typical twenty-one year-old has already played about ten thousand hours video games, read and sent more than two hundred thousand emails, talked on mobile phones for close to ten thousand hours and certainly watched TV for more than ten thousands of hours. It is obvious that he/she is socialized in a very different way from the immediately preceding generations. Thus, the main narration of life is no longer provided by school, church or the family but by modern digital technology. These young people are truly natives, indigenous digital natives as Prensky (2001) calls them, of the new digital world. A world in which the older generations can only be digital immigrants or even digital foreigners. E-mails, Blogs, Wikis, Massive Multiplayer Making on Line Games, Podcasts, WebTV, Web Radio are services that embody the concept of convergence in the everyday life of Internet users. Mainly through their experience of video games, many new users have learned to gather information from many sources and take quick decisions, to understand the rules of the game playing, rather than being taught by someone else or reading them from a manual, to devise strategies to overcome barriers to understanding complex systems, and to experiment on these with their cooperators. Therefore, the current situation is mainly based on data communication; its main feature is its communicative and participatory nature. According to Birney and Barry (2006), the publication achieved by creating a blog is due to the fact that users respond personally to the lack of pluralism that characterizes traditional media, which are controlled by a very small number of business groups.

It is understood from the above that traditional theoretical and empirical issues such as the social creation of reality and everyday life, power and society, identity and social dynamics, gain a new momentum, attracting research interest to the examination of the Web and in particular social forums. In this respect, the influence of virtual forums on the on-line Society, along with their legal, social and ethical implications, becomes increasingly interesting and develops into an issue of special attention and study, acquiring central importance in wider issues of Web science.

The chapter aims to study the new social and technological transformations and the ways to "perform" inter-reactions, as well as the conditions of real life in the potential world. The specific research objectives are:

The influence of virtual communities in economic life through the influences exercised or not exercised by the experience of members of the virtual community in using a product or a service.

The effect of the virtual community in the social life of a true community through the interaction of its members in social, economic, and political events (the case of the activation of various Forums, such as Genoa 2004, meetings of the G8, the death of young Alex in Athens and the events following in December 2008, the political revolution in Tunisia, Egypt, Libya and so on through Facebook, etc.).

The influence of forums on the new political reality. The manner of application of the rules of political Marketing and the interaction between groups of politicians and citizens who do not know each other on a personal level (the case of election of President Obama in the U.S. in 2008).

THEORETICAL UNDERPINNINGS OF THE STUDY

Online Communities

An online community or forum website is normally established from a group of persons which release requests to sign up on the website as members to persons of included in their mailing contacts.

This strategy contributes towards the creation of a community of people with similar interests who will trust and act upon the recommendations of others in the group. Most online communities are currently at a relatively early evolutionary stage and have yet to be subjected to serious study, but

from the company perspective the information posted on relevant community sites can provide valuable research data about such issues as product quality, how useful the website content is and how easy it is for customers to navigate and find what they are looking for. Alongside the growth in company or industry-specific communities, more generic online social networking has become hugely popular in recent years. According to Goad and Mooney (2008), the most popular social networks are currently Facebook, and MySpace, with market shares of 38 per cent, and 19 per cent respectively in UK. The authors cite figures showing that a larger proportion of the population have chosen to communicate with friends through social networks than by email services such as Hotmail or Yahoo since October 2007, and that the age profile of users is rising rapidly with strongest growth amongst the over 55s. Businesses are now recognising the potential of these communities for the development of their brands and to build relationships with key customers. Facebook needs to operate as a profitable business of course, but the danger is that over-commercialisation of a social network through advertising may prompt an exodus of users towards the "next big thing," as well as damage relationships between consumers and participating brands if interactions within the network are not managed carefully. At present, Facebook's policy is for users to receive a maximum of two sponsored messages per day. For this reason, requests are the leading motive for websites to obtain new members. Normally forums permit a user to create and sustain a list of contacts for interacting socially or even professionally. In the centre of an online community boards are custom-made member profiles. Members' profiles are typically a mixture of users' profile photos or avatars, favourite music and books, movies selections, and other links to favourite websites. Dissimilar online forums entail various ranks of confidential information like what should be publicized throughout a page profile to non-members. Profile owners obtain new "friends" through a search in the website and requesting to become friends. There are numerous features of online community information, which have superior power on customer behaviour instead of marketer created online information. Initially, this kind of information, which is available inside online forums, is more likely to contain superior trustworthiness than information created from marketers. This is a significant verification that information originated from a source considered to be additional reliable is likely to lead in a superior cogency of that information (Wilson & Sherrell, 1993). The views and opinions of individual product experiences written on Internet forums are probable to come from reliable sources since these writers are also customers and apparently do not have certain benefit from the product and no reasons to influence the person who reads these reviews (Bickart, & Schindler, 2001).

Even though the participants in online forums discussions are possible not to have similarities and way of life related with those of the receiver, they linked with the receivers since they are associate customers. The information they make available is likely to reveal usual product features. This transforms the information as more appropriate to the reader than marketer created information which the creator may have never used the product in real circumstances.

At last, online communities have a superior capability to produce understanding between readers. The stories of individual experiences create the case of what a person is likely to discover on an online forum. Members in Internet forums are in a way acting for many forum members and assistance to the operation of a forum frequently evaluated regarding the member's skill to amuse and teach several members of the forum. Personal stories have the property to affect the person who reads and to sympathize with the story of the member who posted. Sympathy might influence buyer behaviour not directly by underlining to the user of the product the merits, which are being getting pleasure from other users. More often, the

interest of a forum member unfolding the pleasures of a specific item could straight create almost the same views in the minds of the fellow members (Bickart, & Schindler, 2001).

Social Network

When a computer network connects people or organizations, it is a social network. Just as a computer network is a set of machines connected by a set of cables, a social network is a set of people (or organizations or other social entities) connected by a set of social relationships, such as friendship, co-working or information exchange. Social network analysis focuses on patterns of relations among people, organizations, states, etc. The research into how people use computer-mediated communication has concentrated on how individual users interface with their computers, how two persons interact online, or how small groups function online. As widespread communication via computer networks develops, analysts need to go beyond studying single users, two-person ties, and small groups to examining the computer-supported social networks (Wellman & Gulia, 1995).

Social network analysts seek to describe networks of relations as fully as possible, tease out the prominent patterns in such networks, trace the flow of information (and other resources) through them, and discover what effects these relations and networks have on people and organizations. They treat the description of relational patterns as interesting in its own right e.g., is there a core and periphery? Examine how involvement in such *social networks* helps to explain the behavior and attitudes of network members, e.g., do peripheral people send more email and do they feel more involved? They use a variety of techniques to discover a network's densely knit clusters and to look for similar role relations. When social network analysts study two-person ties, they interpret their functioning in the light of the two persons' relations with other network members. This is a quite different approach than the standard Computer-

Media-Communication (CMC) assumption that relations can be studied as totally separate units of analysis. "To discover how A, who is in touch with B and C, is affected by the relation between B and C . . . demands the use of the social network concept" (Barnes, 1972, p. 3).

Social network analysis reflects a shift from the individualism common in the social sciences towards a structural analysis. This method suggests a redefinition of the fundamental units of analysis and the development of new analytic methods. The unit is [now] the relation, e.g., kinship relations among persons, communication links among officers of an organization, friendship structure within a small group. The interesting feature of a relation is its pattern: it has neither age, sex, religion, income, nor attributes; although these may be attributes of the individuals, among whom the relation exists: "A structuralist may ask whether and to what degree friendship is transitive" (Levine & Mullins, 1978, p. 17).

Social network analysts look beyond the specific attributes of individuals to consider relations and exchanges among social actors. Analysts ask about exchanges that create and sustain work and social relationships. The types of resources can be many and varied; they can be tangibles such as goods and services, or intangibles, such as influence or social support. In a Social network context, the resources are those that can be communicated to others via textual, graphical, animated, audio, or video-based media, for example sharing information (news or data), discussing work, giving emotional support, or providing companionship (Haythornthwait, et al., 1995).

Web 2.0

This latest iteration of the Internet has become known as "Web 2.0." The Internet started life as a peer-to-peer communication tool to exchange data among a number of users, allowing members of the scientific community to collaborate and share information easily. Today encourage users

to review services that they have experienced for the benefit of other users who are considering their own possible purchases. These peer reviews are regarded as far more trustworthy than traditional promotional materials that have been produced by the company itself. Many people in Greece now buy a product or service directly because of comments posted on a community by other consumers. Many Web 2.0 websites are based on the idea that the users would greatly increase their social capital simply because they would be able to know almost everyone on this planet within six steps of hops. Guare (1990) popularized the theory of six degrees of separation, having succeeded in the real world, even directly or indirectly motivated the invention of online societies, especially his states that any two random-selected people on this world can get to know each other by no more than six steps of intermediate friend chains.

Businesses cannot expect to keep control of their marketing message while at the same time allowing a transparent voice to their staff and customers through a blog or other online forum. Negative feedback cannot be hidden without exposing the business to charges of censorship, which can destroy its credibility. Companies after lengthy consultation with the authors of successful blogs, then can develop an appropriate blogging policy offer a simple means of overcoming the lack of human contact online and hence meet consumers' social activity needs. Companies identified peoples' vocal critics from postings on the company blog, and in response invited these people to visit the firm where the company demonstrated how the problems they raised had been addressed. Consequently, many of the critics were converted into active and enthusiastic supporters of the brand.

Microblogging

Another form of social networking activity to emerge recently is "microblogging" through sites such as Twitter, and Jaiku. Using these sites, people can communicate with their chosen network in real time, heavily abbreviated content. These messages can reach a wider audience when they are fed through to display in the author's blog or Facebook profile.

Social Technographics

A key issue that businesses had to bear in mind when considering the use of social networks for communicating with customers was the extent to which different segments of their customer base might be receptive to such approaches. Li (2007) has coined the term "social technographics" to describe the different ways in which consumers may behave online, which in turn governs how they will respond to approaches from companies via social networking channels. For example, a person defined as a "critic" is likely to comment on blog postings whereas a "spectator" is not, and someone categorised as "inactive" is unlikely to respond to any type of new media communication. Li highlights the importance for companies of understanding how their customers use new media, establishing what that means for how they should best communicate with them, and agreeing on how best to handle the possible loss of control associated with public display of negativity about the business on social networks. A business seeking to raise awareness of a new electronic product, for example, would be keen to recruit Twitter users to its cause because they are:

- Early adopters of new technologies;
- Well educated, with high profile careers and large salaries;
- Receptive to relevant advertising and are likely to talk about products within the Twitter community to "spread the word"; and
- Very influential within their own community and potentially able to develop the profile of the brand through its endorsement when interacting with their followers.

There are also issues with regard to the privacy of personal data and how it will be used. Networks such as Facebook offer privacy controls where members can specify the extent of the information they wish to receive or share with individual friends. However, it is not always obvious how to set the controls and information may be unwittingly shared with the wrong people. In addition, if social networks become too greedy in terms of the level of advertising that they permit, then users can very quickly shift their allegiance away from the network. Another warning note comes from recent research by the British Library (Manchester, 2008) which found that the skills and enthusiasm for Web 2.0 tools amongst the "Google generation" had been highly overrated, because while the respondents were prepared to use social networks for personal activities, they were sceptical about their relevance to the business world.

VIRTUAL COMMUNITY IN THE ECONOMIC TRADITIONAL SOCIETY

Community is a centre assembly in communal consideration having an affluent background of political, spiritual, intellectual, and well-liked communication. Until now regardless the acknowledgment of the significance of this assembly, no widespread classification of community can be retrieved (Wilson, 1990). In attempts to obtain a widespread determination regarding the expression "community" numerous analysts implicated in the creation of examinations for sociological determinations regarding the assembly. This attempt proposes that the term "community" characterizes a unification including persons, which allocate social communication, regular relationships among them and the rest members of the society, and live in a common location for a considerable period of time. Researchers describe the traditional kind of community even simpler referred to as a unification of persons linked with social ties, such as companionship or knowledge interchange. Granitz and Ward (1996) refer to the community like "a set of interwoven relationships built upon shared interests, which satisfies members' needs otherwise unattainable individually."

Even though many explanations arise for the community as a term, Granitz and Ward (1996) argues that community consists of three main parts. The primary part is regarded as a fundamental link that persons sense between them and the communal feeling like as differentiation from members, which do not belong inside their community. The second component of a community consists of mutual procedures and customs, which continue community traditions and at the same time inspire specific behavioural attitudes and principles. Continuing with the last constituent that is a characteristic of a community and appears to be a kind of ethical liability and compulsion targeted to the general community and also to the distinctive members of it. The meaning a community has as it could be summarized from the above three centre parts is a feeling that members have of belonging, a feeling that members matter to one another and to the group and a shared faith that members' needs will be met through their commitment to be together.

These main values allow communities to relay on sets of relationships and not on a batch of interactions. Relationships involve members to be strongly dedicated to their community while interactions engage essential, non-obligatory interaction. The above mainstay parts of community could not be limited by geographical borders or other boundaries. As Wilson (1990) comments, the development of public transport and public communication unified distributed persons from all over the world having a mutuality of reason and characteristics and therefore broaden the general perception of a community. The notion like the community therefore it is not limited from any geographical attendance of its affiliates instead it is fairly described as "a network of social relations marked by mutuality and emotional bonds" (Bender, 1978).

The above approach regarding the communities is in coordination with another kind of social communities, which make use of information technology and mass communications as a mean to create Internet communities.

Wilson (1990) argues that persons are capable of establishing important relations through the Internet where they share proximity, familiarity and connection like when having an in person discussion. Wilson (1990) also augments, "The distinction between 'virtual' community and 'real' community is unwarranted. The term virtual means something akin to 'unreal' and so the entailments of calling online communities 'virtual' include spreading and reinforcing a belief that what happens online is like a community, but isn't really a community."

The Internet affects business innovation by expanding reach and minimising the time lag to market. Not so long ago the goal of an online marketing campaign might have been to entice the consumer to click-through to a company's website, but now the objective is to create "sustained engagement" with the consumer. The growing popularity of websites such as YouTube and Facebook demonstrates how the Internet is changing; users are no longer simply downloading static data, but are increasingly uploading and sharing content among them, leading to a proliferation of social networks and other user-generated content sites. Li and Bernoff (2008) refer to this fundamental transfer of power from institutions to individuals and communities as "the groundswell." Online communities will play a key role in the future of economy because they replace customer annoyance with engagement, and control with collaboration for those firms brave enough to take the plunge. The businesses that prosper will be those who proactively embrace this new world, because they regard change as an opportunity rather than as a threat to be avoided.

VIRTUAL COMMUNITY IN THE SOCIAL TRADITIONAL SOCIETY

Nowadays communication in its present form can be described as the "third period of digital communication." The reasons for this phenomenon created a series of political, economic and social developments which more or less variant, is acceptable by many countries. Computer networks have introduced and added various new dimensions to life and society, especially in the realm of information. Communication are fundamental to any form of organizing but is pre-eminent in virtual organizations. Virtual organizations are characterized by (a) highly dynamic processes, (b) contractual relationships among entities, (c) edgeless, permeable boundaries, and (d) reconfigurable structures. Relative to more traditional settings, communication processes that occur in virtual contexts are expected to be rapid, customized, temporary, greater in volume, more formal, and more relationship-based. The major aspects of virtual organization design are: (1) communication volume and efficiency, (2) message understanding, (3) virtual tasks, (4) lateral communication, (5) norms of technology use, and (6) evolutionary effects (DeSanctis & Monge, 1998).

Virtual community being formed in information space is a crucial topic for sociological study. However, the term might be a kind of metaphor of "real community," which sociologists have long studied. If so, it is necessary to clarify its nature and difference from that of real world as a theoretical matter in sociology, and to develop appropriate ways of analyzing those phenomena as a new reality (Araki, 1998).

While Aycock and Alan (1995) has argued that computing via the Internet offers a vision of freedom and a shared humanity, others have claimed with equal vehemence that it may become the instrument of global surveillance and personal alienation.

Society and information technology are rapidly co-evolving, and often-in surprising ways.

In this installment of "trends and controversies," we hear three different views on how society and networked information technology are changing one another. Becoming socialized means learning what kinds of behavior are appropriate in a given social situation. The increasing trend of digitizing and storing our social and intellectual interactions opens the door to new ways of gathering and synthesizing information that was previously disconnected. Barry Wellman-a sociologist and an expert in social network theory explains how the structure of social networks affects the ways we live and work. He describes the move away from a hierarchical society into a society in which boundaries are more permeable and pole is members of many loosely knit groups. He introduces the notion of glocalization: simultaneously being intensely global and intensely local. Wellman describes how computer-mediated communication is contributing to this glocalization transition in social habits and infrastructure.

VIRTUAL COMMUNITY IN THE POLITICAL TRADITIONAL SOCIETY

People discuss politics with friends, families, at work, etc. With the rise of computer-mediated communication, it seems logical that people would create new online groups with whom to discuss politics (Gregson, 1998). In the past, political power was drawn alternatively or in combination from the strength of leaders and institutions, the will of the people and/or the support of the country. Today, those pillars of power are being shaken by tectonic shifts that are transforming the very nature of global society. Nations are facing new rivals for power and influence on the global stage. Power itself is being redistributed, taking new forms and new characteristics. The rules of the game in international relations are changing and the origins of an extraordinary number of those changes can be traced to the information revolution. That revolution has begun and its full extent and implications are unclear. However, for the nations, their ability will depend on the ability to recognize the changes transforming the nature of power in the new world environment and adapt to them (Rothkopf, 1998).

In light of recent discussions of the Internet touting "virtual community" and a capacity to enhance citizen power in democracies, a more rigorous understanding of community, suggest that relationships forged with the aid of electronic technology may do more to foster "categorical identities" than they do dense, multiplex, and systematic networks of relationships and an emphasis on community needs to be complemented by more direct attention to the social bases of discursive publics that engage people across lines of basic difference in collective identities. Previous protest movements have shown that communications media have an ambiguous mix of effects. They facilitate popular mobilization, but they also make it easy for relatively ephemeral protest activity to outstrip organizational roots. Further, they encourage governments to avoid concentrating their power in specific spatial locations and thus make revolution in some ways more difficult (Calhoun & Craig, 1998).

Assesses the utility of Computer-Mediated Communication (CMC) and related communications technologies in helping to create democratization. Electronic democratization (i.e., the enhancement of democracy through new communications technologies) increases the political power of those who have been generally silenced within traditional forms of government (Hacker, et al., 1996).

PATH TO ENHANCE SOCIAL NETWORKING

Shape an Online Character Opinion

Social media could be considered as a way of creating online character opinion. Many social

professional networking gears are accessible which allow working together among persons and societies to provide committed customer assistance and generate time enduring relations. Individuals have the option to submit inquiries, write comments, propose ideas or describe a trouble.

Select the Specified Forum Cautiously

Social networks which originally started for individuals have begun now a transition over the professional environment, although especially designed networks are responsible to assist companies expand its online presence and increase their sales volume inside and outside the Internet. These consist of a absolute variety of professional equipment to assist networking, promotion and interaction with consumers, vendors and associates.

Utilization of Video

Video could be considered as a marketing weapon for professionals in the Internet and it has an enormous increase of the Internet video marketplace. A study carried out on the account of BT Tradespace furthermore proves that on a regular pace, persons consume a significant increased time on Internet sites including video content. In addition to enhanced visibility, video is likely to assist prospective consumers to acquire a distinctive viewpoint about the companies and be noticeable from the mass.

Search Engine Optimisation

To productively participate on the Internet, attaining attention from future consumers is the primary move. Customers are progressively more searching to obtain an item online and a large proportion of them make use of search engines to identify new businesses and Internet sites. Social media assist in the determination of search engine position, by including video, frequently updates of content and making use a number of keywords it is more probable all of these to enchase any business' Internet existence. According to Digital Influence Group there has been developed a new innovative method for launching a marketing campaign within social networks, this implicates three different phases observation, engage and measure. In the first phase, there is identification, mapping and analyzing the environment, attitude, and importance of digital discussions related to business objectives. Consequently, this information is used to establish social media strategies. Continuing to the second phase the information collected is now used for the implementation of the marketing campaign. Throughout this phase social media interaction is being developed, production and sharing of content and involvement in online conversations. Measurement is the third phase where programming, combines the outcomes of the influence campaign with the objectives underlined at the beginning of the client interference. In Figure 1,

the main behaviours and actions to be taken on social networks by marketers in order to achieve desirable outcomes are shown (Digital Influence Group, 2010).

As Ivan Croxford general manager of BT states that "the online world has changed, and consumer behaviour has shifted too. Previously it was enough to have a website, telling your story to the world. Now consumers and customers are part of that story, wanting to engage with you online and shaping your reputation via forums, blogs and posting comments on social networks" (IT Now magazine, The British Computer Society, 2010).

CONCLUSION

The results refer to the investigation of a global society and the changes affected to the individual societies.

Figure 1. Entry methods in social networks

	Media Influence: Outputs	Audience Influence: Outtakes	Business Influence: Outcomes
Goals	Reach	Engagement and Influence	Action and Insight
Metrics	• Visits / views • Unique visitors • Pages viewed • Volume of reviews/comments • Navigation paths • Links • Files embedded	• Sentiment of reviews/comments • Brand affinity • Commenter authority/influence • Time spent • Diggs, votes • Favorites / Friends / Fans • Viral forwards • Number of downloads • Opinions expressed • Membership	•Sales inquiries •New business •Customer satisfaction / loyalty •Marketing efficiency •Risk reduction
How Compiled	Free tools: Google, Technorati Social media platform metrics Web analytics	Social media platform metrics Social media analysis tools	Surveys Market Mix Modeling

The influence of virtual forums on the on-line society, along with their legal, social and ethical implications, become an issue of special attention and study and acquire central importance in wider issues of Web science.

This signals the identification and addressing of common problems with the upcoming new generation of digital users of the Internet. Sociologists and psychologists will be better able to understand the power exercised by virtual communities on real communities, and how this happens in EU countries, USA, and Canada.

The effect on less developed countries can be very powerful if there is no resistance that will enable them to face the coming changes in the economic, social, and political environment.

The study shows the new trends that should be taken into account by social scientists, managers and the leaders of the new society.

On balance, we believe that online communities will play a key role in the future because they replace people annoyance with engagement in the social and political circumstances. Also, replace customer annoyance with engagement and control with collaboration for those firms brave enough to take the plunge.

IMPLICATION POSSIBILITIES

The formation of the new social structure as a multi-cultural society, influencing the millennium generation of society (young people age 16 to 25 years old) more by the electronic media and less from traditional means e.g. school, family, church etc. We have to face a new global community, which personified by the virtual community rather than the real community (family, friends). The new writing of "grikelands" (spelling words with Latin characters), the new form of sending sms messages from young people with dotted words is a new form of communication through mobile phones. A method, which is not taught in any school in any family, gaining more and more

followers. All these "events" are part of the new "society 2.0" which receives influences from a virtual community through blogs, forums, etc. The results of our research at the international level will present the latest trends and it should take into account by the sociologists and the opinion leaders of the new society.

The organizations can take into consideration the reactions of the consumers to the new interactive eRA and take their decisions no longer on an one-way communication policy of promoting products through the use of traditional media (TV, Printed press, radio, posters, etc.) but taking seriously the anonymous consumers who transfer their experience through Blogs, Forums, Facebook, etc. In this digital environment each consumer, is able in a free manner to transfer his view to the unnamed "friends" influencing them positively or negatively by the experience of using products or services. For this reason, the most advanced companies have created their own forums and try to predict and evaluate the most of the reactions of their consumers. Political parties will reflect on the traditional approach followed their supporters through political merger, posters, political advertisements, etc., and to create an interactive strategy of political marketing in the age of "Politics 2.0." The present study can help governments to implement new communication policies with citizens and coalitions for a communication strategy with EU citizens.

REFERENCES

Araki, I. (1998). *An approach to the information space and on-the-line interaction*. New York, NY: International Sociological Association (ISA).

Aycock, A. (1995). Technologies of the self: Foucault and internet discourse. *Journal of Computer-Mediated Communication, 1*(2).

Barnes, S., & Greller, L. M. (1994). Computer-mediated communication in the organization. *Communication Education, 43*(2), 129–142. doi:10.1080/03634529409378970

Bickart, B., & Schindler, R. M. (2001). Internet forums as influential sources of consumer information. *Journal of Interactive Marketing, 15*(3). doi:10.1002/dir.1014

Birney, R., & Barry, M. (2006). Blogs: Enhancing the learning experience for technology students. In *Proceedings of ED-MEDIA 2006,* (p. 1042). ED-MEDIA.

Blumber, J., & Kavanagh, D. (1999). The third age of political communication: Influences and features. *Political Communication, 16,* 209–230. doi:10.1080/105846099198596

Calhoun, C. (1998). Community without propinquity revisited: Communications technology and the transformation of the urban public sphere. *Sociological Inquiry, 68*(3), 373–397. doi:10.1111/j.1475-682X.1998.tb00474.x

DeSanctis, G., & Monge, P. (1998). Communication processes for virtual organizations. *Journal of Computer-Mediated Communication, 3*(4).

Creswell, J. (1998). *Qualitative inquiry and research design: Choosing among five traditions.* London, UK: Sage.

Fontana, A., & Frey, J. (2000). The interview. In Denzin, N., & Lincoln, Y. (Eds.), *Handbook of Qualitative Research* (2nd ed., pp. 645–672). London, UK: Sage.

Goad, R., & Mooney, A. (2008). *The impact of social networking in the UK.* London, UK: Hitwise/Experian.

Granitz, N. A., & Ward, J. C. (1996). Virtual community: A sociocognitive analysis. *Advances in Consumer Research. Association for Consumer Research (U. S.), 23.*

Gregson, K. (1998). Conversation and community or sequential monologues: An analysis of politically oriented newsgroups. In *Proceedings of the ASIS Annual Meeting,* (Vol. 35, pp. 531-541). ASIS.

Guare. (1990). *Six degree of separation.* New York, NY: Random House.

Hacker, K. L., & Todino, M. A. (1996). Virtual democracy at the Clinton white house: An experiment in electronic democratisation. *The Public, 3*(1), 71–86.

Haythornthwaite, C., Wellman, B., & Mantei, M. (1995). Work relationships and media use: A social network analysis. *Group Decision and Negotiation, 4*(3), 193–211. doi:10.1007/BF01384688

Kaid, L. L., & Holtz-Bacha, C. (Eds.). (1995). *Political advertising in western democracies.* Thousand Oaks, CA: Sage Publications.

Li, C., & Bernoff, J. (2008). *Groundswell: Winning in a world transformed by social technologies.* Boston, MA: Harvard Business School Press.

McCracken, G. (1988). *The long interview.* London, UK: Sage.

Patton, M. Q. (1990). *Qualitative evaluation and research design.* Newbury Park, CA: Sage.

Prensky, M. (2001). Digital natives, digital immigrants. *On the Horizon, 9*(5).

Rothkopf, D. (1998). Cyberpolitik: The changing nature of power in the information age. *Journal of International Affairs, 51*(2), 325–359.

Swanson, D., & Mancini, P. (1996). Politics, media and modern democracy – An international study of innovations in electoral campaigning and their consequences. London, UK: Westport.

Wellman, B. (1997). The road to utopia and dystopia on the information highway. *Contemporary Sociology, 26*(4), 445–449. doi:10.2307/2655085

Wellman, B., & Gulia, M. (1995). When social networks meet computer networks: The policy implications of virtual communities. New York, NY: American Sociological Association (ASA).

Wilson, E. J., & Sherrell, D. L. (1993). Sources effects in communication and persuasion research: A meta-analysis of effect size. *Journal of the Academy of Marketing Science, 21.*

Chapter 17
Website Performance Measurement:
Process and Product Metrics

Izzat Alsmadi
Yarmouk University, Jordan

ABSTRACT

Some tasks will be easier to implement and test, and others will either be un-applicable or difficult to test and implement in comparison with testing in traditional software development environments. For engineering management, product and process quality evaluation are important assessment tools by which managers can have significant indicators of the evaluated project or product. There are many ways and characteristics by which websites can be evaluated. Quality attributes can be external or internal. They can be measured based on the developed product (i.e. the website) or the developing process. In this chapter, the author describes in detail some of the product and process metrics by which websites can be evaluated. They are described based on the major classification: process and product metrics. In each one of those two major classes, the author describes possible measurements, how they can be evaluated, and examples of attributes and tools used in this measurement. Values of measurements can in combination provide useful information for project management and planning. Focusing on only one or two attributes can possibly be insufficient or misleading.

INTRODUCTION

In engineering projects in general and software projects in particular, project manages need to have the right information at the right time to be able to make proper decisions. In software projects,

project management tasks include: cost estimation, staff selection and allocation, tasks scheduling, quality assessment, budget distribution and allocation, risk analysis and assessment, etc. In this chapter, the focus is on the goals, approaches, tools, and deliverables of the quality assessment and evaluation task.

DOI: 10.4018/978-1-4666-3658-3.ch017

Initially, the word "quality" is large, vague, generic, and complex by nature. No single or even several parameters can be enough to assess the overall quality of a product. Quality can have several types, levels, and attributes. Some of those quality elements can be directly assessed while many others can be unstructured or subjective to the person who evaluates, or to the nature and the environment of the project and many other environmental factors that can affect the value of such quality attribute. In some cases, complex formulas are built to list all or most of the attributes that can impact the product quality along with the level of impact or effect such attribute can have on the product overall quality.

Product quality attributes can be also external visible to users or internal that need a white box access to the product (e.g. software code) to be able to measure it.

Software metrics are units of measurements that describe one or more attributes of the software. An attribute is a property or a characteristic that the software have. For example, size is a software attribute that gives an indication on how large is a code project (i.e. whether it is small, medium, or large). Lines of Code (LOC) metric is a software metric that is used as an indicator or one of the metrics for this size. Metrics in this sense work as units of measure where you can have several different metrics or units of measure as indicators for size where while they can be different in numbers but generally they should have high positive correlation.

Another related term in the software metrics field is "measurement." While some references do not show differences between the terms: "metrics" and "measurements," other references distinguish the "metrics" from "measurements" where metrics represent more complex formulas relative to measurements that may include one or more of the software attributes.

Attributes can be further divided into external and internal attributes. While literature also may include different definitions of those terminologies, however, generally internal attributes are the actual characteristics that a software or website have and in which, a numerical value can be collected or calculated from the software or website for that attribute. On the other hand, external attributes are high-level attributes that can be measured indirectly from one or more internal attributes. In other words, we can say that "external attributes are the goals or what we want to know or measure, while internal attributes is what we can directly measure). For example, website usability is a popular website external attribute. This attribute cannot be measured directly (i.e. we cannot say for example that for website A, usability $= 5$, etc., i.e. a numerical value). On the other hand, several website internal attributes (e.g. time to learn, number of help features, documents, etc.) can be measured from a website, which collectively can be used to assess the website usability.

Website metrics can be collected manually or automatically through tools. Many tools are developed to collect metrics and attributes automatically. On the other hand, due to their subjective natures, some attributes and metrics require the help of surveyed users or testers to give their "personal" opinion on those attributes.

In the domain of websites, we conducted an extensive survey to find all used and described attributes and metrics for websites. The experiment showed that there is an extensive mix in research documents and published articles between websites attributes and metrics. In the next section, we will provide a list of several described attributes and metrics. Later on, we will classify them according to our proposed conceptual model.

In this chapter, the focus will be on major project quality characteristics. Those are: process and product quality metrics. Focusing on websites as the subject products, we will describe several quality attributes for both product and process.

Product Attributes

1. **Websites usability:** Usability is one of the popular software attributes that is extensively investigated in literature. Through usability, we measure how much it is easy for users to use a website and its information or services. There are many internal attributes that can be measured as part of evaluating usability (which is an external attribute). This may include:

   a. Success rate (i.e. whether users can perform the intended tasks or not).

   b. The time each task in the website requires to accomplish. In some cases, a similar metric called "ease of use" is evaluated by users. It can be also measured subjectively through users' response or satisfaction.

   c. The time it takes for users to know, get used, and complete tasks with the website features and services (i.e. training time).

   d. The error average or rate for typical or average users. This can be also measured based on a ratio between successful to failure tasks executed by the users.

   e. User satisfaction.

   f. Number of features or commands used from the website by users (also called usefulness).

   g. The number of available help files, documents and any other features that can help users perform tasks easier and faster.

Other attributes, which are parts or related to usability include: efficiency, effectiveness, user satisfaction, learnability, and memorability. Efficiency includes all previous internal attributes related to the ability of users to perform tasks quickly such as: Time to complete a task, time to learn, time spent on errors, percent or number of errors, frequency of help or documentation use, and number of repetition or failed commands. Effectiveness related attributes include: Percent of tasks completed, ratio of successes to failures, workload, and number of features or commands used. User satisfaction related attributes include: Rating scale for usefulness of the product or service, rating scale for satisfaction with functions and features, number of times user expresses frustration or anger, rating scale for user versus technological control of task, and perception that the technology supports tasks as needed by the user. Learnability indicates how much it is easy for users to accomplish basic tasks the first time they encounter the website features. Memorability shows how much a website help a user remember its features (i.e. from previous visits). In another aspect, memorability can mean the ability of the website to remember the user and accelerate their ability to perform tasks.

2. **Performance metrics:** Performance is usually used to indicate attributes related to correctness and speed. In this scope, we will focus on attributes related to the speed of transactions processing. This means that some of those metrics can be related to usability. Other papers use: the number of people who visit the website to be part of performance metrics. Other classifications consider it under traffic metrics. If by performance we mean "the ability of the website to perform its services correctly and quickly," then such attribute may not be, directly related to traffic metrics such as visits. Some websites gets popular for the type of services they provide for users. Of course if such services are provided inaccurately or in a slow rhythm, this can affect traffic metrics. The only performance metrics we will consider as performance are those related to the speed of downloading pages, images, the response time it takes when a user trigger a transaction and waits

for the website response. Such performance attributes may be time dependent, which means they can be different depending on the time of the day, or at certain days in the year. They can also depend on the number of users currently requesting services from that website. Interactivity and responsiveness are other related metrics that shows the speed and the amount of time it takes for a user to interact with the website services through ping-pong dialogues.

3. **Traffic and usage metrics:** Those metrics are very popular and the subject of many papers, tools, or evaluations. One of the main success metrics for a website is its popularity or the number of people visiting that website. There are several attributes that can be put under the umbrella of website traffic, popularity, or usage metrics. Those include: Web usage, Web and pages' visits, website and Web pages' rank, visitors (unique, repeated, first time visitors), most viewed pages, most requested pages, single access pages, user average time or stay on website or pages, page refreshes, page views per visitor, and top Directories.

Besides the previous attributes, there are some sub attributes related to traffic metrics. For example, stickiness is a traffic metrics that is used to see if a website is interesting to visitors by calculating the ratio of unique visitors to page views. The higher the ratio of unique visitors to page views, the stickier the website is.

Another metric is relevant track, which shows how long visitors stay on a website. The longer they stay, the more track relevant is the website.

There are some metrics that can be put either in customer, or finance metrics or under traffic metrics as they combine aspects from both sides. Examples of those attributes or metrics include: Cost per visitor, cost per lead, cost per customer, bounce rate, revenue per visit, page attrition, path weight and proxy scoring.

4. **Business or financial metrics:** As mentioned earlier, there are many metrics that are common with traffic metrics. Other metrics includes:
   a. **Operational efficiency:** Measuring and quantifying Web Content Management (WCM) benefits.
   b. **Revenue optimization:** Connecting WCM with revenue.
   c. Time to market

5. **Call centre metrics:** Those are usually considered part of performance metrics. Example of such metrics include: cost per case, first call resolution, call volume, individual agent performance, agent productivity, and customers' waiting time.

   They can be classified generally under several sub categories such as: productivity, quality and customer satisfaction metrics. For example, an inbound productivity metric calculate the number of customer calls handled successfully by customer service along with the time required for the problem handling. Other metrics count the number of sale calls and the amount of successful transactions from those initial sales. Quality related attributes in call centre track errors and problems submitted by users and the type of response from the website or employees. Usually, several dashboard gadgets are implemented based on collected historical data that can be used as Key Performance Indicators (KPIs).

6. **Customer centered metrics:** Those metrics are not directly related to the website and its related attributes. Those are related to the services those websites may offer. Examples of such metrics include: reach, product profitability, brand equity, acquisition of services (e.g. cost of acquisition of new service), service penetration, market share, conversion and retention based on customer communication and interaction with the website and its pages and services,

abandonment, attrition and churn describe migration of users (Schonberg, et al., 2000).

7. **Process quality metrics:** Those are related to the activities and processes that occur in those websites. This may include attributes related to memory and maximum memory usage in website activities. It may also include the amount of time required to accomplish certain tasks in the website. Defect and defect tracking metrics can be also categorized under process quality metrics (e.g. defect arrival and defect fix rates and trends).

8. **Structural metrics:** Those metrics can be gathered from the website components. Web site structural attributes include description of numerical values for website structural elements such as number of links (i.e. internal links), Web pages, forms, frames, text boxes, etc. There are some metrics that can be categorized under Web structural metrics. This include: centrality, or compactness, consistency, which are metrics all related to measuring how much a Web site components are consistence with each other and with Web standard design.

9. **Web content metrics:** Those evaluate the format of Web pages and their components and their consistency with Web design standards. Web content metrics track the evolution of the website such as the number and the rates of Web components added weekly, monthly, etc to the website. It also includes information about the website physical files, database, and the type of software applications used in the website.

10. **Complexity metrics (i.e. those that are related to either maintainability or effort) metrics:** In terms of website complexity, there are many previous attributes such as those in structural metrics are used also in complexity metrics. This means that the majority of those attributes are reused from other metrics for different types of metrics.

Examples of complexity and maintenance related metrics include: testability, error proneness, reliability, fault tolerance, etc. Complexity metrics takes into consideration those components that may cause overhead in maintenance or may slow Web page loading. The metric of counting the number of dynamic pages relative to static pages is also another attribute to consider which may cause overhead on website performance and maintainability.

11. **Quality metrics:** There are several metrics and attributes that are related to quality. In this scope, we will consider only those specifically related to errors in pages, failed logins, errors, crashes, etc.

12. **Marketing metrics:** In many references, some of those metrics are mixed with metrics from popularity or traffic metrics. However, many references listed metrics related purely for marketing. This includes:
    a. Brand impact (i.e., increased product or brand awareness, intent or favourability)
    b. Number of impressions
    c. Position of organic and paid listing
    d. Click through and number of clicks by users
    e. Number of pages indexed
    f. Number of overall inbound links
    g. Authoritative citations/links
    h. Referring traffic sources
    i. Cost per lead
    j. Customer acquisition cost
    k. Customer life cycle
    l. Referring search engines
    m. Top keyword referrals
    n. Top keywords by revenue
    o. And many more items found in good analytics packages
    p. Churn rate
    q. Website grade (i.e. through grader tools, e.g. websitegrader.com)
    r. Ratio of new to returning visitors

s.   Amount of increased website traffic

t.   Duration of website visits (i.e. new, returning, etc.)

u.   Amount of increased traffic to physical store

v.   Amount of increased volume to call centre

w.   Number of leads generated for products sold online

x.   Number of leads generated for products sold offline

y.   Number of immediate sales generated for products sold online

13. **Navigability metrics:** Navigability metrics may also interact with structural and popularity metrics. For example, there are two popular metrics usually collected from websites. Those are inlinks and outlinks. Inlinks include the number of links pointing to a website, while outlinks represent links pointing from the website outward, While outlinks metric is usually considered as navigability metric, in-link is usually considered as popularity metric.

Other types of attributes and metrics in the navigability section include: Entry Pages, Entry Paths, Exit Pages, Clicks to Pages, inlinks, outlinks, broken links, Visitors-per-Page, Pages-per-Visitor, Average-Duration-of-visitor-session, Accessibility and environmental metrics, special needs, wider audience, better optimized for search engines (SEO), platform independent, and help options alternatives.

We should distinguish between pages helping the user to find information and navigating the Web site, particularly navigation pages, as opposed to content pages, considered to be sources of information, organization, arrangement, layout, and sequencing.

14. **Accessibility metrics:** Those metrics evaluate the website ability to deal with different types of users, environments. It also measures the website ability to give users alternative options (e.g. short cuts, alt links, images, etc.). Some accessibility attributes and metrics are mixed with usability metrics and attributes. Web Accessibility metrics aim at measuring in a precise way the conformance of a given Web page with respect to a set of accessibility guidelines. Metrics rely on traditional accessibility guidelines such as WCAG or Section 508, targeting all type of users (www.w3.org/TR/WCAG10). Examples of some of the metrics they have:

a.   Provide equivalent alternatives to auditory and visual content.

b.   Design for device independence.

c.   Provide clear navigation mechanism.

15. **Web security metrics:** Recently, is getting more and more focus. Security metrics for websites are based on calculating the number of weakness (pr vulnerabilities) in that website. Those weaknesses can be and abused by several types of attacks such as: SQL injection, session hijacking, Cross Site Scripting (XSS), Cross Site Request Forgery (CSRF, XSRF), Remote File Include (RFI), Denial and Distributed Denial Of Service (DOS, DDOS), IP, content, packet and form spoofing, phishing, code injection, broken access controls, information leakage, insufficient authentication and authorization, etc. Several metrics are proposed based on previously listed threats. This include:

a.   Exposure or discoverability, e.g. number of weaknesses, or number of days a website is exposed per month or year.

b.   Exploitability or threats.

c.   Impact severity.

d.   Annual expected loss or exposure rate

   i.   Process Attributes

Before describing quality aspects in websites project development, we have to describe major activities that occur in the website design project. The typical goal of Webpage design is similar to that of a software design, which is to design a website that will be easy to learn and use, be attractive to visit and use, enable its users to efficiently perform their tasks, and be easy to maintain.

Some designers divide the website design process based on the major elements of a website design which include:

1. Accessibility design
2. Colour design.
3. Image design
4. Navigation design
5. Database or storage design
6. Text design
7. Sound design
8. Look and feed design
9. Content design
10. Behaviour design

For each one of those listed above there are several quality aspects upon which such element can be evaluated in terms of quality. The major difference between those defined here and those that were defined in the product metrics is that those that were defined in the product metrics were defined at the level of the "product as a whole." For example, performance and security are integrated cumulative aspects. Having some parts of the website secure while the rest are not is pointless. It takes only one weak point for a hacker to bring down a website. Same thing can be stated for performance, reliability, etc. On the other hand, aspects of the process metrics mentioned earlier can be measured individually or on the page by page basis rather than for the whole website.

Webpage design process can be also divided based on the development team: User experience team, graphic and animation design team, database design team, system design, design production and documentation teams.

Websites are usually evolutionary by nature. This is why the project of developing a website (especially active websites) can continue even after the launching or use of the website. As such, design quality attributes are important to facilitate the ability to easily: update, improve, maintain, etc all pages of the website. We will describe below some of the characteristics for a good design of a website:

1. **Maintainability:** A good website should be easy to maintain. This means that the website should not be fragile for changes. Changing a small part of the website should cause a limited impact. Coupling is an aspect that refers to the level of connectivity between the different modules components of the website. It refers to the external links between those components where it should be minimal. High coupling may cause ripple effects upon small changes. As mentioned earlier, as websites are evolutionary by nature, maintainability is a very important characteristic of a good website design that measures its ability to accept changes over time. Several other good design characteristics such as complexity, modularity, etc can be directly related to maintainability.

2. **Usability and reusability:** Usability is a design quality to measure the website ability to be useful for its intended audience and customers. In the Internet world, usability is very important where the value and popularity of the website is a direct indicator to the number of users that they find the subject website useful. On the other hand, reusability is related to the development and the maintainability aspects. Reusability is important for website design managements where websites designer companies would like to minimize the effort of building future websites through reusing some components from previous website design projects.

3. **Performance:** We mentioned those three or four good Web design characteristics in the process metrics section although they measure the product since those attributes

evaluate indirectly the goodness of the Web design process. This is to say that a good Web design development process should produce a website that has those good design characteristics. Nonetheless, there are some metrics that can measure the quality of the development process itself. For example, defects/page or per lines of code is a process metric that measures the percentage of defects in the development process. Other examples of process metrics include effectiveness of defect removal, response time of the fix process, adherence to schedule, accuracy of estimation, etc. Based on the major processes in Web or software design, process metrics can be divided into: process quality metrics, process productivity metrics, process error removal effectiveness and process estimation and scheduling effectiveness.

Productivity is a process indicator to show the amount or level of progress in completing project tasks. However, in many cases, productivity is seen as a rival to design quality attributes where if for a particular designer or developer, productivity is significantly higher than all other team members, it is feared that such high productivity value comes at the cost of quality.

LITERATURE SURVEY

Barnes et al. (2003) uses the WebQual tool to evaluate the quality survey of a cross-national e-government Web site provided by the Organization for Economic Cooperation and Development (OECD), before and after a main redesign process. The researchers adopted the WebQual tool. This research study is conducted on the Web site of Forum for Strategic Management Knowledge Exchange (FSMKE). This tool helps its user to assess Web site usability, information quality, and service interaction quality, in other words it assesses user perceptions of the quality of e-commerce and e-government Web sites. The WebQual tool helps to show how the Web site under consideration has been improved, and identify Web pages that need more improvements.

Barnes et al. (2003) is dedicated towards a quality survey of a cross-national e-government Web site provided by the Organization for Economic Cooperation and Development (OECD). The quality of a specific cross-national Web site under consideration is examined twice, before and after a major redesign process using WebQual tool. The metrics of examination before and after a major redesign of this Web site reveals the strengths, weaknesses, besides the different impressions of users in different countries. Cross-national e-Government Web site offerings will benefit from the findings of this study.

An interesting study by Foglia et al. (2008) attempts to explore the relation between Galvanic Skin Response (GSR) signal sensing as a measure for emotions and traditional usability metrics. Foglia et al. (2008) presented main features of physiological signals, and the way they must be treated in order to get beneficial results. This research is based on a prototype e-government website as a test case. Assessment of the feasibility of relating the Animated Face (AF) effects with usability metrics, to GSR signal variations is conducted. The results of the conducted experiments proved that AF has positive effects on users' performances.

Alsoud & Nakata (2010) examines the accessibility, usability, transparency, and responsiveness to the needs of Jordanian citizens of thirty governmental websites, where the researchers conclude that these Web sites are not mature enough to provide fully their users with all their needs supposed to be provided by their counterpart in developed countries. The design of these thirty Web sites of the Jordanian e-government lacks consistency and similar standards and features.

The identification of Web based measurements to evaluate the reliability and maintainability of hypermedia applications is presented in Dhawan and Kumar (2008), in order to create efficient Web applications. Their study tried to accomplish three goals: first it compares the relative importance of different Web-based metrics and methods, second, constructing User Behaviour Model Graphs (UBMGs) to assess the quality of Web design, and finally, Web page replacement algorithms is used to increase the Web site usability index, maintainability, ranking, and reliability.

Quality of different Web sites is essential to the success of these Web sites to accomplish their mission, therefore Jati and Dominic (2009) exhibits a number of criteria (indeed a usability perspective, a user perspective, a content perspective and quality of service perspective) to test the quality of E-Governmental Web sites in five Asian countries. The conducted tests revealed that these Web sites neglect the quality and performance criteria.

Usability of different Web-applications is essential to the success of different Web sites; therefore Tripathi et al. (2010) proposed a number of usability metrics to measure the usability of Web applications for academic establishments, without conducting any tests to evaluate these metrics.

There are many studies that evaluate Web accessibility and concerned with handicapped Web users, and one of these studies is Huang (2003), which presents an evaluation of thirty five official homepages of Taiwanese E-government Web sites. The evaluation process is based on Web Content Accessibility (WCA) guidelines. Huang (2003) concludes with a number of recommendations to improve the Web sites under consideration.

Alagappan et al. (2009) aims to present new Web metrics based on usability and effectiveness for different Web domain users. Their study attempts to establish a matrix model to determine the total number of different domains visitors to get the most recent information. The study concludes that the total number of Web page visitors is significantly affected by utility and quality oriented metrics.

E-Government websites and their usage in developed countries are explored by Nariman (2010), where you can find many of these with huge-contents. Huge content Web sites in developed countries lacks information management skills and human expertise. Few studies explored the services provided by huge-content Web sites, where Nariman (2010) study explores concepts and models by examining the characteristics and motives of their users in browsing these sites to get the information and services they search for.

Wang et al. (2005) exhibits a general theory for evaluation of Web based application, with a test proving the validity of this theory. There are many studies that evaluate Web sites features to enhance its usability, where as there are few studies to evaluate e-government Web sites services. Their study aims to cover both aspects of Web sites evaluation (usability and services), by providing government agencies with a model to evaluate Web-based e-government services and understand the reasons behind the success or failure of the Web site to help their user to find the information they are searching for. This model is beneficial to be used by other governmental agencies to improve their Web based e-government services.

Signore (2005) defined Web site quality measurement criteria and exhibited in his study a quality model and a set of characteristics to relate external to internal quality factors and giving an indication to the potential problems of Web sites. The proposed model cover automated process for the quality evaluation of Web pages and their components. The quality assessment process is performed automatically by checking the source code, which followed by manual evaluation of a number of users.

In their study, Jinling and Guoping (2005) present a novel mathematical model for e-commerce websites comprehensive evaluation, which is based on concordance analysis approach. In addition, a case study is used to test the proposed model. The suggested model is characterized by presenting satisfaction and dissatisfaction, objective and credible results, simplicity and use easiness, Easy to be programmed and more informative than its counterparts.

In order to enhance different Web sites in terms of their contents and structure, it is essential to analyze the evolution of these sites beside the behaviour of its users. It is possible to use to data mining techniques to get the essential statistics and metrics to understand the structure of the Web site and how much it is beneficial to its users, but it is still difficult to understand and interpret. Therefore Pascual and Dursteler (2007) introduce a preliminary prototype of a customizable exploratory search system for Web Usage Mining (WUM) and Web Structure Mining (WSM), in order to provide end users with a set of tools and visual representations allowing them to explore and to decide how to represent the available data. This system will help its users to evaluate the usability of the implemented interactions, in other words these tools guide its users to the best way to visualize Web mining data for future improvements on the system. In addition, these tools show the strong and weak points of the visual metaphors.

A number of design metrics for navigational models have been proposed, discussed and validated using a formal framework (DISTANCE) to define and theoretically validate those metrics by Abrahão et al. (2003) to analyze the quality of Web applications in terms of size and structural complexity. These metrics are valid for Object-Oriented Web Solutions (OOWS). A controlled experiment is conducted to check the validity of using those metrics as early maintainability indicators.

One of the main concepts within information foraging theory is Information scent. Information scent is used to measure the way users interact with information systems in general, but it is difficult to measure Information sent directly. A research study by Saward et al. (2004) aim to examine the usability of Internet information retrieval systems (IIS) which includes e-commerce Web sites, information portals and corporate intranets, therefore an Online shopping application to examine the relationship between information scent and usability. Their study proposes two methods to measure user perceptions of Information scent.

Zhang et al. (2004) introduces five metrics to measure website's navigability which is based on measuring website structural complexity. Navigability metrics merits are summarized by objectiveness and the possibility of using automated tools to assess Web sites regardless of their size. Their study shows that structural complexity plays a crucial role in Web navigability, therefore structural complexity metrics can be used to measure Web navigability indirectly. Their paper refers to the limitations of metrics approach presented in their paper, such as taxonomy within a Web page, and layout of links.

An assessment model for Web-based systems is proposed by Khan (2008), where this model is dedicated to non-functional properties of the Web-based systems such as reliability, usability, security, efficiency ...etc. The proposed model consists of two main parts, in the first part the quality metrics is derived using Goal-Question-Metric (GQM), while in the second part an evaluation of the quality metrics to rank a Web based system using Multi-Element Component Comparison Analysis technique is conducted, where the evaluation of these metrics is assessed against the quality requirements of the evaluator. The evaluation process adopted in their study is characterized by its flexibility.

Alsmadi (2010) discussed and evaluated several website metrics. The focus of the paper was on the Web structural and traffic metrics. The paper case study was on several websites for universities where the paper tried to evaluate the impact of such metrics on the quality of service those websites are providing to their users.

Geczy et al. (2007) presents a novel formal approach to analyze usability which based on navigation space construct within a large corporate Intranet. Tests prove this is a beneficial approach to analyze, model, and design behaviourally centred Web applications and services, since it helps to extract important usability features.

Most of the users of this large corporate Intranet were knowledge workers who efficiently utilized a small portion of the available resources within the intranet, where this frequently accessed portion is easily identified and navigated. The tests show that the infrequently accessed resources need complex navigation.

The evaluation of a number of local government websites in Indonesia is conducted by Hermana and Silfianti (2011). Their study shows that province websites generally are more popular than that of their city or county counterparts, except for Java Island where the district or city websites generally are more popular than that of the province. In addition, their study shows that the all Web measurements of district websites are lower than their counterparts in province and city, beside the evident digital divide between Java Island and other parts of Indonesia.

Petricek et al. (2006) exhibits an introductory study to test the relation between the statistical properties of the structure of websites and the quality of websites. Therefore, a novel method for evaluating E-government websites is presented in their work, and the structures of government audit office Web sites in Canada, USA, UK, New Zealand, and Czech Republic are examined using the suggested method. The test results reveal the existence of some correlation between structure of websites and the quality of websites, and the government audit office websites in Canada, USA provide better navigability and much better than their counterparts in UK. In addition, government audit office websites in UK has the strongest modality.

E-Government portals need a 24/7 constant monitoring for its services performance, therefore Fong and Meng (2009) exhibited a real time Web-based performance monitoring system (WMS) for portals, in order to avoid expensive costs of performance monitoring by a third-party. WMS consists of a number of tools and algorithms for e-Government portals. WMS is characterized by its open-source, interoperable, flexible and distrib-uted, beside Web usage mining and computation of statistics offered by the hosting company of E-Government portal.

In their study, Spremic et al. (2010) presents issues related to Croatian e-government at three levels: Euro level, transition economies level, and developments at national (Croatia) level. In their study, they presented the necessary e-services needed to be added to the current Croatian e-government and used in other European e-governments.

A metric model or framework is presented by Olsina et al. (2003). The model focused on 4 major metrics: resource, process, product and quality in use metrics. They related those metrics with several examples of attributes that can be categorized under those metrics.

In their research, Singh et al. (2011) collected data related to websites for four years (2007-2010) from www.Webbyawards.com, which is a leading international award honouring excellence website on the Internet. Afterward the collected data by Singh et al. is used to classify the websites into two categories (Good and Bad). The classification depends on the 15 assessed metrics. The results of their work can be beneficial to study the Web design quantitatively.

One of the early websites that offer several Web metrics is: www.Webmetrics.com. Succi et al. (1998) discussed the tools and attributes measured in Webmetrics website. This website focuses on the quality and performance aspects for evaluating websites. The website contains several tools built in Java to evaluate speed quality attributes such as those attributes related to load testing and performance monitoring.

It is essential to define information needs before selecting relevant Web metrics that ensures website improvement and optimization. Click stream data is not enough to have a full perception about the effectiveness of any website, therefore Weischedel et al. (2006) present the strengths and weaknesses of adopting Click stream data for Web optimization, beside conducting an empirical study involving many IT and marketing managers. These mangers face the problem of selecting the most appropriate Web metrics, and they are aware of the value of triangulation. To get deeper understanding of online customer behaviour, mangers have to combine Click stream data with customer surveys and external (comparison) data.

The main existing Web accessibility metrics are presented, reviewed and compared with each other by Freire et al. (2008) with an emphasize on the major features according to important attributes and to experimental data. The correlation between these features is also explored. The aim of their study is to facilitate the process of defining and establishing measurement process by those who are interested by in Web accessibility.

WEBSITE DESIGN QUALITY ASSESSMENT

In this section, an ontology is proposed to correlate the main entities that will take roles in the websites' metrics' evaluation.

Web Metrics Ontology

There are several research papers that tried to present ontology for software metrics in general and Web metrics in particular. Olsina and Martin (2004) assembled a software metrics' ontology to assemble all related components along with their relations with each other. The ontology is composed of the following major elements:

- **Metric:** The defined measurement or calculation method and the measurement scale. It can be divided into direct metric and indirect metric.
- **Unit:** Particular quantity defined and adopted by convention, with which other quantities of the same kind are compared in order to express their magnitude relative to that quantity.
- **Entity:** Object that is to be characterized by measuring its attributes.
- **Tool:** A tool that automates partially or totally a measurement or calculation method.
- **Attribute:** A measurable physical or abstract property of an entity.
- **Indicator:** decision criteria in order to provide an estimate or evaluation of a calcu-

Figure 1. A web metrics ontology proposed in this chapter

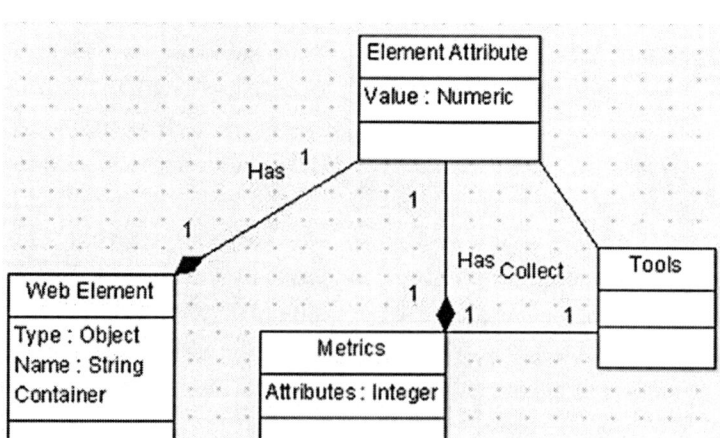

lable concept with respect to defined information needs.

The main entities that will be considered are: website elements, attributes, metrics, and tools. Figure 1 shows a UML diagram representing the relations between those entities. Other less relevant entities like those related to users, and other websites are excluded.

Based on the major entities that we will consider and based on the Web metrics and attributes described in the introduction, we will summarize those components section by section. We will exclude Web components or elements as their contribution will be generic for most Web metrics. Web components are the different types of elements that each website has. This includes container types such as: Web pages, frames, forms, etc. It can also have components such as links, labels, text boxes, images, etc.:

1. **Web usability metrics:** Table 1 shows a summary of Web metrics entities for Web usability. The table shows examples of important usability characteristics and attributes along with examples of possible tools to evaluate them. While in several cases, usability is subjective and can be only measures through a survey or a feedback from users, however, there are several aspects of websites usability that can be collected automatically through tools.

2. **Performance metrics:** The general meaning of the performance quality attribute is that the website should be quick and be able to provide users with the information and services they are requesting quickly. This can be particularly important for e-government websites as it is expected that those websites will have a large number of frequent visitors. In such cases, the ability for the website to provide the service to simultaneous large number of users can be severely degraded. Testing such criteria

Table 1. Usability metric related entities, sample

Metric	Attribute	Tools
Learnability	No of help files	Conceptfeedback.com
Memorability	Time to learn	Chalkmark
Efficiency	User errors	Taskee
Effectiveness	Workload	StomperScrutinizer
User satisfaction		Crazyegg

Table 2. Performance metric related entities, sample

Metric	Attribute	Tools
Interactivity	Speed of response	Showslow
Responsiveness	Images loading time	Yahoo YSlow
Performance	Flash loading time	Pagespeed
	Pages loading time	GZip(GidZip)
		Papiro
		Web Capacity Analysis Tool (WCAT)

should include increasing the load on the website and measure its ability to handle the different volumes of users. Table 2 shows a summary of Web performance metrics. The table shows examples of important performance characteristics and attributes along with examples of possible tools to evaluate them.

3. **Traffic and usage metrics:** Those metrics are usually evaluated to see how much a website is popular and visited by users. It is very important for e-government websites that their main goal is to provide services to citizens and enable them to communicate with the government to assess how much such e-services are used by those citizens. It is also important to see, if their is no high traffic, what kind of possible problems or obstacles that are possible preventing using

and visiting those websites. An initial assessment by authors for e-government websites in Jordan showed that such websites are not visited by the number of visitors it is thought to have. While parts of this problem is due to cultural and society traditions that may change with time, with the Internet large expansion in Jordan, the e-government sponsors should look into the other possible reasons for such issue. Table 3 shows a summary of Web performance metrics. The table shows examples of important traffic characteristics and attributes along with examples of possible tools to evaluate them.

4. **Business or financial metrics:** Those are metrics that are usually measured in e-commerce websites or websites that provide e-services to users. Some of those attributes (e.g. sales trend, bounce rate, etc.) are not applicable for e-government websites that are not revenue oriented. Table 4 shows a summary of Web performance metrics. The table shows examples of important business related characteristics and attributes along with examples of possible tools to evaluate them.

5. **Call centre metrics:** Similar to business metrics, some of call centre attributes are not applicable for e-government websites that are not revenue oriented. Nonetheless, one of the main characteristics of an e-government website is to be able to answer all citizens' queries through the Web or phone services with no need for those users to visit physical governmental offices. Table 5 shows a summary of call centre metrics. The table shows examples of important business related characteristics and attributes along with examples of possible tools to evaluate them.

6. **Customer centred metrics:** Customer metrics can be included in other types of metrics that evaluate users satisfaction (e.g. usability and traffic metrics). Table 6 shows

Table 3. Traffic metric related entities, a sample

Metric	Attribute	Tools
Page rank	Link popularity	www.alexa.com
Site popularity	No of visitors	www.compete.com
Reach	Page visits	Traffic watch
Visitors classes	Unique visitors	Google analytic
Web usage	Repeated visitors	Trafficzap
Page index	Length of visitors stay	http://itools.com/Internet/Web-site-information

Table 4. Business metric related entities, sample

Metric	Attribute	Tools
Operational efficiency	Abandon rate	Webmetrics.com
Revenue optimization	Click rate or visit to purchase	Marketwave
Time to market	Sales trend	ePCR
	Site overlay	iCargo
	Bounce rate	TripViewer

Table 5. Call centre metric related entities, a sample

Metric	Attribute	Tools
first call resolution	cost per case	Force.com Connect CTI Toolkit, SMAC, etc
call volume	Time with customer	Email verification tools
individual agent performance	Time with support	
agent productivity	Customers' waiting time	

examples of customer related metrics and attributes. Attributes column is empty as metrics in the first column represents simple numerical values that can be also considered as attributes. Nonetheless, all listed examples in all tables are not comprehensive and many

Table 6. Customer centre metric related entities, a sample

Metric	Attribute	Tools
Reach		Alexa
product profitability		Compete
brand equity		Google analytics
cost of acquisition of new service		
Market share		

other metrics, attributes and tools can be listed in all tables.

For size purposes, we will list the rest of metrics and tools in the classification through their usage in the case study.

A CASE STUDY

Good design characteristics are not completely absolute. Based on websites classifications, some characteristics can be more important than others. For example, design qualities for an e-commerce website can be significantly different from those deigned as University portals. In this case study, websites from the governmental portals are taken as they can be categorized as websites that need to have a moderate to high level of quality in almost all quality aspects described earlier.

In order to demonstrate the usability of the proposed conceptual Web metrics modeling in this paper, several selected websites from Jordan e-government consortium will be selected. Table 7 shows a summary of those selected websites.

For each one of the listed Web metric sections described above (i.e. 15), we will select one available tool and use the six websites results to analyze the results gathered from the specific tool:

1. Websites usability
2. Performance, efficiency and effectiveness metrics, transaction

3. Traffic and usage metrics
4. Business metrics, financial analysis
5. Call center metrics
6. Customer centered metrics
7. Process quality metrics
8. Structural metrics
9. Web content metrics
10. Complexity (related to either maintainability or effort) metrics
11. Quality metrics
12. Marketing metrics
13. Navigability metrics
14. Accessibility and environmental metrics
15. Web design and reusability metrics

RESULTS AND DISCUSSION

In this section, we provide an overview of the tools and websites used for measuring the required Web site websites attributes and metrics. Several tools and websites are used because we could not measure all the metrics using only one tool. Web Page Analyzer, from website Optimization is a Free website Performance Tool and Web Page Speed Analysis; it calculates page size, composition, and downloads time. The script calculates the size of individual elements and sums up each type of Web page component. Based on these page characteristics the script then offers advice on how to improve page load time (Web Analyzer, 2011). Another tool called W3C Markup Validation Ser-

Table 7. E-government websites selected for the study

No	website	Link
1	Ministry of Information and communication technology	www.moict. gov.jo
2	Ministry of trade	www.mit.gov.jo
3	Ministry of planning	www.mop.gov.jo
4	Civil service Bureau	www.csb.gov.jo
5	Ministry of education	www.moe.gov.jo
6	Ministry of foreign affairs	www.mofa.gov.jo

vice, which checks the markup validity of Web documents in HTML, XHTML, SMIL, MathML, etc. (W3C, 2011). A third tool, which is an online tool, is called WHOIS (Who.is, 2011), WHOIS is a tool that provides valuable domain information such as domain registrar, status, expiration date, and DNS name servers.

Alexa (2011) is the Web information company that provides Free traffic metrics, search analytics, demographics, and more for websites, Alexa is mainly used for measuring metrics related to usability and traffic rank. Another tool is called Xenu's Link Sleuth (TM) (Xenu, 2011), which checks Web sites for broken links. Link verification is done on "normal" links, images, frames, plug-ins, backgrounds, local image maps, style sheets, scripts, and java applets. It displays a continuously updated list of URLs, which you can sort by different criteria. A report can be produced at any time. Another online tool for measuring website metrics is Link Valet (2011) which is a WWW Link checker; when you enter the URL of an HTML page on the Web, it will fetch the page, and print a report on it. Link Valet also spider your site. When a link references another HTML page at the same site and hierarchy as the URL, Link Valet will recursively follow the link and prepare a similar report on the page referenced. (TAW3, 2011), TAW is a tool for the analysis of Web sites, based on the W3C – Web Content Accessibility Guidelines 1.0 (WCAG 1.0). (Page Speed, 2011) is an open-source project started at Google to help developers optimize their Web pages by applying Web performance best practices. Page Speed started as an open-source browser extension, and is now deployed in third-party products such as Webpagetest.org, Show Slow, and Google Webmaster Tools. (Woorank, 2011) is a SEO website Analysis Tool to help developers boosting their client's websites. Other tools include Google chrome, Keywordspy, Yahoosite explorer, and website grader.

Usability Metrics

Table 8 shows the results of evaluating the selected e-government websites for usability.

Table 8 shows the download time required by each domain under test using a connection speed of 56k; the table reveals that the websites of the MOE, MFA, MIT, and CSB required a download time that is almost too high with 227.62s, 216.95s, 201.81s, and 193.28s respectively. On the other hand, the website of the MOICT required less time comparing it with the previous domains with a download time of 55.83s. Finally, the website of the MOP required only 0.23s to be downloaded. Comparing the six websites in term of HTML/XHTML check errors; the CSB website contains only 8 errors followed by the MFA website with 27 errors. On the other hand the websites of MOP, MOE, MIT and MOICT contains lots of errors with 140, 161, 16 and 331 errors, respectively.

Table 9 shows the results for the six websites under test using the Web Page Analyzer, as the table shows the average number of HTML pages found is 1.3 with a minimum number of 1 page in the websites of MOICT, MIT, MOP, CSB and a maximum number of 2 pages in the websites of MOE and MFA. For the total HTML page size,

Table 8. Test results for download time and HTML/XHTML check errors

URLs	Download Time (56K) Second	HTML/XHTML Check Errors
www.moict.gov.jo	55.83	331
www.mit.gov.jo	201.81	166
www.mop.gov.jo	0.23	140
www.csb.gov.jo	193.28	8
www.moe.gov.jo	227.62	161
www.mfa.gov.jo	216.95	27
Download Time (56K): Web Page Analyzer		
HTML/XHTML Check Errors: Markup Validation Service		

Table 9. Results using web page analyzer

URLs	THF	THS	TIS	TNI	TCSS	TCSSS
MOICT	1	54122	161949	25	2	2110
MIT	1	205302	610862	20	16	23935
MOP	1	159	0	0	0	0
CSB	1	200192	532889	34	16	57188
MOE	2	349524	226538	38	10	6129
MFA	2	72212	173875	115	108	61219
Avg.	1.3	146918	284352	38.7	25.3	25096
THF: Total # of HTML Files				TCSS: Total # of CSS Files		
THS: Total HTML Page Size (B)				TCSSS: Total Size of CSS files (KB)		
TIS: Total Size of Images (B)				Results using Web Page Analyzer		
TNI: Total # of Images						

the average size for the six websites is 146918.5B. Regarding images found on the websites the average number of images is 38.7 with an average size of 284352.2B. Finally, the average number of CSS files found is 25.3 with an average size of 25096.8KB.

Performance Metrics

Testing the six websites performance, Table 10 shows the results obtained for the download time required, speed of load, speed percentiles, speed comparison, and page speed score for the websites under the test.

In term of download time, the website MOP achieved the lowest load time followed by the websites of MOICT, CSB, MIT, MFA, and MOE, respectively. The average median load time for the six websites is 2303ms. Comparing the six website speed, the table shows that the website of MOE is the faster followed by the websites of MIT, MFA, CSB, and MOICT; for MOP the speed is unknown. Finally, comparing the six websites based on the page speed score we can reveal that the website of CSB achieved the highest score followed by the websites of MFA, MOICT, MOP, MIT, and MOE, respectively.

In terms of memory usage, Table 11 shows the amount of memory in K that is required by each of the six website. The average memory used by the six website is 17,678K where the website of MOICT required the lowest amount of memory with 8,256 and website of MFA required the highest amount of memory with 43,678K. The average shared memory, private memory, JavaScript memory is 13,315K, 22,483K and 3,467K respectively. No memory is used by the SQLite in every website under the test. Finally, the average image cash and script cash used by all of the six websites is 4,027K and 222K, respectively.

Traffic and Usage Metrics

Table 12 shows results related to traffic metrics. The Table reveals that the website of MOE ranked locally first among the sex websites under test followed by the website of CSB, MIT, MOICT, MFA, and MOP, respectively. On the other hand, the website of CSB ranked globally first among the six websites followed by the websites of MOE, MIT, MFA, MOP, and MOICT, respectively.

Table 13 shows results to usage metrics, as the table demonstrates; CSB achieved the best reach rank among the six websites followed by the websites of MOE, MIT, MOP, MFA, and MOICT. In terms of page views rank; the website of CSB achieved the best rank followed by the websites of MOE, MIT, MOP, MFA, and MOICT, respec-

Table 10. Websites performance attributes

URLs	Download Time (56K) Seconds	Speed - Median Load Time	Speed - %	Speed Comparison (% of sites faster)	Score
MOICT	55.83	3219	14th	86%	57
MIT	201.81	1683	41st	59%	33
MOP	0.23	U	U	U	48
CSB	193.28	2961	17th	83%	71
MOE	227.62	1173	62nd	38%	14
MFA	216.95	4781	18th	82%	59
Download Time (56K): Using Web Page Analyzer					
Load Time, Percentile, Comparison: Using WHOIS					
Speed Score out of 100: using Pagespeed					

Table 11. Memory usage by websites

URLs	Memory Used (K)	Shared Memory (K)	Private Memory (K)	JavaScript Memory (K)	Image Cash (K)	Script Cash (K)
MOI.	8,256	12,808	13,620	1461	1708	2.3
MIT	14,408	13,096	17,728	2418	2920	170
MOP	12,056	12,644	17,360	2952	494	0
CSB	11,408	14,496	16,040	2430	1609	241
MOE	16,688	13,412	21,900	2290	4708	557
MFA	43,252	13,436	48,248	9251	12722	363
Avg.	17,678	13,315	22,483	3,467	4,027	222

tively. For the ratio of reach and page views per million the six websites achieved and average of 2.4 and 0.1 respectively. Finally, the average number of pages viewed by user on all the six websites is 2.7 pages with a number of pages viewed between 2 and 4 pages in all websites.

Business Financial Metrics

Table 14 shows the results of evaluating some selected business and financial metrics.

Examining the selected websites under test we have noticed that none of them have any e-commerce activities. This is why we only selected the overall average price for Web pages of each website. We depend on the average cost per click and the estimation of the markets value for the top ten

keywords from each of the six websites in order to evaluate the business and financial metrics.

As Table 14 demonstrates, the average CPC and Value for the MOICT websites are $0.14 and 8.65. For the website of MIT the averages are $0.42

Table 12. Traffic metrics sample using Alexa

URLs	Global Traffic Rank	Local Traffic Rank	Traffic Rank
MOICT	2,829,133	3,238	4
MIT	1,316,580	1,836	3
MOP	2,137,874	3,977	6
CSB	224,832	195	2
MOE	354,805	185	1
MFA	2,018,028	3,337	5

Table 13. Website usage using WHOIS

URLs	Reach Rank	Page Views Rank	Reach/ Million	Page Views/Million	Page Views/User
MOI.	2,720,561	3,373,305	0.48	0.01	2
MIT	1,218,278	1,697,109	0.02	0.02	2
MOP	2,171,696	2,358,724	0.63	0.01	2.5
CSB	231,021	270,563	7.9	0.23	3.41
MOE	377,016	369,752	4.7	0.16	4
MFA	1,937,667	2,511,039	0.73	0.01	2

and 7.90, MFA website achieved an average values of $0.14 and 7.53, respectively, for both CPC and value. For the website of MOP, the average CPC and value are $0.05 and 2.47 respectively. Finally the websites of MOE and MFA achieved and average CPC and value of $0.23 and 9.82, $0.14 and 7.53, respectively.

Web Structural Metrics

Table 15 shows results related to the structural metrics of the websites. As the demonstrates the total number of pages on the MIT website was the highest one with 3174 pages followed by the websites of MOE, MOP, CSB, MOICT and MFA with 2709, 1977, 1019, 367, and 209 pages, respectively. The total number of Frames/Iframes founded was 0 for each of the MOICT, MIT, MOP and CSB websites, 1 for the MOE website and 2 for the MFA website. In terms of number of text boxes founded on the main page is 1 for all of the websites except the website of the MIT which contains 3 text boxes. The average number of in-links founded on the main page is about 48 links where the average number of in-links founded on the whole website is about 1248 links.

Content Metrics

Table 16 shows the results related to the websites contents metrics. In terms of scripts, the websites of both MOE and MFA contains 11 scripts fol-

lowed by the websites of MIT, CSB and MOICT with 8, 4, 1 scripts, respectively. On the other hand, the website of MOP contains no scripts. Regarding the number of images founded, MFA website contains 118 images followed by the websites of MOE, CSB, MOICT, and MIT with 38, 34, 25, 20 images, respectively; for the MOP websites no Images found. Counting the number of CSS images, the website of MFA contains 99 one followed by the websites of CSB, MIT, MOE, and MOICT with 15, 11, 6, and 1 CSS images; for the MOP website, no CSS images found. For the HTML, all websites contains only 1 except the websites of MOE and MFA which contains 2 HTML objects. Counting the number of CSS objects, MFA websites contains 9 one followed by the websites of MIT, MOE, MOICT, and CSB with 5, 4, 41, 1 CSS objects, respectively; for the MOP websites no CSS objects founded. Finally, the website of CSB contains 10200 indexed pages followed by the websites of MIT, MOE, MFA,

Table 14. Business and financial metrics

Web	CPC	Value %
MOICT	0.14	8.65
MIT	0.42	7.9
MFA	0.14	7.53
MOP	0.05	2.47
MOE	0.23	9.82

Table 15. Website structural attribute

URLs	# of Pages	# of Frames/Iframes	# of Text Boxes	Home Page InLinks	Total # of Inlinks
MOICT	367	0	1	54	441
MIT	3,174	0	3	26	1,472
MOP	1,977	0	1	36	850
CSB	1,019	0	1	29	655
MOE	2,709	1	1	90	2,445
MFA	209	2	1	50	1,626
# of Frames and Iframes: Using Web Page Analyzer					
# of Text Boxes: Manually					
# of InLinks: Using Mozilla Add on					
Total # of Pages, Inlinks: Using Yahoo website Explorer					

Table 16. Websites contents metrics

| URLs | Page Objects | | | | | Indexed Pages |
	Scripts	IMG	CSS IMG	HTML	CSS	
MOICT	1	25	1	1	1	619
MIT	8	20	11	1	5	3570
MOP	0	0	0	1	0	1800
CSB	4	34	15	1	1	10200
MOE	11	38	6	2	4	3120
MFA	11	118	99	2	9	1970
Results using Web Page Analyze						
Indexed Pages Using Woorank						

Table 17. Websites complexity metrics

URLs	# of special components	# InLinks	URL Errors	NO of Page Objects	Indexed Pages
MOICT	2	54	67	130	619
MIT	4	26	1804	1842	3570
MOP	2	36	1248	1296	1800
CSB	2	29	459	492	10200
MOE	4	90	468	587	3120
MFA	5	50	193	299	1970
# of HTML Pages, # of Objects, Frames and Iframes: Using Web Page Analyzer					
# of Text Boxes: Manually					
# of InLinks: Using Mozilla Add on (Inlinks + Oulinks)					
URLs Errors Using Xenu					
Indexed Pages Using Woorank					

MOP, MOE, and MOICT with 3570, 3120, 1970, 1800, and 619 indexed pages respectively.

Complexity Metrics

Table 17 shows the results of evaluating some complexity metrics. Table 17 shows the attributes related to the websites complexity metrics. As the table demonstrates, the total number of HTML pages, Frames, Iframe, and Text Boxes objects founded on the websites of MOICT, MOP, and CSB is 2, where the websites of MIT and MOE contains 4 objects; on the other hand the websites of MFA contains the highest number of objects with 5 objects.

The average number of in-links founded on the main page of the six websites is about 48 links where the average number of URLs error is 707 for all websites, for the number of objects found and number of indexed pages found, the average is 774 object and 3547 indexed pages, respectively.

Quality Metrics

Table 18 shows the results of evaluating quality metrics.

Table 18 shows the quality metrics for the all six websites under test. The table expresses the quality of the websites in term of Web content accessibility guidelines and total number of URL errors. As the table demonstrates the average number of Priority error/human reviews is 524 errors where the average number of priority errors-Automatic is only 124 errors. On the other hand, the average number of URL errors is 187, where the website of MIT contains the highest number of errors and the website of MOICT contains the lowest number of errors.

Marketing Metrics

Table 19 shows the results of evaluating marketing metrics.

Out of the twenty-five marketing metrics we only test the websites under study against four of them namely, website grade, traffic change, indexed pages, and total number of in-links. As the table demonstrates, the websites of the MIT achieved the highest website grade with a value of 90 followed by the websites of MFA, CSB, MOE, MOICT, and MOP with website grades of 88, 85, 84, 72, and 64, respectively. In terms of traffic changes, the website of MOICT, MOP, and

Table 18. Website quality metrics

URLS	Priority Errors-Human Review	Priority Errors-Automatic	URL Errors
MOICT	403	287	67
MIT	2213	987	1804
MOP	N/A	N/A	1248
CSB	1305	68	459
MOE	1471	595	468
MFA	368	78	193
Average	524	124	187
W3C Accessibility Guidelines - Using TAW3			
URL Errors auth required, cancelled/timeout, connection aborted, forbidden request, no connection, no such host, not found, server error, skip type, timeout - Using Xenu			

Table 19. Website marketing metrics

URLs	website Grade	Traffic Change	Indexed Pages	Total # of Inlinks
MOICT	72	-1303224	619	441
MIT	90	245649	3570	1,472
MOP	64	-200168	1800	850
CSB	85	-31834	10200	655
MOE	84	127162	3120	2,445
MFA	88	259399	1970	1,626
website Grade Using website Grader				
Traffic Change over 3 months Using WHOIS (- Rank Increased, + Rank Decreased				
Indexed Pages Using Woorank				
Total # of Pages, Inlinks: Using Yahoo website Explorer				

CSB decreased in traffic ranks where the websites of MIT, MOE, and MFA increased in traffic rank. For the number of indexed pages the website of CSB contains the biggest number of indexed pages followed by the websites of MIT, MOE, MFA, MOP, and MOICT. Finally, the website of MOE contains the biggest number of in-links followed by the websites of MFA, MIT, MOP, CSB, and MOICT.

Navigability Metrics

Table 20 shows the results of evaluating navigability metrics.

As Table 20 demonstrates, the average number of in-links and out-links on the main pages are 48 and 20 respectively. In term of URLs, the average number of URLs with errors is 5 where the average number of redirected URLs is 3, and the average number of changed URLs is 30, none of the websites contains unchanged URLs. Finally, in terms of page views/million and page

Table 20. Website navigability metrics

URLs	InLinks Count	OutLinks Count	Errors	Re-direct	Changed	Other	Page Views/ Million	Page Views/User
MOICT	54	9	11	3	38	49	0.01	2
MIT	26	12	3	2	31	7	0.02	2
MOP	36	12	-	-	-	-	0.01	2.5
CSB	29	4	3	2	26	28	0.23	3.41
MOE	90	29	10	9	48	24	0.16	4
MFA	50	56	5	2	39	54	0.01	2
Avg	48	20	5	3	30	27	0.7	2
InLinks Count OutLinks Count – Mozilla Add on								
Links/Errors, Redirect, Changed, Unchanged, Other - Using Link Valet is a WWW Link checker, All Links Since 1-1-2011 using 2 Recursion Depth								
Page Views/Million Page Views/User - Using WHOIS								

Table 21. Accessibility and environmental metrics

URLs	P1 Errors	P2 Errors	P3 Errors	P1 Errors	P2 Errors	P3 Errors
MOICT	13	246	28	144	214	45
MIT	4	699	284	1,208	474	531
MOP	N/A	N/A	N/A	N/A	N/A	N/A
CSB	0	25	43	551	652	102
MOE	56	410	129	723	505	243
MFA	7	59	12	166	144	58
Average	10.5	82.3	30.7	240.0	216.8	67.2
W3C Accessibility Guidelines: TAW3-Automatic						

views/users the averages are 0.7 and 2, respectively, for the six websites.

Table 21 shows the accessibility of the websites under test. Accessibility tests are done based on the Web Content Accessibility Guidelines (WCAG1.0). As the table demonstrates, the average P1, P2, and P3 errors for MOICT, MIT, CSB, MOE, and MFA websites are 10.5, 82.3 and 30.7 errors respectively using TAW3-Automatic. On the other hand, the average P1, P2, and P3 errors for MOICT, MIT, CSB, MOE, and MFA websites are 240, 216.8, and 67.2 errors, respectively, using TAW3-Human Review.

Web Security Metrics

Table 22 shows the results of security metrics.

In Table 22, all tests are done using the Mozilla Firefox add on. As the table demonstrates

three tests are done to evaluate the security of websites namely; XSS string test, SQL injection string test, and Access Me string test. For the XSS string tests, three websites, namely MOICT, CSB, and MOE, have the same test failures, warning, and passes with 0, 77, and 77, respectively. The website of MIT has 0 failures, 9 warnings and 9 passes. The MOP website contains 11 failures, 154 warnings, and 143 passes. For the SQL injection test the websites of MOICT, MIT, CSB, and MOE contains 15, 17, 14 and 221 failures where the websites of MOP and MFA have no failures reported. All websites under the test have 0 warnings; on the other hand, the SQL injection string test reported too many passes for all websites. Finally, for the access me add on the website of MOICT and MIT reported 0 failures, where the websites of MOP, CSB, MOE and MFA reported 1, 2, 3, and 4 failures. In terms of number of

Table 22. Web security metrics examples

URLs	XSS String Tests			SQL Injection String Tests			Access Me String Tests		
	Failure	Warning	Pass	Failure	Warning	Pass	Failure	Warning	Pass
MOICT	0	77	77	15	0	87705	0	1	1
MIT	0	9	9	17	0	146030	0	4	4
MOP	11	154	143	0	0	43860	1	6	7
CSB	0	77	77	14	0	146186	2	9	10
MOE	0	77	77	221	0	189839	3	11	12
MFA	NA			0	0	102340	4	14	15

warning and passes, MOICT reported 1 warning and 1 pass, MIT have 4 warnings and passes, MOP reported 6 warnings and 7 passes where the CSB website reported 9 warnings and 10 passes. Finally, the websites of MOE and MFA reported 11, 14 warnings and 12, 15 passes, respectively.

CONCLUSION

Evaluating a software product is important and is required for several purposes. Metrics or measurements are used to measure one or more aspects of a software to assess one or more of its aspects. In this chapter, a trial is made to describe all classes of quality attributes that can be measured for software products, in particular, websites. Evaluating one quality aspect alone can be misleading if all other aspects are ignored. For the least, a good collection of such quality metrics should be selected and assessed.

E-government websites are examples of multipurpose information sensitive websites that should have advanced levels of quality for several quality attributes such as: performance, security, reliability, usability, etc. In this chapter, we presented a comprehensive description, evaluation, and classification for website metrics and tools based on the different characteristics of those websites. The goal is to understand those different evaluation concerns and know their relations with each other. Through using tools and metrics to assess websites under study, we tried also to study the type of information related to quality attributes that we can get from each one of those metrics. A case study of several e-government websites in Jordan is selected. In each metric and website quality attribute, we used some of the available tools. Then we evaluate and compare those websites relative based on the selected attributes and metrics. websites evaluators should consider all those different quality attributes into consideration when evaluating any website specially as in many cases some of those quality attributes

can have conflicting or trade off requirements among each other.

As it is impractical to be able to build a website with all quality attributes as optimal, it is important for website designers and project managers to understand the need for the particular website to be able to prioritize and give more focus to a selection of quality metrics and optimize them while getting a satisfied level for the rest of the quality metrics.

ACKNOWLEDGMENT

This chapter is an expanded version of a previously published article ("A Conceptual Organization for Websites Metrics: e-Government Websites – A Case Study") in the IGI journal, *International Journal of Information Communication Technologies and Human Development* (IJICTHD), Volume 4, Issue 1, 2012.

REFERENCES

Abrahao, S., Condori-Fernandez, N., Olsina, L., & Pastor, O. (2003). Defining and validating metrics for navigational models. In *Proceedings of the 9th International Software Metrics Symposium (METRICS 2003)*, (pp. 200-210). Sydney, Australia: METRICS.

Alagappan, B., Alagappan, M., & Danishkumar, S. (2009). Web metrics based on page features and visitor's web behavior. In *Proceedings of the 2nd International Conference on Computer and Electrical Engineering (ICCEE 2009)*, (vol. 2, pp. 236-241). Dubai, UAE: ICCEE.

Alexa. (2011). *The web information company.* Retrieved from http://www.alexa.com/

Alhenshiri, A., & Duffy, J. (2010). User studies in web information retrieval: User-centered measures in web IR evaluation. In *Proceedings of the 2010 IEEE 24th International Conference on Advanced Information Networking and Applications Workshops (WAINA)*, (pp. 632-637). IEEE Press.

Alsmadi, I. (2010). The automatic evaluation of website metrics and state. *International Journal of Web-Based Learning and Teaching Technologies*, 5(4), 1–17. Retrieved from http://www.igi-global.com/bookstore/article.aspx?TitleId=52596 doi:10.4018/jwltt.2010100101

Alsoud, A., & Nakata, K. (2010). Evaluating e-government websites in Jordan: Accessibility, usability, transparency and responsiveness. In *Proceedings of 2010 IEEE International Conference on Progress in Informatics and Computing (PIC 2010)*, (pp. 761-765). Shanghai, China: IEEE Press.

Barnes, S., & Vidgen, R. T. (2003). Assessing the quality of a cross-national e-government web site: A study of the forum on strategic management knowledge exchange. In *Proceedings of the 36th Annual Hawaii International Conference on System Sciences (HICSS 2003)*. Hawaii, HI: IEEE.

Dhawan, S., & Kumar, R. (2008). Analyzing performance of web-based metrics for evaluating reliability and maintainability of hypermedia applications. In *Proceedings Third International Conference on Broadband Communications, Information Technology & Biomedical Applications (BroadCom 2008)*, (pp. 376–383). Pretoria, South Africa: BroadCom.

Foglia, P., Prete, C. A., & Zanda, M. (2008). Relating GSR signals to traditional usability metrics: A case study with an anthropomorphic web assistant. In *Proceedings of 2008 IEEE International Instrumentation & Measurement Technology Conference (I2MTC 2008)* (*Vol. 1*, pp. 1814–1819). Victoria, Canada: IEEE Press. doi:10.1109/IMTC.2008.4547339

Fong, S., & Meng, H. S. (2009). A web-based performance monitoring system for e-government services. In *Proceedings of the 3rd International Conference on Theory and Practice of Electronic Governance (ICEGOV 2009)*, (pp. 74-82). Bogota, Colombia: ICEGOV.

Freire, A., Fortes, R. P. M., Turine, M. A. S., & Paiva, D. M. B. (2008). An evaluation of web accessibility metrics based on their attributes. In *Proceedings of the 26th Annual ACM International Conference on Design of Communication*, (pp. 73-79). Lisbon, Portugal: ACM Press.

Geczy, P., Izumi, N., Akaho, S., & Hasida, K. (2007). Navigation space based intranet usability analysis. In *Proceedings of the IEEE Symposium on Computational Intelligence and Data Mining (CIDM 2007)*, (pp. 147-154). Honolulu, HI: IEEE Press.

Grader, W. (2011). *Website*. Retrieved from http://websitegrader.com/

Hermana, A., & Silfianti, W. (2011). Evaluating e-government implementation by local government: Digital divide in internet based public services in Indonesia. *International Journal of Business and Social Science*, 2(3), 156–163.

Horan, T., Abhichandani, T., & Rayalu, R. (2006). Assessing user satisfaction of e-government services: Development and testing of quality-in-use satisfaction with advanced traveler information systems (ATIS). In *Proceedings of the 39th Annual Hawaii International Conference on System Sciences (HICSS 2006)*. Hawaii, HI: IEEE.

Huang, C. J. (2003). Usability of e-government web-sites for people with disabilities. In *Proceedings of the 36th Annual Hawaii International Conference on System Sciences (HICSS 2003)*. Washington, DC: IEEE Press.

Jati, H., & Dominic, D. D. (2009). Quality evaluation of e-government website using web diagnostic tools: Asian case. In *Proceedings of the 2009 International Conference on Information Management and Engineering*, (pp. 85-89). Kuala Lumpur, Malaysia: IEEE.

Jinling, C., & Guoping, X. (2005). Comprehensive evaluation of e-commerce web site based on concordance analysis. In *Proceedings of the 2005 IEEE International Conference on e-Business Engineering (ICEBE 2005)*, (pp. 179-182). Beijing, China: IEEE Press.

Khan, K. (2008). Assessing quality of web based systems. In *Proceedings of the 2008 IEEE/ACS International Conference on Computer Systems and Applications (AICCSA 2008)*, (pp. 763-769). Doha, Qatar: IEEE Press.

Link, V. (2011). *A WWW link checker*. Retrieved from http://htmlhelp.com/tools/valet/

Nariman, A. (2010). E-government websites evaluation using correspondence analysis. In *Proceedings of the 2010 International Conference on Complex, Intelligent and Software Intensive Systems (CISIS)*, (pp. 1147-1152). Krakow, Poland: CISIS.

Olsina, L., Lafuente, G., & Pastor, O. (2002). Towards a reusable repository for web metrics. *International Journal of Web Engineering*, *1*(1), 61–73.

Olsina, L., & Maria, M. (2004). Ontology for software metrics and indicators: Building process and decisions taken. [ICWE.]. *Proceedings of ICWE*, *2004*, 176–181.

Page Analyzer, W. - 0.98. (2011). *A free website performance tool and web page speed analysis*. Retrieved from http://www.websiteoptimization.com/services/analyze/

Pascual, V., & Dursteler, J. C. (2007). WET: A prototype of an exploratory search system for web mining to assess usability. In *Proceedings of the 11th International Conference of Information Visualization*, (pp. 211-215). Zurich, Switzerland: IEEE.

Petricek, V., Escher, T., Cox, I. J., & Margetts, H. (2006). The web structure of e-government: Developing a methodology for quantitative evaluation. In *Proceedings of the 15th International World Wide Web Conference (WWW 2006)*, (pp. 669-678). Edinburgh, UK: IEEE.

Qi, S., Ip, C., Leung, R., & Law, R. (2010). A new framework on website evaluation, e-business and e-government. In *Proceedings of the International Conference on E-Business and E-Government (ICEE 2010)*, (pp. 78-81). Guangzhou, China: ICEE.

Saward, G., Hall, T., & Barker, T. (2004). Assessing usability through perceptions of information scent. In *Proceedings of the 10th International Symposium on Software Metrics*, (pp. 337-346). Chicago, IL: IEEE.

Schonberg, E., Cofino, T., Hoch, R., Podlaseck, M., & Spraragen, S. L. (2000). Measuring success. *Communications of the ACM*, *40*(3), 53–57. doi:10.1145/345124.345142

Signore, O. (2005). A comprehensive model for web sites quality. In *Proceedings of the Seventh IEEE International Symposium on Web Site Evolution (WSE 2005)*, (pp. 30-36). Budapest, Hungary: IEEE Press.

Singh, Y., Malhotra, R., & Gupta, P. (2011). Empirical validation of web metrics for improving the quality of web page. *International Journal of Advanced Computer Science and Applications*, *2*(5), 22–28. doi:10.5120/2414-3226

Speed, P. (2011). *Website*. Retrieved from http://code.google.com/speed/page-speed/

Spremic, M., Šimurina, J., Jaković, B., & Ivanov, M. (2010). E-government in transition economies. *International Journal of Human and Social Sciences*, *5*(2), 82–90.

Storm, K., Kraemer, E., Aurrecoeche, C., Heiges, M., Pennington, C., & Kissinger, J. (2009). Web site evolution: Usability evaluation using time series analysis of selected episode graphs. In *Proceedings of the 11th IEEE International Symposium on Web Systems Evolution (WSE), 2009*, (pp. 27-36). IEEE Press.

Succi, G., Benedicenti, L., Bonamico, C., & Vernazza, T. (1998). The webmetrics project - Exploiting software tools on demand. In *Proceedings of the 1998 World Multiconference on Systemics, Cybernetics, and Informatics*. Orlando, FL: IEEE.

TAW3. (2011). *A tool for the analysis of web sites, based on the W3C - Web content accessibility guidelines 1.0 (WCAG 1.0)*. Retrieved from http://www.tawdis.net/ingles.html

Taksa, I., & Spink, A. (2007). Evaluating usability of a long query meta search engine. In *Proceedings of the 40th Annual Hawaii International Conference on System Sciences, 2007*, (p. 82). IEEE Press.

Tripathi, P., Pandey, M., & Bharti, D. (2010). Towards the identification of usability metrics for academic web-sites. In *Proceedings of the 2nd International Conference on Computer and Automation Engineering (ICCAE)*, (vol. 2, pp. 393-397). Singapore, Singapore: ICCAE.

W3C. (2011). *W3C markup validation service.* Retrieved from http://validator.w3.org/

Wang, L., Bretschneider, S., & Gant, J. (2005). Evaluating web-based e-government services with a citizen-centric approach. In *Proceedings of the 38th Annual Hawaii International Conference on System Sciences (HICSS 2005)*. Hawaii, HI: IEEE Press.

Weischedel, A., & Huizingh, E. K. R. E. (2006). Website optimization with web metrics: A case study. In *Proceedings of the 8th International Conference on Electronic Commerce - ICEC 2006*, (pp. 463-470). Fredericton, Canada: ICEC.

Who.Is. (2011). *Website.* Retrieved from http://whois.is/

Woorank. (2011). *A SEO website analysis tool.* Retrieved from http://www.woorank.com

Xenu. (2011). *Link sleuth.* Retrieved from http://home.snafu.de/tilman/xenulink.html#Description

Zhang, Y., Zhu, H., & Greenwood, S. (2004). Web site complexity metrics for measuring navigability. In *Proceedings of the Fourth International Conference on Quality Software*, (pp. 172-179). Braunschweig, Germany: IEEE.

Chapter 18
Measuring the Conceptual Variables for E–Services Acceptance:
A Descriptive Statistical Analysis

Kamaljeet Sandhu
University of New England, Australia

ABSTRACT

This case study examines the Web Electronic Service framework for a University in Australia. The department is in the process of developing and implementing a Web-based e-service system. The user experience to use e-services requires insight into the attributes that shape the experience variable. The descriptive data about the attributes that form the experience variable is provided in this study.

INTRODUCTION

Descriptive statistics can offer powerful insights into the factors involved in e-Services system acceptance by analysing students' responses from the survey. A large volume of descriptive data can highlight the role of these factors in influencing the international student's use of the e-Services system on the University of Australia (not the real name) website. Analysing descriptive statistics is particularly effective for illuminating the inner workings of the constructs and ascertaining the strength of their effects in the international student's e-Services acceptance process. The international students were from first, second, and third year degree courses including commerce, arts, science, and nursing. They were studying full time. Both male and female international students participated in the survey.

DOI: 10.4018/978-1-4666-3658-3.ch018

Data Collection

The data for the survey was collected from the university's international student population. An invitation to the international students to participate in the survey was sent to them by an email. In addition, the researcher also visited computer laboratories and distributed leaflets to international students requesting them to participate in the survey. This approach is believed to have been effective as the international students who were in the computer laboratories and in front of the computers responded immediately. An invitation was also sent in the international student's fortnightly newsletter email, which is sent by the international students union. Participants could complete the survey online at anytime. The total numbers of responses received was 403. All responses received electronically were complete and without any errors. This resulted in eliminating incomplete survey responses, which has its limitations in a paper-based survey.

Research Project Background

This chapter is part of a larger study investigating the critical success factors in e-service user acceptance. The aim of the case study is to investigate the acceptance and use of e-services system amongst student users. The case study examines the Web Electronic Service framework of the University of Australia (not the real name). The department is in the process of developing and implementing Web-based e-service system. International students have the option to lodge the admission application through either of any: Web-based e-service system on the World Wide Web, phone, fax, or in person. On receiving the application, a decision is made by the staff on the admission status. The department is implementing the electronic delivery of its services on the website. The Web electronic service is believed to be in use for approximately last two and half years. The e-service process involves students making the application and the staff processing application on the website.

The items that are measured in this study are adopted for building and testing the user experience variable for studying the user acceptance of e-services. The items are also tested for is effectiveness and performance for the measurement of the variables.

THE INTERNATIONAL STUDENT EXPERIENCE WITH USING THE E-SERVICES SYSTEM

The international students were asked to respond to seven category items about the extent of their experience with the e-Services system. The seven items that measured international students experience were: a) their skills; b) finding information on the e-Services system; c) knowing about using the e-Services system; d) finding the e-Services system; e) movement within the e-Services system; f) awareness when interacting with the e-Services system; and g) confidence in using the e-Services system.

Skills

Taylor and Todd (1995) and Yaobin and Tao (2007) argue that the user's experience and skills influence IT usage. In this study, fifty-one percent of international students agreed that they had become skilled at using the university's e-Services system. A further 28.8% strongly agreed with the same statement. Nine percent of international students disagreed and another 11% were unsure. The data suggests that international students' skills had improved, and that they strongly believed that the e-Services system had assisted them in their university work. The descriptive data is shown in Table 1.

Table 1. Students skills in using the e-services system

		Frequency	Percent	Valid Percent	Cumulative Percent
Valid	Strongly Disagree	2	.5	.5	.5
	Disagree	35	8.7	8.7	9.2
	Not Sure	44	10.9	10.9	20.1
	Agree	205	50.9	51.0	71.1
	Strongly Agree	116	28.8	28.9	100.0
	Total	402	99.8	100.0	
Missing	Not Applicable	1	.2		
Total		403	100.0		

The connection between improved skills and using the university's e-Services system may explain the role of the international student's experience in enhancing their acceptance of e-Services system.

Finding Information on the E-Services System

Ramaswami et al. (2001) demonstrated that user adoption of on-line services takes a path of gaining familiarity and comfort before people use them for other purposes. Traditional consumer behaviour literature suggests that high involvement of users on e-Services system generates intense efforts by the user to attend to and search out sources of information (Ramaswami, et al., 2001; Huang, et al., 2007). In this study, the international students agreed (53.6%) that they knew where to find in-formation on the university e-Services system. A further 20.8% strongly agreed. The data is shown in Table 2. Given that 75% of the international students indicated they knew where to find what they wanted on the university e-Services system, it follows that they probably had a clear idea of the purpose of their activity within the e-Services system. International students who were not sure of where to find what they wanted on the university e-Services system accounted for 13.6%. Nearly 10% (9.9%) of international students disagreed with the statement.

Understanding About Using the E-Services System

D'Ambra and Rice (2001) argue that technology usage is dependent on user's perceptions of the impact of the e-Services system on their tasks, as

Table 2. Students ability to find information on the e-Services system

		Frequency	Percent	Valid Percent	Cumulative Percent
Valid	Strongly Disagree	8	2.0	2.0	2.0
	Disagree	40	9.9	9.9	11.9
	Not Sure	55	13.6	13.6	25.6
	Agree	216	53.6	53.6	79.2
	Strongly Agree	84	20.8	20.8	100.0
	Total	403	100.0	100.0	

well as on a host of social and contextual factors. A total of 50.9% of the international students reported that they were not sure if they knew more about using the university e-Services system than most other international students. These international students may not have discussed their experience with others, or may have felt either apprehensive or hesitant about the consequences of disclosing such information.

The self-evaluation of their use of the e-Services system by international students may require specific understanding of the student-specific experiences that may drive the use of e-Services across different student groups. Twenty-two percent of the international students agreed that they knew more about using the university e-Services system than most other international students (see Table 3). These international students responded that they had from intermediate to advanced experience in using the university's e-Services system. Fifteen percent of international students disagreed about knowing more about using the university e-Services system than most other international students.

Finding the E-Services System

Cockburn and McKenzie (2000) show that e-Services system revisitation is a prevalent activity amongst most users. Forty-eight percent of the international students in this study agreed that the university e-Services system was easy to locate as it was on the homepage on all university computers. Sometimes international students had difficulty in relocating previously visited websites or sub-websites on the e-Services system that required remembering a website address or other information that would assist the international students in returning to that website.

If relocating the e-Services system is made easy, it is reasonable to expect an increase in continued use of e-Services by international students. This prediction is based on the understanding that international students relocating the e-Services system may display enhanced experience on continued visits. The evidence is made stronger by the finding that another 38% of international students strongly agreed that the e-Services system was easy to locate (see Table 4). The ever-increasing number of websites on the e-Services system will influence an international student's experience in terms of finding website addresses that are difficult to remember.

The e-Services system website address often marks the starting point of an international student's evaluation of using e-Services.

Table 3. A student knowing more about using the e-services system than most other students

		Frequency	Percent	Valid Percent	Cumulative Percent
Valid	Strongly Disagree	10	2.5	2.5	2.5
	Disagree	62	15.4	15.4	17.9
	Not Sure	205	50.9	51.0	68.9
	Agree	89	22.1	22.1	91.0
	Strongly Agree	36	8.9	9.0	100.0
	Total	402	99.8	100.0	
Missing	Not Applicable	1	.2		
Total		403	100.0		

Table 4. E-services system is easy to locate by students

		Frequency	Percent	Valid Percent	Cumulative Percent
Valid	Strongly Disagree	4	1.0	1.0	1.0
	Disagree	15	3.7	3.8	4.8
	Not Sure	31	7.7	7.8	12.5
	Agree	195	48.4	48.9	61.4
	Strongly Agree	154	38.2	38.6	100.0
	Total	399	99.0	100.0	
Missing	Not Applicable	4	1.0		
Total		403	100.0		

Movement within the E-Services System

Hoffman and Novak (1995), Heijden (2000), and Koufaris (2002) demonstrate that user's experience improves with sustained movement within e-Services system website. E-Services system websites that are difficult to navigate may have low usage (Heijden, 2000). The majority of international students in this study (50.6%) agreed that it was easy to move between different sections of the university e-Services system. If the international student's navigation between different sections of the e-Services system is straightforward, it will probably influence e-Services usage.

The role of flow in a user's experience in moving from one section to another is important in understanding use of different features of the e-Services system (Zeithaml, et al., 2000; Chea & Lou, 2008; Chellappan, 2008). In this study, the evidence is made stronger with a further 23.6% of international students strongly agreeing that it was easy to move within the university e-Services system (see Table 5). The level of e-Services delivered to international students then, it can be argued, had a degree of easy accessibility and reach. Students indicated that they could reach the points of each of the e-Services without difficulty.

Awareness when Interacting with the E-Services System

Webster et al. (1993) argue that users enter a state of intense focus when interacting with e-Services system websites and thus merge their actions with awareness. Hoffman and Novak (1995) argue further that the user's experience consists of skills and challenges, and that their focused attention must be present in e-Services system interaction for successful use. In this study, only 28.8% of the international students reported that they were unaware of potential distractions around them when interacting with the e-Services system. The evidence subjectively points to very intense student concentration when in that state of focus and a decreased awareness of other things happening around them. Another 21.8% of international students agreed that they were less aware of other things when interacting with the university e-Services system.

A significant percentage of international students (38% in Table 6) disagreed that they were less aware of other things around them when interacting with e-Services. The students would have had some degree of awareness of what was happening in their surroundings, and it may be expected that there would be a decreased focus on

Table 5. Students ease of movement within the e-services system

		Frequency	Percent	Valid Percent	Cumulative Percent
Valid	Strongly Disagree	7	1.7	1.7	1.7
	Disagree	46	11.4	11.5	13.2
	Not Sure	49	12.2	12.2	25.4
	Agree	204	50.6	50.9	76.3
	Strongly Agree	95	23.6	23.7	100.0
	Total	401	99.5	100.0	
Missing	Not Applicable	2	.5		
Total		403	100.0		

their work and a lower level of experience as their attention was being drawn elsewhere (Webster, et al., 1993).

E-Services flow then may require complete user attention while interacting with a e-Services system. Distractions could potentially affect their experience and acceptance process (Csikszentmihalyi, 1990).

Confidence in Using the E-Services System

Ramaswami et al. (2001) and Morris and Turner (2001) argue that users who use the on-line channel for information search have a greater tendency to use it for other transactions. Obtaining information gives users an opportunity to experience the range of e-Services and obtain a certain level of comfort before they are ready to use it for other purposes (Ramaswami, et al., 2001; Chellappan, 2008). This analysis anticipates that the international student's confidence level is expected to increase with every e-Services based activity in which they are successful. A study by Gardner et al. (1993) show that the more people use computers, the more their self-confidence with respect to computer increases.

The majority of international students in this study (51.6%) believed that when they were successful in using the e-Services system for one task, they felt confident about using it for other tasks. The evidence is strengthened by a further 17% of international students strongly agreeing to the item. The data shows that a high number of international students were willing to try to test the use of e-Services in more than one e-Services

Table 6. Students awareness when interacting with the e-services system

		Frequency	Percent	Valid Percent	Cumulative Percent
Valid	Strongly Disagree	22	5.5	5.5	5.5
	Disagree	150	37.2	37.6	43.1
	Not Sure	116	28.8	29.1	72.2
	Agree	88	21.8	22.1	94.2
	Strongly Agree	23	5.7	5.8	100.0
	Total	399	99.0	100.0	
Missing	Not Applicable	4	1.0		
Total		403	100.0		

activity. This suggests that in this case international student usage grows with increased experience.

Hoffman and Novak (1995) and Zeithaml et al. (2000) argue that users experience a number of positive consequences of flow such as: increased learning, increased perceived behavioural control, and increased exploratory and participatory behaviour, when interacting with e-Services system websites. Wanting to continue to use an e-Services system then may stem from the interactivity involved, which may positively or negatively influence client's perceptions, and such perceptions favourably affect the use of different types of e-Services. In this study, 24% of the international students in this study were not sure whether, if successful in one type of task, they would feel confident using the e-Services system for other tasks (see Table 7).

The international student's experience involves direct interaction with the university e-Services system. User's skills should provide them with experience in using different e-Services features that may facilitate their interaction with a system (Koufaris, 2002; Chea & Lou, 2008; Chellappan, 2008). van Riel et al. (2001) and Chea and Lou (2008) found information on an e-Services system should provide a clear understanding to the user about where to look for that information rather than wasting time and effort in searching for information in the wrong place. However, some

users may not have a very clear perception of knowing more compared to other users, which has been demonstrated in studies by Taylor and Todd (1995) and Morris and Turner (2001). It may be that every user's experience is unique and may differ from that of another user (Morris & Turner, 2001). In this study, 68% of international students had confidence in using and then re-using the e-Services system.

CONCLUSION

The descriptive statistical data shows that student user experience was an important factor in deciding the acceptance, use, and continued use of the e-Services System. The empirical evidence shows the importance of their skills, ability to find information on the e-Services system, ease of movement within the e-Services system, and confidence in using the e-Services system in accepting and using the system.

The initial acceptance of e-Services is the starting point where the international students evaluate the different features and functions of e-Services based on individual experiences. The international student's assessment of the e-Services application is formed on the basis of whether it is easy to follow the e-Services, whether the application is used in their work, and whether information is quicker to

Table 7. When successful in a task, students feel confident using e-services for other tasks

		Frequency	Percent	Valid Percent	Cumulative Percent
Valid	Strongly Disagree	4	1.0	1.0	1.0
	Disagree	22	5.5	5.6	6.6
	Not Sure	95	23.6	24.0	30.6
	Agree	208	51.6	52.5	83.1
	Strongly Agree	67	16.6	16.9	100.0
	Total	396	98.3	100.0	
Missing	Not Applicable	7	1.7		
Total		403	100.0		

find. Such aspects of e-Services form an underlying measurement related to how international students form their beliefs about using e-Services. After initial acceptance, the international student makes careful consideration about using e-Services on a continuous basis in their work.

REFERENCES

Chea, S., & Lou, M. (2008). Post-adoption behaviors of e-service customers: The interplay of cognition and emotion. *International Journal of Electronic Commerce, 12*(3), 29–56. doi:10.2753/JEC1086-4415120303

Chellappan, C. (2008). *E-services: The need for higher levels of trust by populace.* Chennai, India: Anna University.

Cockburn, A., & McKenzie, B. (2000). What do web users do? An empirical analysis of web use. *International Journal of Human-Computer Studies, 54*(6), 903–922. doi:10.1006/ijhc.2001.0459

D'Ambra, J., & Rice, R. E. (2001). Emerging factors in user evaluation of the world wide web. *Information & Management, 38*, 373–384. doi:10.1016/S0378-7206(00)00077-X

Heijden, H. (2000). *Using the technology acceptance model to predict website usage: Extension and empirical test.* Research Memorandum. Amsterdam, The Netherlands: Vrije Universiteit.

Hoffman, D. L., & Novak, T. P. (1995). *Marketing in hypermedia computer-mediated environments: Conceptual foundations.* Working Paper No. 1. Project 2000: Research Program on Marketing in Computer-Mediated Environments.

Huang, C.-Y., Shen, Y.-C., Chiang, I.-P., & Lin, C.-S. (2007). Concentration of web users' online information behavior. *Information Research, 12*(4). Retrieved from http://InformationR.net/ir/12-4/paper324.html

Koufaris, M. (2002). Applying the technology acceptance model and flow theory to online consumer behaviour. *Information Systems Research, 13*(2), 205–223. doi:10.1287/isre.13.2.205.83

Morris, M. G., & Turner, J. M. (2001). Assessing users' subjective quality of experience with the world wide web: An exploratory examination of temporal changes in technology acceptance. *International Journal of Human-Computer Studies, 54*, 877–901. doi:10.1006/ijhc.2001.0460

Ramaswami, S. N., Strader, T. J., & Brett, K. (2001). Determinants of on-line channel use for purchasing financial products. *International Journal of Electronic Commerce, 5*(2), 95–118.

Taylor, S., & Todd, P. A. (1995). Understanding information technology usage: A test of competing models. *Information Systems Research, 6*(2), 144–176. doi:10.1287/isre.6.2.144

van Riel, A. C. R., Liljander, V., & Jurriëns, P. (2001). Exploring consumer evaluations of e-services: A portal site. *International Journal of Service Industry Management, 12*(40), 359–377. doi:10.1108/09564230110405280

Webster, J., Trevino, L. K., & Ryan, L. (1993). The dimensionality and correlates of flow in human computer interactions. *Computers in Human Behavior, 9*(4), 411–426. doi:10.1016/0747-5632(93)90032-N

Yaobin, L., & Tao, Z. (2007). A research of consumers' initial trust in online stores in China. *Journal of Research and Practice in Information Technology, 39*(3). Retrieved from http://www.jrpit.acs.org.au/jrpit/JRPITVolumes/JRPIT39/JRPIT39.3.167.pdf

Zeithaml, V. A., Parasuraman, A., & Malhotra, A. (2000). *A conceptual framework for understanding e-service quality: Implications for future research and managerial practice.* Working paper, report no. 00-115. Marketing Science Institute.

Chapter 19
Empirical Analysis for E–Services Acceptance Model:
Important Findings

Kamaljeet Sandhu
University of New England, Australia

ABSTRACT

This study investigates factors that influence the acceptance and use of e-Services. The research model includes factors such as user experience, user motivation, perceived usefulness, and perceived ease of use in explaining the process of e-Services acceptance, use, and continued use. The two core variables of the Technology Acceptance Model (TAM), perceived usefulness and perceived ease of use, are integrated into the Electronic Services Acceptance Model (E-SAM).

INTRODUCTION

E-Services system continuance at the individual user level is central to the survival of many electronic commerce firms, such as Internet Service Providers (ISPs), online retailers, online banks, online brokerages, online travel agencies, and the like (Bhattacherjee, 2001). Information, if not available to individual users in online and offline environments, directs the user in adopting the search process to meet their information require-

ments. If users are successful in the search process, it forms a positive perception and becomes part of user experience.

E-Services system on website is generally perceived as being successful, but insufficient evaluation has been conducted on how well the websites meets their user's primary information requirements (D'Ambra & Rice, 2001; Chea & Lou, 2008; Chellapan, 2008). Accordingly they argues that in the pre-adoption process users evaluate the extent to which their expectations were met and in the post-adoption process, they make a decision about continuing or discontinuing use of the system.

DOI: 10.4018/978-1-4666-3658-3.ch019

An individual's intention to adopt (or continue to use) Information Technology (IT), including e-Services is determined by two basic sets of factors: one set reflects personal interests and the other reflects social influence (Karahanna, et al., 1999; Liu & Khalifa, 2002; Xue, et al., 2004; Hsu & Chu, 2004). With regard to personal factors, attitudes toward adopting (or continuing to use) an IT system or innovation reflects an individual's positive and negative evaluations of it (i.e. experience, motivation, perceived usefulness, and perceived ease of use).

THE RESEARCH MODEL: ELECTRONIC SERVICES ACCEPTANCE MODEL (E-SAM)

The research model includes e-Services acceptance and continued use of e-Services as dependent variables. The dependent variable, that is acceptance of e-Services, relates to: 1) the intention to navigate e-Services system, which makes way for user interaction; 2) the intention to use the e-Services system more for work, indicating that user interaction is facilitated; and 3) the likelihood of finding information quickly and so meets with user objectives. The other dependent variable is

'continued use' of the e-Services system in relation to international students' use in areas other than work, success or failure in work, task focus, screen design personalisation, effects of incomplete information and unfriendly features. All of the variables were measured on six point Likert scales (see Figure 1).

Ease of Use Factor

Ease of use (Factor 1) was determined by international student response to items about how to use the e-Services system, learning to use those service, the flexibility in using the e-Services system, the level of effort that was required, the ease to use the e-Services system, navigation of the e-Services system, movement that related to accessibility of Web pages on the e-Services system, and information on the e-Services systems Web pages.

Usefulness Factor

Usefulness (Factor 2) was determined from measures of e-Services system support, ability to do work quickly, work performance, work productivity, work effectiveness, ease of use, usefulness, and work usage of the e-Services system.

Figure 1. Electronic services acceptance model (E-SAM)

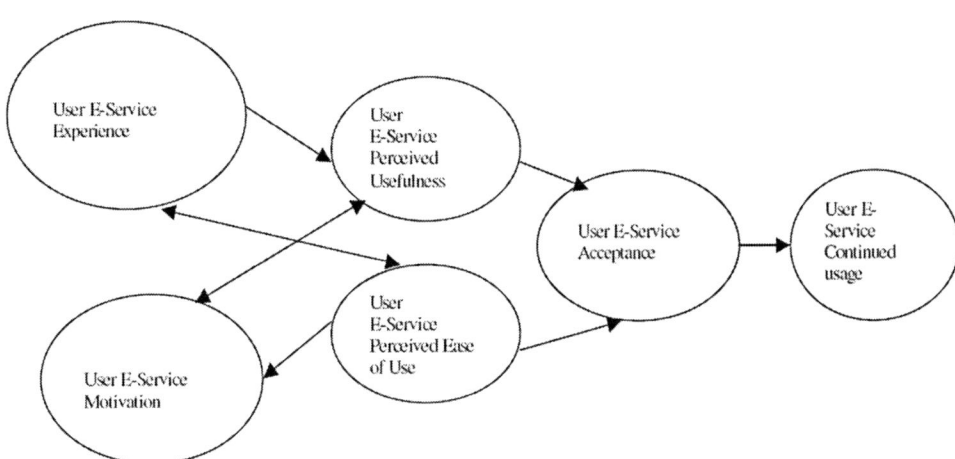

Experience Factor

International student experience (Factor 3) was determined by measures of e-Services system control when interacting with the e-Services system, feelings of confusion in use, feelings of calmness when interacting with the e-Services system, feelings of frustration when faced with problems when using the system, ability to locate the e-Services system, the process of becoming skilled when interacting with the e-Services system, and knowing about e-Services system availability on the university's website.

Motivation Factor

Motivation (Factor 4) was determined from measures of student inspiration, encouragement to use e-Services based on past and present interactions, the level of expectations in using the e-Services system, and peer influence. The acceptance of e-Services (Factor 5) was determined by measures about monitoring time when using the e-Services system, money spent when using the system, and about proposed benefits that positively or negatively affect acceptance of the e-Services system.

Acceptance and Continued Usage Factor

The continued usage of e-Services (Factor 6) was determined by measures that focused on e-Services system features that the international students used when revisiting the e-Services system. These were work accomplishment in terms of the task they were able to do, perception of the level of support, feelings of independence when using the e-Services system, e-Services system screen displays, e-Services system personalised welcome, levels of non-work usage, and concentration levels when interacting with the e-Services system. The continued use of e-Services was assumed to have been developed over a period of time when students have formed a perception of their successful

and unsuccessful activities on e-Services system, on the concentration needed to interact with the e-Services system, on the state of the customised screens and on other features of the e-Services system that assisted them in their work.

HYPOTHESES

From the analysis of previous research, it was hypothesized that in the in the acceptance, use, and continued use of the e-Services system:

H1: User experience positively affects perceived usefulness,

H2: User experience positively affects perceived ease of use,

H3: Perceived usefulness positively affects user motivation,

H4: Perceived ease of use positively affects user motivation.

The Technology Acceptance Model (TAM) provides a useful comparison between perceived usefulness and perceived ease of use. Based on usefulness of the e-Services system the user will have a favorable perception to accept e-Services if it results in some form of improvement or increase in work productivity that is directly related to the use of the e-Services. The useful outcomes from using e-Services are dependent on e-Services acceptance. Similarly, if the user's perceptions of the e-Services system are that it is easy to use it would reflect in the level of acceptance of the system. Therefore, it was also hypothesized that in the acceptance, use and continued use of the e-Services system:

H5: Perceived usefulness positively affects user acceptance,

H6: Perceived ease of use positively affects user acceptance,

H7: User acceptance positively affects continued use of e-Services.

The continued use of e-Services is formed on the basis that the user has accepted the system as a productive tool in their work.

DATA ANALYSIS

For the data analysis, the study used a two-step estimation approach, in which the first step analyses the measurement model and the second estimates a structural relationship. The two-step estimation details the distinctive characteristics of the relationship between indicators (observed variables) and the underlying variables. After completing an estimation of the measurement model through an exploratory factor analysis, the next step involves a path analysis (structural model) using Partial Least Squares (Chin, 1998).

To conduct a factor analysis it is important to test the data initially for the adequacy of the data and the degree of relatedness of the variables. The Kaiser-Meyer-Olkin Measure of Sampling Adequacy is a statistic which indicates the proportion of variance in the variables which is common variance, i.e. which might be caused by underlying factors (SPSS, 2003). Table 1 shows the empirical value for Kaiser-Meyer-Olkin Measure of Sampling is 0.915, which explains that a factor analysis will be useful with the data (SPSS, 2003). Bartlett's test of sphericity indicates whether the correlation matrix is an identity matrix, which would indicate that the variables are unrelated (SPSS, 2003). The Bartlett's test of sphericity is 0.000. Very small values (less than .05) indicate that there are probably significant relationships among the variables (SPSS, 2003).

The next step was to do a factor analysis using SPSS (2003).

FACTOR ANALYSIS

An exploratory factor analysis was conducted with Varimax rotation to test the construct validity of the survey questions. Rummel (1970) suggests that a hypothesis regarding dimensions of user attitude can be measured using factor analysis because meaning which is associated with 'dimension' is that of a cluster or group of highly inter-correlated characteristics or behaviour for which factor analysis may be used to test for their empirical existence. The author further suggests that those characteristics or behaviours which should, in theory, be related to a particular dimension, can be postulated in advance and statistical tests of significance can be applied to the factor analysis results. Another technique for testing the hypotheses in this study is 'factor analysis' to test patterns within the variables. The testing of the hypotheses is intended to demonstrate further validation of the instrumentation, as discussed earlier. If the variables perform as predicted by theory, then we can infer that the measurement of the variables is nomologically valid (Koufaris, 2002).

For scale assessment, a combination of confirmatory factor analysis and reliability analysis is used. Confirmatory factor analysis is used to assess variable validity for the variables considered in

Table 1. Kaiser-Meyer-Olkin measure and Bartlett's test

Kaiser-Meyer-Olkin Measure of Sampling Adequacy.		0.915
Bartlett's Test of Sphericity	Approx. Chi-Square	5515.282
	df	703
	Sig.	.000

this research. Follow-up reliability analysis is used to further assess the stability of the scales used.

To test the reliability of the measures, Cronbach's (1971) alpha values for the multiple items were calculated. All variables (factors) that emerged from the factor analysis showed high Cronbach's (except for continued use), establishing the reliability of the instrument. Perceived usefulness had a Cronbach .908, perceived ease of use .899, motivation .731, experience .7213, e-Services acceptance .735, and e-Services continued usage .519. This analysis helps reinforce the validity and reliability of the scales used in this study. According to Nunnally (1978), a reliability value of 0.8 or greater is desired, but 0.7 or greater is acceptable. Thus, the reliability for all the measures is acceptable.

Two sets of Factor Analysis were conducted (Factor Analysis 1 and 2). The first set included only the TAM variables of usefulness and ease of use (see Tables 3 and 4). The purpose of first conducting a factor analysis with TAM variables was to analyse how the TAM items in a factor perform. The second factor analysis includes other variables such as international students experience, their motivation, e-Services acceptance, and continued usage of e-Services as well as the TAM variables of usefulness and ease of use (see Table 3 and Table 4). The comparison of two sets of Factor analysis provides information about the relationship of TAM variables to other variables that have been introduced in this study and enable examination of the performance of TAM variables that were adopted into the Electronic Services Acceptance Model (E-SAM).

FACTOR ANALYSIS 1 FINDINGS

Table 2 shows the total variance explained for the TAM model factors, perceived ease of use and perceived usefulness. Only two factors have been extracted with eigenvalues above 1, in total the two factors account for 60.351% of the variance in the items and therefore appear to be a good representation of the original data set. Factor 1- perceived ease of use explains 30% of the variance and Factor 2- perceived usefulness also explains 30% of the variance.

The factor scores for each item measuring these two factors, ease of use and usefulness, are shown in Table 3. To test the construct validity of ease of use and usefulness, factor loadings were assessed. Loadings for items above .50 are show. A high loading value indicates that a strong relationship exists between that item and either ease of use or usefulness. Factor 1 relates to ease of use and Factor 2 to usefulness.

Ease of Use Factor

The first factor consists of items that measured: understandability about how to use the system, learning to use the e-Services system, the flexibility in using the e-Services system, the level of effort that is required to use e-Services system, the ease to use the e-Services system, navigation of the e-Services system on the Web, movement that related to accessibility of Web pages on the e-Services system, and finding information on the e-Service system. The result of the analysis

Table 2. Total variance explained extraction method: principal component analysis

Factors	Initial Eigenvalues			Extraction Sums of Squared Loadings			Rotation Sums of Squared Loadings		
	Total	% of Var	Cumul %	Total	% of Var	Cumul %	Total	% of Var	Cumul %
1	7.390	46.187	46.187	7.390	46.187	46.187	4.845	30.279	30.279
2	2.266	14.164	60.351	2.266	14.164	60.351	4.812	30.073	60.351
3	.744	4.653	65.004						

Table 3. Rotated component matrix for students user group extraction method: principal component analysis, rotation method: varimax with Kaiser normalization. Rotation converged in 3 iterations.

Factors	Factor 1 Ease of use	Factor 2 Usefulness
Ease of use		
Understandability	.772	.215
Learning	.779	.148
Flexible	.716	.217
Effort	.750	.250
Easy to use	.818	.249
Navigation	.769	.259
Movement	.711	.095
Find	.580	.275
Usefulness		
e-Services system support	.065	.639
Quicker	.330	.728
Performance	.185	.849
Productivity	.234	.841
Effectiveness	.179	.842
Easier	.290	.751
Useful	.249	.679
Work usage	.225	.593

shows that most items load highly on their associated factors.

Usefulness Factor

The second factor consists of items that measured: the level of support on the e-Services system, impact on working quicker, effect on work in terms of performance, productivity, and effectiveness, usage of the system, ease of use of the system, and usefulness in work. All items in the 'usefulness' factor had loadings above 0.8. The items student's work performance (.849), work productivity (.841), and work effectiveness (.842) emerged with the highest measures. The results reveal that these variables are valid and strongly influence the usefulness factor.

The higher loading items in the ease of use variable are the ease of use of the e-Services system (.818), learning to use the e-Services system (.779), and in understanding how to use the system (.772). These measures are valid and

have a strong influence on the ease of use Factor 1 (see Table 3). These items in the ease of use Factor 1 are important and form a reliable scale for this variable. For both factors, the composite reliability coefficients values exceeded 0.7 and are thus considered valid according to Nunnally (1978). Thus, the measures of these variables are reliable.

The results from the factor analysis (see Table 3) support the positioning of TAM variables (i.e. perceived ease of use and perceived usefulness) in the Electronic Services Acceptance Model (E-SAM). The consistent data pattern suggests that the measurement scale has high validity and reliability.

FACTOR ANALYSIS 2 FINDINGS

A second factor analysis was then undertaken which included the other factors: student experience and motivation, acceptance and continued

Table 4. Rotated component matrix for students user group

Factors	1 Ease of use	2 Usefulness	3 Experience	4 Motivation	5 Accept	6 Cont. usage	7 Cont. usage	8 Cont. usage	9 Cont. usage
Ease of use									
Understandability	**.686**	.205	.322	.080	-.085	-.062	.095	.041	.010
Learning	**.770**	.135	.141	.043	-.033	.049	.209	.065	-.019
Flexible	**.672**	.220	.125	.177	-.007	-.122	-.046	-.010	.016
Effort	**.766**	.246	.138	.039	.071	.099	.022	.025	.014
Easy to use	**.713**	.250	.309	.032	-.028	-.099	.021	.085	.141
Navigation	**.704**	.244	.332	.054	.070	.010	-.007	.086	.129
Movement	**.627**	.068	.191	.142	.018	-.197	.073	-.022	.124
Find	**.635**	.187	.244	.137	.146	-.134	-.087	.000	-.016
Usefulness									
Website support	.065	**.536**	.040	.235	.204	-.095	.231	.137	.026
Quicker	.287	**.721**	.240	.111	.104	.048	-.041	.020	-.063
Performance	.107	**.826**	.179	.130	.088	-.021	.017	.118	.084
Productivity	.206	**.840**	.109	.138	.078	.015	.063	.015	-.007
Effectiveness	.144	**.842**	.126	.139	.060	.060	-.024	.044	-.021
Easier	.255	**.742**	.123	.160	.077	-.029	.121	-.055	.014
Useful	.248	**.704**	.084	.086	-.003	.009	.105	-.056	.045
Work usage	.179	**.609**	.101	.036	.067	-.042	.014	.199	.276
Experience									
Control	.456	.161	**.531**	.101	-.003	.024	.050	.115	.048
Confused	-.308	-.131	**-.730**	-.077	.066	.101	-.082	.010	.023
Calm	.345	.086	**.546**	.114	.030	-.058	-.062	.233	-.105
Frustrated	-.274	-.152	**-.731**	-.079	-.012	.153	-.006	.014	.112
Skilled	.283	.263	**.590**	-.069	.043	.124	.031	-.038	.160
Know	.306	.229	**.583**	.125	.016	-.001	.065	-.029	.284
Motivation									
Inspiration	.372	.360	.119	**.523**	.090	-.057	.062	.062	.001
Encouragement	.165	.148	.120	**.693**	.014	.336	-.014	-.081	.061
Expectations	.203	.288	.073	**.689**	-.037	.055	.163	.082	.052
Peers influence	.039	.199	.046	**.650**	.151	-.087	-.058	.140	.145
Acceptance									
Monitor time	.096	.102	-.027	.036	**.864**	-.015	-.020	.074	.011
Dollar spending	-.072	.088	.158	-.001	**.787**	.082	-.017	.063	.203
Proposed benefits	.038	.217	-.137	.134	**.688**	.071	.166	-.060	-.122
Continued usage									
Website features	-.120	-.018	-.098	-.070	.132	**.683**	.081	.273	-.091
Accomplishment	-.139	-.017	-.062	.185	.002	**.759**	.091	-.026	.086
Service perception	.163	.113	.040	.164	.128	-.028	**.731**	.118	-.212
Inadequate support	.026	.145	.062	-.073	-.014	.235	**.773**	.028	.200
Independent	.310	.207	.076	.035	.064	.067	**.653**	-.167	-.078
Screens	.009	.098	.047	.009	.019	.363	-.011	**.748**	-.027
Welcome	.139	.098	.048	.142	.060	-.066	.144	**.731**	.143
Non-work usage	.243	.051	-.010	-.017	.161	.098	-.195	.083	**.609**
Concentration	-.027	.081	.047	.274	-.050	-.076	.169	.030	**.611**

Extraction Method: Principal Component Analysis. Rotation Method: Varimax with Kaiser Normalization.
Rotation converged in 9 iterations.

usage of the e-Services system along with the TAM factors - perceived usefulness and ease of use. Table 4 displays the loadings in a matrix for the international student user group. Factor 1 represents ease of use, Factor 2 usefulness, Factor 3 student experience, Factor 4 student motivation, Factor 5 acceptance of e-Services, and Factors 6, 7, 8, and 9 relate to continued usage of the e-Services system.

Factorial validity is concerned with whether the ease of use and usefulness form distinct variables (Davis, 1989). Table 4 shows the factors, loadings, and resultant scale reliabilities for the ease of use, usefulness, experience, motivation, acceptance, and continued use of the e-Services system. As expected, the ease of use and usefulness factors measured strongly in this study.

Ease of Use Factor

The factors, ease of use, tests the student's abilities in learning how to use the e-Services system.

Learning to use the e-Services system seems reasonable for student user interaction to begin the use with the e-Services system. Similar findings about the learning to use the system have been positively reported by Dillon and Gabbard (1998), Chea and Lou (2008), and Chellapan (2008). This study has arrived at similar findings that the users perceive the e-Services system to provide some form of understanding about how to use the e-Services system, which may form a basic criteria determining user-e-Services system interaction. This finding has also consolidated previous research findings (e.g. Csikszentmihalyi, 1975; Venkatesh & Spier, 2000; Venkatesh, 1999) that the user in a task environment associates with their willingness to greater learning on how to use those systems.

Usefulness Factor

Factor 2 measured usefulness of the e-Services. The most influential item in this factor is student user's work effectiveness. It is reasonable to argue that the usefulness of a system, amongst other things, may be determined on the basis of how effective an e-Services system is in user work. This item has been similarly recorded as the highest measuring item in Davis's study (1989) when he first theorized TAM. This item is a fundamental structure that forms the usefulness of the systems. The findings of user work effectiveness in perceived usefulness factor has been demonstrated to be equally validated here in this study. This study has also demonstrated the robustness of this item and its application and validation for the e-Services system and has also tapped into an important aspect of usefulness of the e-Services system.

Experience Factor

Factor 3 reports on the student users experience to use the e-Services system. The influential item captured in this factor is user's feelings of frustra-

tion. The feelings of frustration in using the system scored in the top number three items in Davis' study (1989) and has occupied number one place in this study. This item has been considered as an important aspect of the user's expressing their experience with the e-Services system. However, this item was measured in Davis' study (1989) as a perceived ease of use factor item and not as user experience, as has been adopted in this study. Nevertheless, the ranking of this item in the number one place in the formation of the user experience factor is considered as influential in e-Services systems use. The two items having negative values in Factor 3 of user experience are users feeling confused when using the e-Services system and users feeling frustrated. These are shown in Table 4. These two items were rotated in SPSS to attain a positive valuation in order to be included in the experience factor. One other interesting finding that emerged in the data is that the item, users feeling confused using the system, scored on the seventh rank in the Davis study (1989), but scored at the number two rank for the items in this study. The scores and rankings of these two items, user frustration and user confusion in using the e-Services system, have improved when adopted in the user experience factor, rather than in the perceived ease of use factor.

Motivation Factor

Factor 4 reports on student motivation to use the e-Services system. Previous research in technology adoption by Venkatesh (1999) suggests that user's motivation includes the desires to perform an activity based on perceptions of past and present efforts. The item that is user encouragement to use e-Services based on past and present interactions is the most influential item in the motivation factor of this study. Bandura (1986) suggests two types of expectations, efficacy, and outcome that best determine user behaviour. An outcome expectancy is defined as a "person's estimate that a given behaviour will lead to certain outcomes"

(Bandura, 1986). Therefore users understanding of interaction with the e-Services system can largely be based on this assumption that users are encouraged to use system and are driven by the outcomes of those perceptions and which are formed from past and present outcomes that the user had in using the system. Koufaris (2002) and Heijden (2000) found that perceived enjoyment on websites that is associated with user motivation is influential in driving the continued use of, and visits to, the websites.

Acceptance of E-Services Factor

Factor 5 reports on acceptance of the e-Services system. The influential item in this factor is users monitoring time. This item in Davis (1989) study was dropped after it was placed in the eleventh rank of the item list for perceived usefulness. Though the wordings of this item has been improved in this research to—users monitoring time from the Davis (1989) study that was phrased as—Saves me time. This item was adopted in the user acceptance factor compared to Davis' (1989) study that adopted this item as a perceived usefulness item. The findings of this item on the first rank of the user acceptance factor also demonstrates that this item performs well when adopted in the user acceptance of e-Services and better explains the user perception towards e-Services systems use.

Continued E-Services Usage Factor

Factors 6, 7, and 8 have reported on the continued e-Services usage. The item users inadequate support measured as the most influential item in continued use of the e-Services system. This item has performed well when adopted in the continued use factor compared to Davis' (1989) study which was ranked number twelve on the item list and not adopted in the perceived ease of use factor. Though this item has been shown to perform well in continued use of the e-Services system in this study, it was subsequently dropped

from the perceived ease of use in the Davis study (Davis, 1989). The wordings of this item has been improved to—users feeling inadequate support on the e-Services system, which in Davis' (1989) study was phrased as—provides guidance. This item after scoring number one on the item list also demonstrates that it best fits in the continued use factor in this study.

In order to retain an item, it must have a significant factor loading. Hair et al. (1995) suggests that a factor loading value of 0.3 should be considered a significant value while a factor loading value of 0.4 can be considered to be more significant value. A factor loading value that exceeds 0.5 is considered to be very significant value for convergent validity (Hair, et al., 1995). All the items in the survey used in this study achieved convergent validity.

The items that had a loading value below .05 are e-Services system enjoyment (0.466), confidence to use the e-Services system (0.442), improved self-service to use the e-Services system (0.383), finding support on the e-Services (0.425), motivation by improved work performance (0.413), training to use the e-Services system (0.383), and new features of the e-Services system (0.463). These items were dropped to improve the factor scores. The result of the new factor analysis is shown in Table 4. Eigen values above the threshold level (1.0) were used as the criterion for the factor analysis and the factor analysis now explains 63.315% variance among items.

As shown in Table 5, the independent variables converged well into four factors, perceived ease of use (Factor 1) explains 14% of the variance, perceived usefulness (Factor 2) explains 14% of the variance, user experience (Factor 3) explains 8% of the variance, and user motivation (Factor 4) explains 5% of the variance, verifying that each item measuring one of the four variables representing that specific factor.

Similarly, factor analysis of the dependent variables converged well into five factors, user acceptance which is Factor 5 explains 5% of the variance, continued use which consists of Factors 6, 7, 8, and 9 together explains only 3% of the variance shown in Table 5.

The data shows that the independent variables, i.e. perceived ease of use, perceived usefulness, and user experience have a greater impact on the factor score affecting the dependent variables, i.e. user acceptance and use of e-Services and that the independent variables, i.e. user motivation and continued use has the least impact on the dependent variables, i.e. user acceptance and use of the e-Services system.

CONCLUSION

The evaluation of the factors is reported in detail. However, six key conclusions have emerged. The variables perceived ease of use and perceived usefulness are significant in the E-SAM model in explaining those factors, which influence the acceptance, use and continued use of the e-Services

system by international students. The data shows that the most impact on user acceptance, use and continued use of the e-Services system comes from perceived ease of use, perceived usefulness, and user experience, and perceived usefulness is more significant than perceived ease of use in this study. Student experience and motivation to use e-Services are also significant factors in terms of influencing the acceptance, use, and continued use of e-Services. This also suggests that perceived usefulness is an important guiding factor to user acceptance of the e-Services system. Student evaluation of e-Services acceptance is based on time they spent in doing work on the e-Services system; money (dollars) spent in using the e-Services, and proposed benefits in student's work to use e-Services. Finally, it can also be concluded that continued use of e-Services is primarily influenced by e-Services system features, work accomplishment that assists the user in continuously using them in their work and determine their importance based on meeting their work

Table 5. Total variance explained

Factors	Initial Eigenvalues			Extraction Sums of Squared Loadings			Rotation Sums of Squared Loadings		
	Total	% of Var	Cumul %	Total	% of Var	Cumul %	Total	% of Var	Cumul%
1. Ease of use	10.64	28.78	28.780	10.64	28.78	28.780	5.256	14.206	14.206
2. Usefulness	3.255	8.798	37.578	3.255	8.798	37.578	5.248	14.185	28.390
3.Experience	1.876	5.071	42.649	1.876	5.071	42.649	3.052	8.249	36.639
4. Motivation	1.720	4.648	47.297	1.720	4.648	47.297	2.135	5.770	42.410
5. Acceptance	1.430	3.865	51.162	1.430	3.865	51.162	2.069	5.591	48.000
6. Continued use	1.303	3.521	54.684	1.303	3.521	54.684	1.558	4.210	52.210
7. Continued use	1.139	3.079	57.763	1.139	3.079	57.763	1.468	3.969	56.179
8. Continued use	1.044	2.820	60.583	1.044	2.820	60.583	1.414	3.822	60.001
9. Continued use	1.011	2.732	63.315	1.011	2.732	63.315	1.226	3.314	63.315
10	.961	2.598	65.913						

requirements. Because the validity and reliability of the measures are within acceptable levels, the next step was to perform a path analysis, using PLS (Cin, 1998) to test the formulated research hypotheses and evaluate some degree of causality.

Factor analysis is not explanatory and only an indication of the factors. The factor analysis shows the validity of the factors that have been measured on the scale, but it tells little about the working structure of the E-SAM model. Then, to test the factors in a model, PLS was used to determine how the factors (i.e. the factors derived from factor analysis such as user experience, user motivation, perceived ease of use, perceived usefulness, user acceptance, and continued use of e-Services) behave in the E-SAM model. And to know the influence of those factors (i.e. user experience, user motivation, perceived ease of use, perceived usefulness) in user acceptance, use and continued use of e-Services. The PLS provides the fundamental basis to understand the relationship of the factors in the model and to distinguish from important factors to unimportant factors.

The evidence from the PLS analysis supports the E-SAM model variables for user experience, perceived ease of use, and user acceptance and that they are important in explaining acceptance, use and continued use of the e-Services System. The consistent pattern suggests that the measurement scale has high validity and reliability. In addition, student acceptance of e-Services system is likely to increase if students perceive significant ease of use within the e-Services system. Such features en-

hance the ability of students, endowing them with higher flexibility within the e-Services system.

In contrast, experience does not appear to influence perceived usefulness. This is perhaps supported by e-Services system characteristics being different from a student's perspective. It is proposed that student experience affects the usage pattern for e-Services system. Better control features, self-service ability and good support within the e-Services system are likely to increase usage frequency. It is interesting to note that a high correlation exists between student experience (0.572) and perceived ease of use (0.715), which can further enhance student acceptance of e-Services.

Table 6 shows a summary of hypotheses tested in this study. Student experience does not affect perceived usefulness to use e-Services whereas perceived ease of use does. Perceived usefulness affects student motivation but it does not affect acceptance of e-Services, which is rather surprising since there is no correlation between the perceived motivation and acceptance of e-Services.

There is no obvious link in this data between perceived usefulness, motivation, and acceptance. Rather the link exists only between perceived usefulness and motivation. It is more surprising to note that perceived ease of use affects acceptance of e-Services but it does not affect student motivation. There is no evidence to support the notion that e-Services acceptance will lead to continued use of e-Services.

Table 6. Summary of hypotheses tests

Hypotheses	Supported
H1: Student experience positively affects perceived usefulness to use e-Services	No
H2: Student experience positively affects perceived ease of use to use e-Services	Yes
H3: Perceived usefulness positively affects student motivation to use e-Services	Yes
H4: Perceived ease of use positively affects student motivation to use e-Services	No
H5: Perceived usefulness positively affects student acceptance of e-Services	No
H6: Perceived ease of use positively affects student acceptance of e-Services	Yes
H7: Student acceptance positively affects continued use of e-Services	No

REFERENCES

Bandura, A. (1986). *Social foundations of thought and action: A social cognitive theory*. Englewood Cliffs, NJ: Prentice-Hall.

Bhattacherjee, A. (2001). Understanding information systems continuance. *Management Information Systems Quarterly, 25*(3), 351–370. doi:10.2307/3250921

Bianchinia, D., Antonellisa, V., Pernicib, B., & Plebanib, P. (2006). Ontology-based methodology for e-service discovery. *Information Systems, 31*, 361–380. doi:10.1016/j.is.2005.02.010

Chea, S., & Lou, M. (2008). Post-adoption behaviors of e-service customers: The interplay of cognition and emotion. *International Journal of Electronic Commerce, 12*(3), 29–56. doi:10.2753/JEC1086-4415120303

Chellappan, C. (2008). *E-services: The need for higher levels of trust by populace*. Chennai, India: Anna University.

Chin, W. W. (1998). *The partial least squares approach for structural equation modelling*. Hoboken, NJ: Lawrence Erlbaum Associates.

Cronbach, L. J. (1971). *Test validation: Education measurement*. Washington, DC: American Council on Education.

Csikszentmihalyi, M. (1975). *Beyond boredom and anxiety*. San Francisco, CA: Jossey-Bass.

D'Ambra, J., & Rice, R. E. (2001). Emerging factors in user evaluation of the world wide web. *Information & Management, 38*, 373–384. doi:10.1016/S0378-7206(00)00077-X

Davis, F. D. (1989). Perceived usefulness, perceived ease of use, and user acceptance of information technology. *Management Information Systems Quarterly, 13*(2), 319–340. doi:10.2307/249008

Dillon, A., & Gabbard, R. (1998). Hypermedia as an educational technology: A review of the quantitative research literature on learner comprehension, control, and style. *Review of Educational Research, 68*(3), 322–349.

Gelderman, M. (1998). The relation between user satisfaction, usage of information systems and performance. *Information & Management, 34*, 11–18. doi:10.1016/S0378-7206(98)00044-5

Hair, J. F., Anderson, R. E., Tatham, R. L., & Black, W. C. (1995). *Multivariate data analysis with reading*. Englewood Cliffs, NJ: Prentice Hall.

Heijden, H. (2000). *Using the technology acceptance model to predict website usage: Extension and empirical test*. Research Memorandum. Amsterdam, The Netherlands: Vrije Universiteit.

Hsu, M., & Chu, C. (2004). Predicting electronic service continuance with a decomposed theory of planned behaviour. *Behaviour & Information Technology, 23*(5), 359–373. doi:10.1080/01449290410001669969

Karahanna, E., Straub, D. W., & Chervany, N. L. (1999). Information technology adoption across time: A cross-sectional comparison of pre-adoption and post-adoption beliefs. *Management Information Systems Quarterly, 23*(2), 183–213. doi:10.2307/249751

Khalifa, M., & Liu, V. (2003). Satisfaction with internet-based services: The role of expectations and desires. *International Journal of Electronic Commerce, 7*(2), 31–49.

Koufaris, M. (2002). Applying the technology acceptance model and flow theory to online user behaviour. *Information Systems Research, 13*(2), 205–223. doi:10.1287/isre.13.2.205.83

Nunnally, J. C. (1978). *Psychometric theory*. New York, NY: McGraw-Hill.

Rummel, R. J. (1970). *Applied factor analysis.* Evanston, IL: Northwestern University Press.

Venkatesh, V. (1999). Creation of favourable user perceptions: Exploring the role of intrinsic motivation. *Management Information Systems Quarterly, 23*(2), 239–260. doi:10.2307/249753

Venkatesh, V., & Spier, C. (2000). Computer technology training in the workplace: A longitudinal investigation of the effect of mood. *Organizational Behavior and Human Decision Processes, 79*(1), 1–28. doi:10.1006/obhd.1999.2837

Xue, M., Harker, P. T., & Heim, G. R. (2004). *Incorporating the dual customer roles in e-service design.* Philadelphia, PA: The Wharton Financial Institutions Center.

Compilation of References

Abrahamson, E. (2000). Change without pain. *Harvard Business Review*, 78(4), 75–79.

Abrahao, S., Condori-Fernandez, N., Olsina, L., & Pastor, O. (2003). Defining and validating metrics for navigational models. In *Proceedings of the 9th International Software Metrics Symposium (METRICS 2003)*, (pp. 200-210). Sydney, Australia: METRICS.

Adair, J. (1985). *Management decision-making*. Aldershot, UK: Gower Publishing.

Adams, W. C. (1986, Spring). Whose lives count? TV coverage of natural disasters. *The Journal of Communication*. doi:10.1111/j.1460-2466.1986.tb01429.x

Agile Manifesto. (2012) *Website*. Retrieved from http://www.agilemanifesto.org

Ahmadi, R. H., Bagchi, U., & Roemer, T. A. (2005). Coordinated scheduling of customer orders for quick response. *Naval Research Logistics*, 52, 493–512. doi:10.1002/nav.20092

Ahonen, J. J., & Savolainen, P. (2010). Software engineering projects may fail before they are started: Post-mortem analysis of five cancelled projects. *Journal of Systems and Software*, 83, 2175–2187. doi:10.1016/j.jss.2010.06.023

Alagappan, B., Alagappan, M., & Danishkumar, S. (2009). Web metrics based on page features and visitor's web behavior. In *Proceedings of the 2nd International Conference on Computer and Electrical Engineering (ICCEE 2009)*, (vol. 2, pp. 236-241). Dubai, UAE: ICCEE.

Alavi, A., & Leidner, D. (2001). Review: Knowledge management and knowledge management systems: Conceptual foundations and research issues. *Management Information Systems Quarterly*, 25(1), 107–136. doi:10.2307/3250961

Alderfer, C. (2011). *The practice of organizational diagnosis: Theory and methods*. Oxford, UK: Oxford University Press.

Alexa. (2011). *The web information company*. Retrieved from http://www.alexa.com/

Alhenshiri, A., & Duffy, J. (2010). User studies in web information retrieval: User-centered measures in web IR evaluation. In *Proceedings of the 2010 IEEE 24th International Conference on Advanced Information Networking and Applications Workshops (WAINA)*, (pp. 632-637). IEEE Press.

Ali, A. (1989). Decision styles and work satisfaction of Arab executives: A cross-national study. *International Studies of Management and Organization*, 19(2), 22–37.

Allwood, C. M., & Selart, M. (2001). *Decision-making: Social and creative dimensions*. Dordrecht, The Netherlands: Kluwer.

Al-Neimat, T. (2005). *Why IT projects fail*. Retrieved from www.projectperfect.com.au

Alqarni, A. (2003). *The managerial decision styles of Florida's State University libraries' mangers*. (Unpublished Dissertation). Florida State University. Tallahassee, FL.

Alsmadi, I. (2010). The automatic evaluation of website metrics and state. *International Journal of Web-Based Learning and Teaching Technologies*, 5(4), 1–17. Retrieved from http://www.igi-global.com/bookstore/article.aspx?TitleId=52596doi:10.4018/jwltt.2010100101

Alsoud, A., & Nakata, K. (2010). Evaluating e-government websites in Jordan: Accessibility, usability, transparency and responsiveness. In *Proceedings of 2010 IEEE International Conference on Progress in Informatics and Computing (PIC 2010)*, (pp. 761-765). Shanghai, China: IEEE Press.

Al-Tmeemy, S. M., Abdul-Rahman, H., & Harun, Z. (2011). Future criteria for success of building projects in Malaysia. *International Journal of Project Management, 29*(3), 241–356. doi:10.1016/j.ijproman.2010.03.003

American National Standard Institute. (2011). *ANSI/EIA-649-B-2011 configuration management standard.* Washington, DC: ANSI.

Araki, I. (1998). *An approach to the information space and on-the-line interaction.* New York, NY: International Sociological Association (ISA).

Arroba, T. (1977). Styles of decision-making and their use: an empirical study. *British Journal of Guidance & Counselling, 5*, 149–158. doi:10.1080/03069887708258110

Asrilhant, B., Meadows, M., & Dyson, R. G. (2004). Exploring decision support and strategic project management in the oil and gas sector. *European Management Journal, 22*(1).

Attewell, P., & Rule, J. (1984). Computing and organizations: What we know and what we don't know. *Communications of the ACM, 27*, 1184–1192. doi:10.1145/2135.2136

Auf der Heide, E. (1989). *Disaster response: Principles of preparation and coordination.* St. Louis, MO: CV Mosby.

Aycock, A. (1995). Technologies of the self: Foucault and internet discourse. *Journal of Computer-Mediated Communication, 1*(2).

Ba, S., Lang, K. R., & Whinston, A. B. (1997). Enterprise decision support using intranet technology. *Decision Support Systems, 20*(2), 99–134. doi:10.1016/S0167-9236(96)00068-1

Bahgat, G. (2010). United States' energy security: challenges and opportunities. *The Journal of Social, Political, and Economic Studies, 35*(4), 409–425.

Baietti, A., Shlyakhtenko, A., La Rocca, R., & Patel, U. (2012). *Green infrastructure finance: Leading initiatives and research.* Washington, DC: World Bank Publications.

Bakker, K. D., Boonstra, A., & Wortmann, H. (2010). Does risk management contribute to IT project success? A meta-analysis of empirical evidence. *International Journal of Project Management, 28*, 493–503. doi:10.1016/j.ijproman.2009.07.002

Balogun, J., & Hope Hailey, V. (2004). *Exploring strategic change* (2nd ed.). London, UK: Prentice-Hall.

Bandura, A. (1986). *Social foundations of thought and action: A social cognitive theory.* Englewood Cliffs, NJ: Prentice-Hall.

Banks, J. (1998). *Handbook of simulation: Principles, methodology, advances, applications, and practice.* New York, NY: John Wiley & Sons Inc.

Barnes, S., & Greller, L. M. (1994). Computer-mediated communication in the organization. *Communication Education, 43*(2), 129–142. doi:10.1080/03634529409378970

Barnes, S., & Vidgen, R. T. (2003). Assessing the quality of a cross-national e-government web site: A study of the forum on strategic management knowledge exchange. In *Proceedings of the 36th Annual Hawaii International Conference on System Sciences (HICSS 2003).* Hawaii, HI: IEEE.

Barney, M. (2002). Motorola's second generation. *Six Sigma Forum Magazine, 1*(3), 13–16.

Bayraktara, E., Lenny Koh, S. C., Gunasekaranc, A., Sarid, K., & Tatoglu, E. (2008). The role of forecasting on bullwhip effect for E-SCM applications. *International Journal of Production Economics, 113*, 193–204. doi:10.1016/j.ijpe.2007.03.024

Beck, K. (2000). *Extreme programming explained: Embrace change.* Reading, MA: Addison-Wesley.

Beckhard, R., & Harris, R. (1987). *Organizational transitions: Managing complex change* (2nd ed.). Reading, MA: Addison-Wesley.

Bedingfield, J. D., & Thal, A. E. (2008). Project manager personality as a factor for success. In *Proceedings of Portland International Conference on Management of Engineering & Technology 2008 (PICMET 2008),* (pp. 1303-1314). Cape Town, South Africa: PICMET.

Beer, M., & Noria, N. (2000). Cracking the code of change. *Harvard Business Review, 78*(3), 133–141.

Benson, B. E. (1986). *Self-reported decision styles for chief nurses and assistant chief nurses in veterans administration field hospitals.* (Unpublished Doctoral Dissertation). Kansas State University. Lawrence, KS.

Bhattacherjee, A. (2001). Understanding information systems continuance. *Management Information Systems Quarterly, 25*(3), 351–370. doi:10.2307/3250921

Bhutta, K. (2003). *Taguchi approach to design of experiments*. Houston, TX: Southwest Decision Sciences Institute.

Bianchinia, D., Antonellisa, V., Pernicib, B., & Plebanib, P. (2006). Ontology-based methodology for e-service discovery. *Information Systems, 31*, 361–380. doi:10.1016/j.is.2005.02.010

Bickart, B., & Schindler, R. M. (2001). Internet forums as influential sources of consumer information. *Journal of Interactive Marketing, 15*(3). doi:10.1002/dir.1014

Billinton, R., & Allan, R. (1992). *Reliability evaluation of engineering systems: Concepts and techniques*. New York, NY: Plenum Press.

Bischofberger, W. R., & Pomberger, G. (1992). *Prototyping-Oriented software development–Concepts and tools*. Berlin, Germany: Springer-Verlag. doi:10.1007/978-3-642-84760-8

Black, J., & Gregersen, H. (2002). *Leading strategic change: Breaking through the brain barrier*. Upper Saddle River, NJ: Prentice-Hall.

Blackburn, J. L., & Domin, T. J. (2007). *Protective relaying principles and applications*. Boca Raton, FL: Taylor & Francis Group.

Blocher, J. D., & Chhajed, D. (1998). The customer order lead-time problem on parallel machines. *Naval Research Logistics, 43*(5), 629–654. doi:10.1002/(SICI)1520-6750(199608)43:5<629::AID-NAV3>3.0.CO;2-7

Blocher, J. D., Chhajed, D., & Leung, M. (1998). Customer order scheduling in a general job shop environment. *Decision Sciences, 29*(4), 951–981. doi:10.1111/j.1540-5915.1998.tb00883.x

Blumber, J., & Kavanagh, D. (1999). The third age of political communication: Influences and features. *Political Communication, 16*, 209–230. doi:10.1080/105846099198596

Bodenhausen, G. H. C. (1968). *Guide to the application of the Paris convention for the protection of industrial property as revised at Stockholm in 1967*. New York, NY: World Intellectual Property.

Boehm, B., & Turner, R. (2003). *Balancing agility and discipline: A guide for the perplexed*. Boston, MA: Addison-Wesley.

Boehm, B. W. (1988). A spiral model of software development and enhancement. *IEEE Computer, 21*(5), 61–72. doi:10.1109/2.59

Bolles, D. (2004). *A guide to the project management body of knowledge* (3rd ed.). Newton Square, PA: Project Management Institute.

Booch, G., Maksimchuk, R., Engle, M., Young, B., Conallen, J., & Houston, K. (2007). *Object-oriented analysis and design with applications* (3rd ed.). Reading, MA: Addison-Wesley Professional.

Bozovsky, I. (1961). *Reliability theory and practice*. Englewood Cliffs, NJ: Prentice-Hall.

Burke, W. (2008). *Organizational change: theory and practice* (2nd ed.). Thousand Oaks, CA: Sage Publications.

By, R. (2005). Organizational change management: A critical review. *Journal of Change Management, 5*(4), 369–380. doi:10.1080/14697010500359250

Cachon, G., & Fisher, M. (2000). Supply chain inventory management and the value of shared information. *Management Science, 46*, 1032–1048. doi:10.1287/mnsc.46.8.1032.12029

Cachon, G., & Terwiesch, C. (2009). *Matching supply with demand: an introduction to operations management* (2nd ed.). New York, NY: McGraw-Hill/Irwin.

Cai, J., Ghali, S., Giannelia, M., Hughes, A., Johnson, A., & Khoo, T. (2003). *Identifying best practices in information technology project management*. Retrieved from http://www.pdfport.com/view/134033-identifying-best-practices-in-information-technology-project.html

Calhoun, C. (1998). Community without propinquity revisited: Communications technology and the transformation of the urban public sphere. *Sociological Inquiry, 68*(3), 373–397. doi:10.1111/j.1475-682X.1998.tb00474.x

Calisir, F., & Gumussoy, C. A. (2005). Determinants of budget overruns on IT projects. *Technovation, 25*(6), 631–636. doi:10.1016/j.technovation.2003.10.011

Carrillo, R. (2004). Managing knowledge: Lessons from the oil and gas sector. *Construction Management and Economics, 22*(6).

Carter, S. (2012). *Get bold: Using social media to create a new type of social business*. Boston, MA: Pearson Education, Inc.

Case, H. B. S. (1991). *Digital equipment corporation: Complex order management*. Boston, MA: Harvard Business School.

Cerpa, N., & Verner, J. M. (2009). Why did your project fail? *Communications of the ACM, 52*(12), 130–134. doi:10.1145/1610252.1610286

Chao, H. P. (2011). Demand response in wholesale electricity markets: The choice of customer baseline. *Journal of Regulatory Economics, 39*(1), 68–88. doi:10.1007/s11149-010-9135-y

Chea, S., & Lou, M. (2008). Post-adoption behaviors of e-service customers: The interplay of cognition and emotion. *International Journal of Electronic Commerce, 12*(3), 29–56. doi:10.2753/JEC1086-4415120303

Chellappan, C. (2008). *E-services: The need for higher levels of trust by populace*. Chennai, India: Anna University.

Chen, R., Lu, F., Li, Q., Fu, G., Wang, H., Chai, Z., & Xu, Z. (2009). Failure analysis on foreign object damage of aero-engine compressor blade. In *Proceedings of 8ᵗʰ International Conference on Reliability, Maintainability and Safety,* (pp.1085-1088). IEEE.

Chin, W. W. (1998). *The partial least squares approach for structural equation modelling*. Hoboken, NJ: Lawrence Erlbaum Associates.

Chiong, J. (2008). *Predictors of project success: A Singapore study*. (Dissertation). University of Western Australia. Crawley, Australia.

Chou, S.-W. (2005). Knowledge creation: Absorptive capacity, organizational mechanisms, and knowledge storage/retrieval capabilities. *Journal of Information Science, 31*(6), 453–465. doi:10.1177/0165551505057005

Chrysler Corporation. Ford Motor Company, & General Motors Corporation. (1995). *Advanced product quality planning and control plan (APQP) and control plan: Reference manual*. Detroit, MI: Chrysler Corporation.

Clark, K. B., Chew, W. B., & Fujimoto, T. (1987). Product development in the world auto industry. *Brookings Papers on Economic Activity, 3*, 729–771. doi:10.2307/2534453

Clark, R. B. (2010). *Information technology project management*. Helena, MT: State Government Publication.

Cockburn, A. (2008). *Crystal methodologies*. Retrieved from: http://alistair.cockburn.us/Crystal+methodologies

Cockburn, A., & McKenzie, B. (2000). What do web users do? An empirical analysis of web use. *International Journal of Human-Computer Studies, 54*(6), 903–922. doi:10.1006/ijhc.2001.0459

Cohen, D., & Prusak, L. (2001). *In good company: How social capital makes organizations work*. Boston, MA: Harvard Business School Press. doi:10.1145/358974.358979

Colenso, M. (2000). *Kaizen strategies for successful organizational change: Enabling evolution and revolution within the organization*. London, UK: Pearson Education Ltd.

Cooper, R. G., Edgett, S. J., & Kleinschmidt, E. J. (2012). *Portfolio management: Fundamental for new product success*. Retrieved from http://www.stage-gate.net/downloads/working_papers/wp_12.pdf

Coram, M., & Bohner, S. (2005). *The impact of agile methods on software project management*. Washington, DC: IEEE Press.

Cornish, K. (2011). Communications technology for the smart grid. *Electric Light and Power, 89*(3), 62–63.

Crawford, L. (2005). Senior management perceptions of project management competence. *International Journal of Project Management, 23*(1), 7–16. doi:10.1016/j.ijproman.2004.06.005

Crawford, L., & Nahmias, A. (2010). Competencies for managing change. *International Journal of Project Management, 28*(4), 405–412. doi:10.1016/j.ijproman.2010.01.015

Crawford-Mason, C. (Producer), (1980, June 24). *If Japan can …Why can't we?*. New York, NY: NBC.

Creswell, J. (1998). *Qualitative inquiry and research design: Choosing among five traditions*. London, UK: Sage.

Cronbach, L. J. (1971). *Test validation: Education measurement*. Washington, DC: American Council on Education.

Crow, K. (2001). *Product development strategic orientation*. Retrieved from http://www.npd-solutions.com/strategy.htm

Csikszentmihalyi, M. (1975). *Beyond boredom and anxiety*. San Francisco, CA: Jossey-Bass.

D'Ambra, J., & Rice, R. E. (2001). Emerging factors in user evaluation of the world wide web. *Information & Management*, *38*, 373–384. doi:10.1016/S0378-7206(00)00077-X

Davenport, T. H. (1993). *Process innovation: Reengineering work through information technology*. Boston, MA: Harvard Business School.

Davis, F. D. (1989). Perceived usefulness, perceived ease of use, and user acceptance of information technology. *Management Information Systems Quarterly*, *13*(2), 319–340. doi:10.2307/249008

Davis, P. (2011). Smart grid, evolution of DR and the impact of FERC 745. *Electric Light and Power*, *89*(3), 46–47.

de Wit, A. (1988). Measurement of project success. *International Journal of Project Management, 6*(3).

Dearlove, D. (1998). *Key management decisions: Tools and techniques of the executive decision-maker*. London, UK: Pitman Publishing.

Dearlove, D. (2001). *The ultimate book of business thinking: Harnessing the power of the world's greatest business ideas*. Oxford, UK: Capstone Publishing.

D'Emilia, P. (2011). *Tsunami nucleare, I 30 giorni che sconvolsero il giappone*. Rome, Italy: Il Manifesto Libri.

Deming, W. (1982). *Out of the crisis*. Cambridge, MA: MIT Press.

Dennis, P. (2002). *Lean production simplified: A plain-language guide to the world's most powerful production system*. New York, NY: Productivity Press.

Denny, F. I., & Dismukes, D. E. (2002). *Power system operations and electricity markets*. Boca Raton, FL: CRC Press.

Department of Defence. (2002). *MIL-HDBK-61B military handbook configuration management guidance*. Retrieved from http://www.everyspec.com

Desanctis, G., & Gallupe, R. B. (1987). A foundation for the study of group decision support systems. *Management Science*, *33*(5). doi:10.1287/mnsc.33.5.589

DeSanctis, G., & Monge, P. (1998). Communication processes for virtual organizations. *Journal of Computer-Mediated Communication, 3*(4).

Dhawan, S., & Kumar, R. (2008). Analyzing performance of web-based metrics for evaluating reliability and maintainability of hypermedia applications. In *Proceedings Third International Conference on Broadband Communications, Information Technology & Biomedical Applications (BroadCom 2008)*, (pp. 376–383). Pretoria, South Africa: BroadCom.

Dillon, A., & Gabbard, R. (1998). Hypermedia as an educational technology: A review of the quantitative research literature on learner comprehension, control, and style. *Review of Educational Research*, *68*(3), 322–349.

Ding, R., & Wang, Y. (2008). An empirical study on critical success factors based on governance for IT projects in China. In *Proceedings of the 4th International Conference on Wireless Communications, Networking and Mobile Computing, 2008*. WiCOM.

Driver, M. J., Brousseau, K. E., & Hunsaker, P. L. (1993). *The dynamic decision maker*. San Francisco, CA: Jossey-Bass Publishers.

Driver, M. J., Svensson, K., Amato, R. P., & Pate, L. E. (1996). A human information- Processing approach to strategic change. *International Studies of Management and Organization*, *26*(1), 41–58.

Duhigg, C. (2012). *The power of habit: Why we do what we do in life and business*. New York, NY: Random House, Inc.

Earl, M. (1996, Spring). The risks of outsourcing IT. *Sloan Management Review*, 26–32.

Earle, J. H. (2008). *Engineering design graphics with AutoCAD 2007*. Upper Saddle River, NJ: Prentice Hall, Inc.

Elam, J. J., & Leidner, D. G. (1995). EIS adoption, use and impact: The executive perspective. *Decision Support Systems*, *14*(2), 89–103. doi:10.1016/0167-9236(94)00004-C

El-khattam, W., & Salama, M. M. A. (2004). Distributed generation technologies, definitions and benefits. *Electric Power Systems Research*, *71*(2), 119–128. doi:10.1016/j.epsr.2004.01.006

Erel, E., & Ghosh, J. B. (2007). Customer order scheduling on a single machine with family setup times: Complexity and algorithms. *Applied Mathematics and Computation*, *185*, 11–18. doi:10.1016/j.amc.2006.06.086

Evans, M. W., Abela, A. M., & Beltz, T. (2002, April). Seven characteristics of dysfunctional software projects. *The Journal of Defense Software Engineering, CrossTalk*, 16-20.

Fagrell, H., & Ljungberg, F. (2000). A field study of news journalism: Implications for knowledge management systems. In *Proceedings of PDC 2000*. PDC.

Fagrell, H., Kristoffersen, H. S., & Ljungberg, F. (1999). Exploring support for knowledge management in mobile work. [Berlin, Germany: Springer.]. *Proceedings of ECSCW, 1999*, 259–275. doi:10.1007/978-94-011-4441-4_14

Fan, D. (2010). Analysis of critical success factors in IT project management. In *Proceedings of the 2nd International Conference on Industrial and Information Systems*, (pp. 487-490). IEEE Press.

Feil, P., Menzel, W., Nguyen, T. P., Pichot, C., & Migliaccio, C. (2008). Foreign objects debris detection (FOD) on airport runways using a broadband 78 GHz sensor. In *Proceedings of 38th European Microwave Conference*, (pp. 1608-1611). IEEE.

Fendt, J. (2006, Winter). Are you promoting change—Or hindering it?. *Harvard Management Communication Newsletter*, 1-6.

Fisher, R. (1935). *The design of experiments*. Edinburgh, UK: Oliver & Boyd.

Fitzgerald, S. P. (2002). *Decision making*. Oxford, UK: Capstone Publishing.

Fitzpatrick, G. (2002). Bootstrapping expertise sharing. In Ackerman, M., Pipek, V., & Wulf, V. (Eds.), *Sharing Expertise: Beyond Knowledge Management* (pp. 81–110). Cambridge, MA: MIT Press.

Foglia, P., Prete, C. A., & Zanda, M. (2008). Relating GSR signals to traditional usability metrics: A case study with an anthropomorphic web assistant. In *Proceedings of 2008 IEEE International Instrumentation & Measurement Technology Conference (I2MTC 2008)* (*Vol. 1*, pp. 1814–1819). Victoria, Canada: IEEE Press. doi:10.1109/IMTC.2008.4547339

Fong, S., & Meng, H. S. (2009). A web-based performance monitoring system for e-government services. In *Proceedings of the 3rd International Conference on Theory and Practice of Electronic Governance (ICEGOV 2009)*, (pp. 74-82). Bogota, Colombia: ICEGOV.

Fontana, A., & Frey, J. (2000). The interview. In Denzin, N., & Lincoln, Y. (Eds.), *Handbook of Qualitative Research* (2nd ed., pp. 645–672). London, UK: Sage.

Fox, T. L., & Spence, J. W. (1999). An examination of the decision styles of project managers: Evidence of significant diversity. *Information & Management, 36*, 313–320. doi:10.1016/S0378-7206(99)00025-7

Framinan, J. M., Gonzales, P. L., & Ruiz-Usano, R. (2003). The conwip production control system: Review and research issues. *Production Planning and Control, 14*(3), 255–265. doi:10.1080/0953728031000102595

Freire, A., Fortes, R. P. M., Turine, M. A. S., & Paiva, D. M. B. (2008). An evaluation of web accessibility metrics based on their attributes. In *Proceedings of the 26th Annual ACM International Conference on Design of Communication*, (pp. 73-79). Lisbon, Portugal: ACM Press.

Galbraith, J. (1977). *Organization design*. Reading, MA: Addison-Wesley Publishing Inc.

Galer, M., Harker, S., & Ziegler, J. (1992). *Methods and tools in user-centered design for information technology (human factors in information technology)*. London, UK: Elsevier Science Ltd.

Geczy, P., Izumi, N., Akaho, S., & Hasida, K. (2007). Navigation space based intranet usability analysis. In *Proceedings of the IEEE Symposium on Computational Intelligence and Data Mining (CIDM 2007)*, (pp. 147-154). Honolulu, HI: IEEE Press.

Gelderman, M. (1998). The relation between user satisfaction, usage of information systems and performance. *Information & Management, 34*, 11–18. doi:10.1016/S0378-7206(98)00044-5

Gibson, J. L., Ivancevich, J. M., & Donnelly, J. H. Jr. (1994). *Organizations* (8th ed.). Burr Ridge, IL: Richard D. Irwin.

Goad, R., & Mooney, A. (2008). *The impact of social networking in the UK*. London, UK: Hitwise/Experian.

Goodyear, R. (1987). *A descriptive correlational study of the decision-making patterns of nurse practitioners in primary care.* (Unpublished Doctoral Dissertation). University of San Diego. San Diego, CA.

Gore, W. J. (1964). *Administrative decision-making: A heuristic model.* New York, NY: John Wiley and Sons.

Gottschalk, P., & Karlsen, J. T. (2005). A comparison of leadership roles in internal IT projects versus outsourcing projects. *Industrial Management & Data Systems, 105*(9), 1137–1149. doi:10.1108/02635570510633220

Grader, W. (2011). *Website.* Retrieved from http://websitegrader.com/

Granitz, N. A., & Ward, J. C. (1996). Virtual community: A sociocognitive analysis. *Advances in Consumer Research. Association for Consumer Research (U. S.), 23.*

Grant, K. A., & Qureshi, U. (2006). Knowledge management systems - Why so many failures? In *Proceedings of the Innovations in Information Technology Conference, 2006.* IEEE Press.

Greenberg, J., & Baron, A. (1993). *Behavior in organizations* (4th ed.). Boston, MA: Allyn and Bacon.

Gregson, K. (1998). Conversation and community or sequential monologues: An analysis of politically oriented newsgroups. In *Proceedings of the ASIS Annual Meeting,* (Vol. 35, pp. 531- 541). ASIS.

Greneker, G. (1999). *Radar to detect foreign object ingestion by a jet engine.* Retrieved from http://digitalcommons.unl.edu/birdstrike1999/17

Guare. (1990). *Six degree of separation.* New York, NY: Random House.

Gurumurthy, A., & Kodali, R. (2008). A multi-criteria decision-making model for the justification of lean manufacturing systems. *International Journal of Management Science and Engineering Management, 3,* 100–118.

Hacker, K. L., & Todino, M. A. (1996). Virtual democracy at the Clinton white house: An experiment in electronic democratisation. *The Public, 3*(1), 71–86.

Haidt, J. (2006). *The happiness hypothesis: Finding modern truth in ancient wisdom.* New York, NY: Basic Books.

Hair, J. F., Anderson, R. E., Tatham, R. L., & Black, W. C. (1995). *Multivariate data analysis with reading.* Englewood Cliffs, NJ: Prentice Hall.

Hammond, J. S. (1999). *Smart choices.* Boston, MA: Harvard Business School Press.

Harper, P. (2011, March 21). *Media slammed for Japan crisis coverage.* Retrieved from http://www.nzherald.co.nz/japan-tsunami/news/article.cfm?c_id=1503051&objectid=10713939

Harren, V. A. (1979). A model of career decision-making for college students. *Journal of Vocational Behavior, 14,* 119–133. doi:10.1016/0001-8791(79)90065-4

Harris, C. (2006). *Electricity markets pricing, structures, and economics.* Hoboken, NJ: John Wiley & Sons.

Harry, M., & Lawson, R. (1992). *Six sigma producibility analysis and process characterization.* Reading, MA: Addison-Wesley Publishing Company, Inc.

Harry, M., & Schroeder, R. (2000). Six sigma – The break-through management strategy revolutionizing the world's top corporations. *Soundview Executive Book Summaries, 22*(11).

Harry, M. J. (2000). A new definition aims to connect quality with financial performance. *Quality Progress, 33*(1), 64–66.

Harry, M. J., & Schroeder, R. (2002). *Six sigma: The breakthrough management strategy revolutionizing the world's top corporations.* New York, NY: Doubleday.

Hartman, F., & Ashrafi, R. (2004). Development of the SMART project planning framework. *International Journal of Project Management, 22,* 499–510. doi:10.1016/j.ijproman.2003.12.003

Hatfield, M. (2011). Meter data management systems and the paradigms of time. *Utility Products, 15*(7), 48–50.

Haythornthwaite, C., Wellman, B., & Mantei, M. (1995). Work relationships and media use: A social network analysis. *Group Decision and Negotiation, 4*(3), 193–211. doi:10.1007/BF01384688

Hazır, Ö., Günalay, Y., & Erel, E. (2008). Customer order scheduling problem: A comparative metaheuristics study. *International Journal of Advanced Manufacturing Technology, 37,* 589–598. doi:10.1007/s00170-007-0998-8

Heath, C., & Heath, D. (2010). *Switch: How to change things when change is hard*. New York, NY: Broadway Books.

Heijden, H. (2000). *Using the technology acceptance model to predict website usage: Extension and empirical test*. Research Memorandum. Amsterdam, The Netherlands: Vrije Universiteit.

Herman v. Youngstown Car Mfg. (1911) Co., 191 F. 579, 584-85, 112 CCA 185 (6th Cir. 1911)

Hermana, A., & Silfianti, W. (2011). Evaluating e-government implementation by local government: Digital divide in internet based public services in Indonesia. *International Journal of Business and Social Science*, 2(3), 156–163.

Hersey, P., Blanchard, K. H., & Johnson, D. E. (1996). *Management of organizational behavior: Utilizing human resources* (7th ed.). Upper Saddle River, NJ: Prentice Hall.

Hersey, P., Blanchard, K., & Johnson, D. (2001). *Management of organizational behavior: Leading human resources* (8th ed.). Upper Saddle River, NJ: Prentice Hall.

Hicks, B. J. (2007). Lean information management: Understanding and eliminating waste. *International Journal of Information Management*, 27, 233–249. doi:10.1016/j.ijinfomgt.2006.12.001

Hidding, G. J., & Nicholas, J. (2009). Reducing IT project management failures: A research proposal. In *Proceedings of the 42nd Hawaii International Conference on System Sciences – 2009*, (pp. 1-10). IEEE Press.

Highsmith, J. (2000). *Adaptive software development*. New York, NY: Dorset House.

Hill, P. H. (1979). *Making decisions: A multidisciplinary introduction*. London, UK: Addison-Wesley Pub. Co.

Hoffman, D. L., & Novak, T. P. (1995). *Marketing in hypermedia computer-mediated environments: Conceptual foundations*. Working Paper No. 1. Project 2000: Research Program on Marketing in Computer-Mediated Environments.

Holland, W. (2000). *Change is the rule: Practical actions for change: On target, on time, on budget*. Chicago, IL: Dearborn Financial Publishing, Inc.

Holsapple, C. W., & Whinston, A. B. (1996). *Decision support systems: A knowledge based approach*. New York, NY: West Publishing Company.

Holweg, M. (2007). The genealogy of lean production. *Journal of Operations Management*, 25, 420–437. doi:10.1016/j.jom.2006.04.001

Hopp, W. J., & Spearman, M. L. (2004). To pull or not to pull: What is the question? *Manufacturing and Service Operations Management*, 6, 133–148. doi:10.1287/msom.1030.0028

Horan, T., Abhichandani, T., & Rayalu, R. (2006). Assessing user satisfaction of e-government services: Development and testing of quality-in-use satisfaction with advanced traveler information systems (ATIS). In *Proceedings of the 39th Annual Hawaii International Conference on System Sciences (HICSS 2006)*. Hawaii, HI: IEEE.

Hoverstadt, P. (2004). Mosaic transformations in organizations. *Journal of Organizational Transformation and Social Change*, 1(2-3), 163–177. doi:10.1386/jots.1.2.163/0

Hsu, M., & Chu, C. (2004). Predicting electronic service continuance with a decomposed theory of planned behaviour. *Behaviour & Information Technology*, 23(5), 359–373. doi:10.1080/01449290410001669969

Hsu, S.-Y., & Liu, C.-H. (2009). Improving the delivery efficiency of the customer order scheduling problem in a job shop. *Computers & Industrial Engineering*, 57, 856–866. doi:10.1016/j.cie.2009.02.015

Huang, C. J. (2003). Usability of e-government websites for people with disabilities. In *Proceedings of the 36th Annual Hawaii International Conference on System Sciences (HICSS 2003)*. Washington, DC: IEEE Press.

Huang, C.-Y., Shen, Y.-C., Chiang, I.-P., & Lin, C.-S. (2007). Concentration of web users' online information behavior. *Information Research, 12*(4). Retrieved from http://InformationR.net/ir/12-4/paper324.html

Huang, Z., Poli, M., & Mithiborwala, H. S. (2009). Project strategy: Success themes for strategic projects. In *Proceedings of Portland International Conference on Management of Engineering & Technology 2008 (PICMET 2009)*, (pp. 1282-1289). Portland, OR: PICMET.

Hughes-White, S. (2011). Fleet efficiency in a wireless world. *Utility Products, 15*(6), 10–14.

Humphrey, W. (1989). *Managing the software process.* Reading, MA: Addison-Wesley.

Huysman, M., & Wulf, V. (2004). *Social capital and information technology.* Cambridge, MA: MIT-Press.

Huysman, M., & Wulf, V. (2006). IT to support knowledge sharing in communities: Towards a social capital analysis. *Journal of Information Technology, 1*(21), 40–51. doi:10.1057/palgrave.jit.2000053

Imamoglu, O., & Gozlu, S. (2008). The sources of success and failure of information technology projects: Project managers' perspective. In *Proceedings of Portland International Conference on Management of Engineering & Technology 2008 (PICMET 2008),* (pp. 1430-1435). Cape Town, South Africa: PICMET.

International Intellectual Property Alliance. (2012). *Article 27.1.* Retrieved from http://www.iipa.com

International Organization for Standardization. (2003). *ISO 10007:2003 quality management systems -- Guidelines for configuration management.* Retrieved from http://www.iso.org

IPO. (2011). *Manual of patent office practice and procedure, version 01.11.* Retrieved from http://www.ipindia.nic.in

Iveroth, E. (2010). Inside ericsson: A framework for the practice of leading global IT-enabled change. *California Management Review, 53*(1), 136–153. doi:10.1525/cmr.2010.53.1.136

Jackson, T. (1995). *Implementing a lean management system.* Portland, OR: Productivity Press.

Jalote, P., & Agarwal, N. (2006). Using defect analysis feedback for improving quality and productivity in iterative software development. In *Proceedings of ITI 3rd International Conference on Information and Communications Technology: Enabling Technologies for the New Knowledge Society,* (pp. 703-713). ITI.

Jati, H., & Dominic, D. D. (2009). Quality evaluation of e-government website using web diagnostic tools: Asian case. In *Proceedings of the 2009 International Conference on Information Management and Engineering,* (pp. 85-89). Kuala Lumpur, Malaysia: IEEE.

Jill, M. J. (2006). *Relationship between principals' decision making styles and technology acceptance & use.* (Dissertation). University of Pittsburgh. Pittsburgh, PA.

Jinling, C., & Guoping, X. (2005). Comprehensive evaluation of e-commerce web site based on concordance analysis. In *Proceedings of the 2005 IEEE International Conference on e-Business Engineering (ICEBE 2005),* (pp. 179-182). Beijing, China: IEEE Press.

Joia, A. L. (2007). Knowledge management strategies: Creating and testing a measurement scale. *International Journal of Learning and Intellectual Capital, 4*(3), 203–221. doi:10.1504/IJLIC.2007.015607

Jugdev, K., & Muller, R. (2005). A retrospective look at our evolving understanding of project success. *Project Management Journal, 36*, 19–31.

Julien, F. M., & Magazine, M. J. (1990). Scheduling customer orders: An alternative production scheduling approach. *Journal of Manufacturing Operations, 3*, 177–199.

Kaid, L. L., & Holtz-Bacha, C. (Eds.). (1995). *Political advertising in western democracies.* Thousand Oaks, CA: Sage Publications.

Karahanna, E., Straub, D. W., & Chervany, N. L. (1999). Information technology adoption across time: A cross-sectional comparison of pre-adoption and post-adoption beliefs. *Management Information Systems Quarterly, 23*(2), 183–213. doi:10.2307/249751

Karsak, E. E., & Tolga, E. (2001). Fuzzy multi-criteria decision-making procedure for evaluating advanced manufacturing system investments. *International Journal of Production Economics, 69*(1), 49–64. doi:10.1016/S0925-5273(00)00081-5

Kavanagh, M., & Ashkanasy, N. (2006). The impact of leadership and change management strategy on organizational culture and individual acceptance of change during a merger. *British Journal of Management, 17*, 81–103. doi:10.1111/j.1467-8551.2006.00480.x

Keen, P. (1981). Information systems and organizational change. *Communications of the ACM, 24*(1), 24–33. doi:10.1145/358527.358543

Keil, M., & Mann, J. (1997). Understanding the nature and extent of IS project escalation: Results from a survey of IS audit and control professionals. In *Proceedings of the 30th Hawaii International Conference on System Sciences (HICSS)*, (Vol. 3, pp. 139-148). IEEE.

Keil, M., Rai, A., Mann, J., & Zhang, G. (2003). Why software projects escalate: The importance of project management constructs. *IEEE Transactions on Engineering Management*, *50*(3), 251–261. doi:10.1109/TEM.2003.817312

Kelton, W. D., Sadowski, R. P., & Sturrock, D. T. (2007). *Simulation with arena* (4th ed.). New York, NY: McGraw-Hill Publishing Co.

Khalifa, M., & Liu, V. (2003). Satisfaction with internet-based services: The role of expectations and desires. *International Journal of Electronic Commerce*, *7*(2), 31–49.

Khan, K. (2008). Assessing quality of web based systems. In *Proceedings of the 2008 IEEE/ACS International Conference on Computer Systems and Applications (AICCSA 2008)*, (pp. 763-769). Doha, Qatar: IEEE Press.

Khan, N., & Manarvi, I. A. (2011). Identification of delay factors in C-130 aircraft overhaul and finding solutions through data analysis. In *Proceedings of 2011 IEEE Aerospace Conference*, (pp. 1-8). Big Sky, MT: IEEE.

Khan, R. H., Aditi, T. F., Sreeram, V., & Iu, H. H. C. (2010). A prepaid smart metering scheme based on WiMAX prepaid accounting model. *Smart Grid and Renewable Energy*, *1*(2), 63-69. Doi:10.4236/sgre/2010/12010doi:10.4236/sgre.2010.12010

Khandani, S. (2005). *Engineering design process*. Retrieved from http://www.iisme.org/etp/HS%20Engineering-%20Engineering.pdf

Kirschen, D., & Strbac, G. (2004). *Fundamentals of power system economics*. Chichester, UK: Wiley. doi:10.1002/0470020598

Knight, S. (2011). Italian journalist denounces media's disaster coverage. *Asahi.com*. Retrieved from http://www.asahi.com/english/TKY201107050261.html

Kolb, D., & Frohman, A. (1970). An organizational development approach to consulting. *Sloan Management Review*, *12*, 51–65.

Kotter, J., & Cohen, C. (2002). *The heart of change: Real-life stories of how people change their organizations*. Boston, MA: Harvard Business School Press.

Kotter, J., & Schlesinger, L. (2008, July-August). Choosing strategies for change. *Harvard Business Review*, 130–139.

Koufaris, M. (2002). Applying the technology acceptance model and flow theory to online consumer behaviour. *Information Systems Research*, *13*(2), 205–223. doi:10.1287/isre.13.2.205.83

KPGMS. (2003). *KPMG's international 2002-2003 programme management survey*. Retrieved from www.transformed.com.au/_.../Reports_-_Programm

Kroenke, D., & Hatch, R. (1994). *Management information systems*. New York, NY: McGraw-Hill.

Kueneman, R. M., & Wright, J. E. (1975). New policies of broadcast stations for civil disturbances and disasters. *The Journalism Quarterly*, *52*(4). doi:10.1177/107769907505200409

Kumar, U. D., Croker, J., & Knezrvic, J. (1999). Evolutionary maintenance for aircraft engines. In *Proceedings of Reliability and Maintainability Symposium*, (pp. 62-68). IEEE.

Laplante, P. (2003, January-February). Remember the human element in IT project management. *IT Pro*, 46-50.

Larman, C., & Basili, R. V. (2003). Iterative and incremental development: A brief history. *IEEE Computer*, *36*(6), 47–56. doi:10.1109/MC.2003.1204375

Lasagni, M., & Richeri, G. (2004). *Un anno di temi sociali nella programmazione RAI*. Retrieved from http://www.segretariatosociale.rai.it/palinsesto/indice_dati_program_soc.html

Latendresse, P., & Chen, J. (2003). *The information age and why it projects must not fail*. Retrieved from www.sbaer.uca.edu/research/swdsi/2003/.../045.pdf

Latorella, K. A., & Prabhu, P. V. (2000). A review of human error in aviation maintenance and inspection. *International Journal of Industrial Ergonomics*, *1*(2), 133–161. doi:10.1016/S0169-8141(99)00063-3

Leonard, N. H., Scholl, R. W., & Kowalski, K. B. (1999). Information processing style and decision making. *Journal of Organizational Behavior, 20*, 407–420. doi:10.1002/(SICI)1099-1379(199905)20:3<407::AID-JOB891>3.0.CO;2-3

Leung, J. Y.-T., Li, H., & Pinedo, M. (2006). Scheduling orders for multiple product types with due date related objectives. *European Journal of Operational Research, 168*, 370–389. doi:10.1016/j.ejor.2004.03.030

Leung, J. Y.-T., Li, H., & Pinedo, M. (2008). Scheduling orders on either dedicated or flexible machines in parallel to minimize total weighted completion time. *Annals of Operations Research, 159*, 107–123. doi:10.1007/s10479-007-0270-5

Levinston, E. (1964). System reliability analysis, introduction. In Rothbart, E. (Ed.), *Mechanical Design and System Handbook*. New York, NY: McGraw-Hill.

Lewin, K. (1951). *Field theory in social science: Selected theoretical papers*. New York, NY: Harper and Brothers.

Li, C., & Bernoff, J. (2008). *Groundswell: Winning in a world transformed by social technologies*. Boston, MA: Harvard Business School Press.

Li, C.-L., & Ou, J. (2007). Coordinated scheduling of customer orders with decentralized machine locations. *IIE Transactions, 39*(9), 899–909. doi:10.1080/07408170600972990

Li, C., & Vairaktarakis, G. (2007). Coordinating production and distribution of jobs with bundling operations. *IIE Transactions, 39*, 203–215. doi:10.1080/07408170600735561

Li, H. (2005). *Order scheduling in dedicated and flexible machine environments*. (Ph.D. Thesis). New Jersey Institute of Technology. Newark, NJ.

Library of Congress. (2009). *A century of lawmaking for a new nation: U.S. congressional documents and debates*. Washington, DC: Law Library of Congress.

Liker, J. K. (2004). The Toyota way: 14 management principles from the world`s greatest manufacturer. *Business Book Review, 12*, 1–11.

Link, V. (2011). *A WWW link checker*. Retrieved from http://htmlhelp.com/tools/valet/

Liu, C.-H. (2009). Lot streaming for customer order scheduling problem in job shop environments. *International Journal of Computer Integrated Manufacturing, 22*(9), 890–907. doi:10.1080/09511920902866104

Liu, C.-H. (2010). A coordinated scheduling system for customer orders scheduling problem in job shop environments. *Expert Systems with Applications, 37*(12), 7831–7837. doi:10.1016/j.eswa.2010.04.055

Lombardi, M. (1993). *Tsunami: 'Crisis management' della comunicazione*. Milan, Italy: Vita & Pensiero.

Lombardi, M. (2005). *Comunicare nell'emergenza*. Milan, Italy: Vita & Pensiero.

Lombardi, M. (2005). *La comunicazione dei rischi internazionali: Un confronto internazionale*. Milan, Italy: Vita & Pensiero.

Lubani, E., & Qirjo, M. (2002). *Developing skills for NGOS: Project management*. Szentendre, Hungary: The Regional Environmental Center For Central And Eastern Europe.

Lucy, J. (2008). Why is the failure rate for organizational change so high? *Management Services, 52*(4), 10–18.

Lui, T. J., Stirling, W., & Marcy, H. O. (2010). Get smart. *Power and Energy Magazine, 8*(3), 66–78. doi:10.1109/MPE.2010.936353

Lyons, J. (2009). The impact of leadership on change readiness in the U.S. military. *Journal of Change Management, 9*(4), 459–475. doi:10.1080/14697010903360665

Macdonald, S. (1995). Learning to change: An information perspective on learning in the organization. *Organization Science, 6*(5), 557–568. doi:10.1287/orsc.6.5.557

MacLeod, C. (2002). *Inventing the industrial revolution: The english patent system, 1660-1800*. Cambridge, UK: Cambridge University Press.

Manarvi, I. A., & Umer, W. (2010). Analyzing the defects of C-130 Aircraft through maintenance history. In *Proceedings of 2010 IEEE Aerospace Conference (AERO 2010)*, (pp 1-7). Big Sky, MT: IEEE Press.

Marakas, G. M. (2003). *Decision support systems in the 21st century* (2nd ed.). Upper Saddle River, NJ: Prentice Hall.

Maris, G. M., & Robert, M. D. (2007). Strategic decision making and support systems: Comparing American, Japanese and Chinese management. *Decision Support Systems*, *43*, 284–300. doi:10.1016/j.dss.2006.10.005

Mason, F. A., Kraus, D. C., Johnson, W. B., & Watson, J. (2001). *Reducing foreign object damage through improved human performance: Best practices*. Washington, DC: Galaxy Scientific Corporation.

Massachusetts Institute of Technology. (1980). *International motor vehicle program (IMVP)*. Retrieved from http://web.mit.edu/ctpid/www/imvp/

McCarty, T., Jordan, M., & Probst, D. (2011). *Six sigma for sustainability*. New York, NY: McGraw-Hill Professional.

McConnell, S. (1995). *Rapid development*. Redmond, WA: Microsoft Press.

McCracken, G. (1988). *The long interview*. London, UK: Sage.

McDonald, J. D. (2007). *Electric power substations engineering*. Boca Raton, FL: Taylor & Francis Group. doi:10.1201/9781420007312

McKenney, J. L., & Keen, P. G. W. (1974). How manager's minds work. *Harvard Business Review*, *52*(3), 79.

McNurlin, B. C., & Sprague, R. H. (2004). *Information systems management in practice* (6th ed.). Englewood Cliffs, NJ: Prentice Hall.

Mech, T. F. (1993). The managerial decision styles of academic library director. *College & Research Libraries*, *54*(5), 375–386.

Meredith, J. R., & Mantel, S. L. (2003). *Project management-A managerial approach* (5th ed.). New York, NY: John Wiley & Sons.

Merrill, D. W., & Reid, R. H. (1981). *Personal styles and effective performance: Making your style work for you*. Radnor, PA: Chilton Book Co.

Miles, R. (2010). Accelerating corporate transformations (don't lose your nerve!): Six mistakes that can derail your company's attempts to change. *Harvard Business Review*, *88*(1/2), 68–75.

Miller, R. H., & Malinowski, J. H. (1994). *Power system operation*. Boston, MA: McGraw-Hill.

Mirvis, P., & Berg, D. (1977). *Failures in organization development and change: Cases and essays for learning*. New York, NY: John Wiley & Sons.

Mohan, T. (1992). Deep inside the black box: Early case study findings on technology planning in product development projects. In Feldman, Hustad, & Page (Eds.), *Managing Product Development: Winning in the 90s: Proceedings of the PDMA International Conference*, (pp. 57-65). Chicago, IL: PDMA.

Morgan, J., & Liker, J. (2006). *The Toyota product development system*. Portland, OR: Productivity Press.

Morgenshtern, O., Raz, T., & Dvir, D. (2007). Factors aVecting duration and eVort estimation errors in software development projects. *Information and Software Technology*, *49*, 827–837. doi:10.1016/j.infsof.2006.09.006

Morris, M. G., & Turner, J. M. (2001). Assessing users' subjective quality of experience with the world wide web: An exploratory examination of temporal changes in technology acceptance. *International Journal of Human-Computer Studies*, *54*, 877–901. doi:10.1006/ijhc.2001.0460

Morsink, J. (1999). *The universal declaration of human rights: Origins, drafting, and intent*. Philadelphia, PA: University of Pennsylvania Press.

Mossoff, A. (2001). Rethinking the development of patents: An intellectual history, 1550-1800. *The Hastings Law Journal*, *52*, 1255.

Muller, R., & Turner, T. (2007). The influence of project managers on project success criteria and project success by type of project. *European Management Journal*, *25*(4), 298–309. doi:10.1016/j.emj.2007.06.003

Munns, A. K., & Bjeirmi, B. F. (1996). The role of project management in achieving project success. *International Journal of Project Management*, *14*(2), 81–87. doi:10.1016/0263-7863(95)00057-7

Nariman, A. (2010). E-government websites evaluation using correspondence analysis. In *Proceedings of the 2010 International Conference on Complex, Intelligent and Software Intensive Systems (CISIS)*, (pp. 1147-1152). Krakow, Poland: CISIS.

NASA. (2008). *NASA-STD-0005 NASA configuration management (CM) standard.* Washington, DC: NASA. Retrieved from http://www.standards.nasa.gov/documents/viewdoc/3315133/3315133

Netjasov, F., & Janic, M. (2008). A review of research on risk and safety modelling in civil aviation. *Journal of Air Transport Management, 1*(4), 213–220. doi:10.1016/j.jairtraman.2008.04.008

Ng, C., Cheng, T., & Yuan, J. (2003). Concurrent open shop scheduling to minimize the weighted number of tardy jobs. *Journal of Scheduling, 6,* 405–412. doi:10.1023/A:1024284828374

Night, C., Jervis, E., & Herd, G. (1955). Terms of interest in the study of reliability. *I.R.E. Transactions on Reliability and Quality Control, 5,* 34–35. doi:10.1109/IRE-PGRQC.1955.5007222

Nohria, N., & Khurana, R. (1993). *Executing change: Seven key considerations.* Harvard Business School Note 9-494-038. Boston, MA: Harvard Business School Press.

North American Electric Reliability Corporation. (2010). *Reliability functional model.* Retrieved March 10, 2012 from http://www.nerc.com/files/Functional_Model_V5_Final_2009Dec1.pdf

Nulden, U. (1996). Escalation in IT projects: Can we afford to quit or do we have to continue. In *Proceedings of the Information Systems Conference of New Zealand,* (pp. 136-142). Palmerston North, New Zealand: IEEE Computer Society Press.

Nulden, U. (1996). *Failing projects: Harder to abandon than to continue.* Bayonne, France: Projectics.

Nunnally, J. C. (1978). *Psychometric theory.* New York, NY: McGraw-Hill.

Nutt, P. C. (1990). Strategic decision made by top executive and middle managers with data and process dominant styles. *Journal of Management Studies, 27*(2), 172–194. doi:10.1111/j.1467-6486.1990.tb00759.x

Ohno, T. (1988). *Toyota production system: Beyond large scale production.* Portland, OR: Productivity Press.

Olberding, J., Williams, B., Schreiner, A., & Paulsen, J. (2009). *Robust engineering.* Iowa City, IA: Quality Control Mini Culture.

Olsina, L., Lafuente, G., & Pastor, O. (2002). Towards a reusable repository for web metrics. *International Journal of Web Engineering, 1*(1), 61–73.

Olsina, L., & Maria, M. (2004). Ontology for software metrics and indicators: Building process and decisions taken. [ICWE.]. *Proceedings of ICWE, 2004,* 176–181.

Osmanbhoy, M. Z., Runo, S., & Mallasch, P. (2010). Development of fault detection and reporting for non-central maintenance aircraft. In *Proceedings of 2010 IEEE Aerospace Conference (AERO 2010),* (pp. 1-7). Big Sky, MT: IEEE Press.

Owotoki, P., & Mayer-Lindenberg, F. (2005). Comprehensible hierarchical intelligent (CHI) framework for monitoring and preventive maintenance of aircraft systems. In *Proceedings of Third International Workshop on Intelligent Solutions in Embedded Systems,* (pp. 175-184). IEEE.

Oxford. (2007). *Advanced learner's dictionary* (7th ed). Oxford, UK: Oxford University Press.

Page Analyzer, W. - 0.98. (2011). *A free website performance tool and web page speed analysis.* Retrieved from http://www.websiteoptimization.com/services/analyze/

Palmer, S. R., & Felsing, J. M. (2002). *A practical guide to feature-driven development.* Upper Saddle River, NJ: Prentice Hall International.

Pant, I., & Baroudi, B. (2008). Project management education: The human skills imperative. *International Journal of Project Management, 26,* 124–128. doi:10.1016/j.ijproman.2007.05.010

Paris Convention. (2011). *World intellectual property organization (WIPO).* Retrieved from http://www.wipo.int/

Parker, E. C. (1980). What is right and wrong with media coverage of disaster? In *Disasters and the Mass Media.* Washington, DC: The National Research Council.

Pascal, D. (2002). *Lean production simplified: A plain-language guide to the world's most powerful production system.* New York, NY: Productivity Press.

Pascual, V., & Dursteler, J. C. (2007). WET: A prototype of an exploratory search system for web mining to assess usability. In *Proceedings of the 11th International Conference of Information Visualization,* (pp. 211-215). Zurich, Switzerland: IEEE.

Patton, M. Q. (1990). *Qualitative evaluation and research design*. Newbury Park, CA: Sage.

Paulk, M. (2001, November-December). Extreme programming from a CMM perspective. *IEEE Software*, 19–26. doi:10.1109/52.965798

Paynter, H. M. (1990). The first patent. *Invention & Technology*. Retrieved from http://www.me.utexas.edu/~longoria/paynter/hmp/The_First_Patent.html

Pearlson, K. E., & Saunders, C. S. (2010). *Managing and using information systems*. Hoboken, NJ: John Wiley & Sons.

Petricek, V., Escher, T., Cox, I. J., & Margetts, H. (2006). The web structure of e-government: Developing a methodology for quantitative evaluation. In *Proceedings of the 15th International World Wide Web Conference (WWW 2006)*, (pp. 669-678). Edinburgh, UK: IEEE.

Pipek, V., Hinrichs, J., & Wulf, V. (2003). Sharing expertise: Challenges for technical support. In Ackerman, M., Pipek, V., & Wulf, V. (Eds.), *Sharing Expertise: Beyond Knowledge Management* (pp. 111–136). Cambridge, MA: MIT Press.

Plevyak, H. M., Jr., & Pistolessi, J. F. (1994). *USAF the F-22 integrated product development (IPD) implementation improvement plan*. Retrieved from http://www.mitre.org/work/sepo/toolkits/ippd/examples/F22_IPD_ImpTemplate.doc

PMI. (2004). *A guide to the project management body of knowledge (PMBOK® guide)*. PMI.

Polanyi, M. (1966). *The tacit dimension*. London, UK: Routledge and Kegan.

Poppendieck, M. (2001, June). Lean programming: Part 2. *Software Development*, 71-75.

Poppendieck, M., & Poppendieck, T. (2006). *Implementing lean software development - From concept to cash*. Reading, MA: Addison-Wesley Professional.

Prensky, M. (2001). Digital natives, digital immigrants. *On the Horizon, 9*(5). Birney, R., & Barry, M. (2006). Blogs: Enhancing the learning experience for technology students. In *Proceedings of ED-MEDIA 2006*, (p. 1042). ED-MEDIA.

Pressman, R. S. (1996). *A manager's guide to software engineering*. New York, NY: McGraw-Hill.

Pugh, S. (1996). *Creating innovative products using total design*. Reading, MA: Addison-Wesley Publishing Company.

Puksic, M., & Goricanec, D. (2005). Increasing quality and economic efficacy of health institutions in public and private sectors in Slovenia. In *Proceedings of the 5th WSEAS International Conference on Distance Learning and Web Engineering*, (pp. 59-64). Corfu, Greece: WSEAS.

Qi, S., Ip, C., Leung, R., & Law, R. (2010). A new framework on website evaluation, e-business and e-government. In *Proceedings of the International Conference on E-Business and E-Government (ICEE 2010)*, (pp. 78-81). Guangzhou, China: ICEE.

Qi, Y., Lu, Z., & Song, B. (2002). New concept for aircraft maintenance management: New cognition for aircraft maintenance study in R&M field of China. In *Proceedings of Reliability and Maintainability Symposium*, (pp. 401–405). IEEE.

Qi, Y., Lu, Z., & Song, B. (2005). New concept for aircraft maintenance management: The "dolphin curve life cycle model" of a typical repairable system. In *Proceedings of Reliability and Maintainability Symposium*, (pp. 533-538). IEEE.

Quarantelli, E. L. (1991). *Lessons from research: Findings on mass communications system behavior in the pre, trans and postimpact periods*. Newark, NJ: Disaster Research Center.

Quarantelli, E. L., & Dynes, R. (1972). When disaster strikes: It isn't much like what you've heard and read about. *Psychology Today, 5*(9).

Ramaswami, S. N., Strader, T. J., & Brett, K. (2001). Determinants of on-line channel use for purchasing financial products. *International Journal of Electronic Commerce, 5*(2), 95–118.

Ranatunga, R. A. S. K., Annakkage, U. D., & Kumble, C. S. (2003). Algorithms for incorporating reactive power into market dispatch. *Electric Power Systems Research, 65*(3), 179–186. doi:10.1016/S0378-7796(02)00217-1

Rassouli, H. (2011). *GD&T - Turning concepts into products*. Retrieved from http://www.brighthub.com/engineering/mechanical/articles/36159.aspx

Rassouli, H. (2011). *Quality function deployment*. Retrieved from http://www.brighthub.com/engineering/mechanical/articles/94890.aspx

Ravicher, D. B. (2008). *Protecting freedom in the patent system: The public patent foundation's mission and activities*. Retrieved from http://events.stanford.edu/events/50/5004/

Rehman, A., & Hussain, R. (2009). Software project management methodologies/frameworks dynamics: A comparative approach. In *Proceedings of the International Conference on Information and Emerging Technologies*. ICIET.

Reich, B. H. (2007). Managing knowledge and learning in IT projects: A conceptual framework and guideline for practices. *Project Management Institute, 38*(2), 5–17.

Remawi, H., Bates, P., & Dix, I. (2011). The relationship between the implementation of a safety management system and the attitudes of employees towards unsafe acts in aviation. *Journal of Safety Science, 1*(5), 625–632. doi:10.1016/j.ssci.2010.09.014

Repiso, R. L., Setchi, R., & Salmeron, J. L. (2007). Modelling IT projects success with fuzzy cognitive maps. *Expert Systems with Applications, 32*, 543–559. doi:10.1016/j.eswa.2006.01.032

Repiso, R. L., Setchi, R., & Salmeron, J. L. (2007). Modelling IT projects success: Emerging methodologies reviewed. *Technovation, 27*, 582–594. doi:10.1016/j.technovation.2006.12.006

Reyes, V. (2011). Human resources transformation for power and utilities Companies. *Electric Light and Power, 89*(3), 20–23.

Robbins, S. P. (1998). *Organizational behavior* (8th ed.). Upper Saddle River, NJ: Prentice Hall.

Robbins, S. P. (1999). *Management* (6th ed.). Englewood Cliffs, NJ: Prentice Hall.

Robey, D., & Taggart, W. (1981). Measuring managers' minds: The assessment of style in human information processing. *Academy of Management Review, 6*, 375–383.

Robillard, P. N. (1999). The role of knowledge in software development. *Communications of the ACM, 42*(1), 87–92. doi:10.1145/291469.291476

Rodrigue, J. P. (2009). *The geography of transport systems* (2nd ed.). Abingdon, UK: Routledge.

Rogers, E. M., & Sood, R. (1981). *Mass media operations in a quick-onset natural disaster: Hurricane David in Dominica*. Boulder, CO: Natural Hazards Research and Applications Information Center.

Rosenberg, R. (2003, Summer). The eight rings of organizational influence: How to structure your organization for successful organizational change. *Journal for Quality and Participation*, 30–34.

Rothkopf, D. (1998). Cyberpolitik: The changing nature of power in the information age. *Journal of International Affairs, 51*(2), 325–359.

Rowe, A. J., & Boulgarides, J. D. (1992). *Managerial decision making: A guide to successful business decisions*. New York, NY: McMillan.

Rowe, A. J., & Mason, R. O. (1987). *Managing with style: A guide to understanding, assessing, and improving decision making*. San Francisco, CA: Jossey Bass.

Rowley, J. (2002). Reflections on customer knowledge management in e-business. *Qualitative Market Research, 5*(4), 268–280. doi:10.1108/13522750210443227

Royce, W. (1970). Managing the development of large software systems. *Proceedings of the IEEE, 26*, 1–9.

Rummel, R. J. (1970). *Applied factor analysis*. Evanston, IL: Northwestern University Press.

Saeed, S., Pipek, V., Rohde, M., & Wulf, V. (2010). Managing nomadic knowledge: A case study of the European social forum. In *Proceedings of the 28th International Conference on Human Factors in Computing Systems*. Atlanta, GA: IEEE.

Saeed, S., Reichling, T., & Wulf, V. (2008). Applying knowledge management to support networking among NGOs and donors. In *Proceedings of IADIS International Conference on E-Society*. Algarve, Portugal: IADIS.

Saksvik, P. (2007). Developing criteria for healthy organizational change. *Work and Stress, 21*(3), 243–263. doi:10.1080/02678370701685707

Saward, G., Hall, T., & Barker, T. (2004). Assessing usability through perceptions of information scent. In *Proceedings of the 10th International Symposium on Software Metrics*, (pp. 337-346). Chicago, IL: IEEE.

Scanlon, J. (1998). The search for non-existent facts in the reporting of disaster. *Journalism and Mass Communication Educator, 53*(2). doi:10.1177/107769589805300205

Scanlon, T. J. (1992). *Disaster preparedness some myths and misconceptions*. Easingwold, UK: The Emergency Planning College.

Scanlon, T. J., & Alldred, S. (1982). Media coverage of disasters: The same old story. In Jones, B. G., & Tomazevic, M. (Eds.), *Social and Economic Aspects of Earthquakes*. Ithaca, NY: Cornell University.

Scanlon, T. J., Dixon, K., & McClellan, S. (1982). *The Miramichi earthquakes: The media respond to an invisible emergency*. Ottawa, Canada: Emergency Communications Research Unit.

Schein, E. (1987). Process consultation: *Vol. II. Lessons for managers and consultants*. Reading, MA: Addison-Wesley Publishing Company, Inc.

Schmidt, K. (1991). Cooperative work: A conceptual framework. In Rasmussen, J., Brehmer, B., & Leplat, J. (Eds.), *Distributed Decision Making: Cognitive Models for Cooperative Work*. New York, NY: John Wiley & Sons Ltd.

Schonberg, E., Cofino, T., Hoch, R., Podlaseck, M., & Spraragen, S. L. (2000). Measuring success. *Communications of the ACM, 40*(3), 53–57. doi:10.1145/345124.345142

Schuh, P. (2005). *Integrating agile development in the real world*. Hingham, MA: Charles River Media.

Schwaber, K., & Beedle, M. (2002). *Agile software development with scrum*. Upper Saddle River, NJ: Prentice-Hall.

Securities and Exchange Commission. (2010). *Website*. Retrieved from www.sec.gov/news/press/2010/2010-15.htm

Seymour, T., & Kadrmas, W. (2011). Smart grid U.S. transmission grid: Issues and opportunities. *The Review of Business Information Systems, 15*(3), 1–7.

Shah, R., & Ward, P. T. (2007). Defining and developing measures of lean production. *Journal of Operations Management, 25*, 785–805. doi:10.1016/j.jom.2007.01.019

Shapiro, R., & Pham, N. (2007). *Economic effects of intellectual property-intensive manufacturing in the United States*. Retrieved from http://www.sonecon.com/docs/studies/0807_thevalueofip.pdf

Shapiro, B. P., Rangan, V. K., & Sviokla, J. J. (1992). Staple yourself to an order. *Harvard Business Review, 70*(4), 113–121.

Sherman, B., & Bently, L. (1999). *The making of modern intellectual property law: The British experience, 1760-1911*. Cambridge, UK: Cambridge University Press.

Shou, Y., & Ying, Y. (2005). Critical failure factors of information system projects in Chinese enterprises. In *Proceedings of International Conference on Services Systems and Services Management,* (pp. 823-827). IEEE Press.

Signore, O. (2005). A comprehensive model for web sites quality. In *Proceedings of the Seventh IEEE International Symposium on Web Site Evolution (WSE 2005)*, (pp. 30-36). Budapest, Hungary: IEEE Press.

Simchi-Levi, D., Kaminsky, P., & Simchi-Levi, E. (2007). *Designing and managing the supply chain* (3rd ed.). New York, NY: McGraw-Hill/Irwin.

Singh, Y., Malhotra, R., & Gupta, P. (2011). Empirical validation of web metrics for improving the quality of web page. *International Journal of Advanced Computer Science and Applications, 2*(5), 22–28. doi:10.5120/2414-3226

Skarke, G. (1999). *The change management toolkit: A step-by-step methodology for successfully implementing dramatic organizational change* (2nd ed.). Houston, TX: Winhope Press.

Slater, R. (1999). *Jack Welch & the GE way: Management insights and leadership secrets of the legendary CEO*. New York, NY: McGraw-Hill.

Smith, C. (1998). Visual evidence in environmental catastrophe TV stories. *Journal of Mass Media Ethics, 13*(4). doi:10.1207/s15327728jmme1304_4

Smith, D., Passos, J., & Isaacs, R. (2010). How IT project managers cope with stress. In *Proceedings of the 48th Annual Conference on Computer Personnel Research*, (pp. 15-24). ACM Press.

Smith, J. S. (2003). Survey on the use of simulation for manufacturing system design and operation. *Journal of Manufacturing Systems*, 22(2), 157. doi:10.1016/S0278-6125(03)90013-6

Smollan, J., & Sayers, J. (2009). Organizational culture, change and emotions: A qualitative study. *Journal of Change Management*, 9(4), 435–457. doi:10.1080/14697010903360632

Sood, R., Stockdale, G., & Rogers, E. M. (1987). How the news media operate in natural disasters. *The Journal of Communication*, 37.

Spear, S., & Bowen, H. K. (1999). Decoding the DNA of the Toyota production system. *Harvard Business Review*. Retrieved from http://hbr.org/1999/09/decoding-the-dna-of-the-toyota-production-system/ar/1

Spearman, M. L., Woodruff, D. L., & Hopp, W. J. (1990). CONWIP: A pull alternative to Kanban. *International Journal of Production Research*, 28(5), 879–894. doi:10.1080/00207549008942761

Speed, P. (2011). *Website*. Retrieved from http://code.google.com/speed/page-speed/

Spender, J. C. (1996). Making knowledge the basis of a dynamic theory of the firm. *Strategic Management Journal*, 17, 45–62.

Spremic, M., Šimurina, J., Jaković, B., & Ivanov, M. (2010). E-government in transition economies. *International Journal of Human and Social Sciences*, 5(2), 82–90.

Standing, C., Guilfoyle, A., Lin, C., & Love, P. E. (2006). The attribution of success and failure in IT projects. *Industrial Management & Data Systems*, 106(8), 1148–1165. doi:10.1108/02635570610710809

Standish Group International. (2009). *Website*. Retrieved from http://blog.standishgroup.com/

Steinberg, P. W. (2003). *Decision making styles within different hierarchical levels in the South African military health service*. (Thesis). Technikon Pretoria. Pretoria, South Africa.

Stensakar, I. (2003). *Excessive change: Unintended consequences of strategic change*. Briarcliff Manor, NY: Academy of Management Press.

Stiglitz, J. (2006). *Making globalization work*. New York, NY: W.W. Norton and Company.

Storm, K., Kraemer, E., Aurrecoeche, C., Heiges, M., Pennington, C., & Kissinger, J. (2009). Web site evolution: Usability evaluation using time series analysis of selected episode graphs. In *Proceedings of the 11th IEEE International Symposium on Web Systems Evolution (WSE), 2009*, (pp. 27-36). IEEE Press.

Storm, P., & Savelsbergh, C. (2005). *Lack of managerial learning as a potential cause of project failure*. Retrieved from http://www.ou.nl/Docs/Faculteiten/MW/Congres%20Papers/2005/18-20%20aug%202005.pdf

Streufert, S., & Streufert, S. (1978). *Behavior in the complex environment*. Washington, DC: Winston-Wiley.

Su, L. -.H., Ping-Shun Chen, P.-S., & Chen, S.-Y. (2012). Scheduling on parallel machines to minimise maximum lateness for the customer order problem. *International Journal of Systems Science*. doi:doi:10.1080/00207721.2011.649366

Succi, G., Benedicenti, L., Bonamico, C., & Vernazza, T. (1998). The webmetrics project - Exploiting software tools on demand. In *Proceedings of the 1998 World Multiconference on Systemics, Cybernetics, and Informatics*. Orlando, FL: IEEE.

Sundari, R. T., Barwal, P. N., Prakash, R., Yadav, R., Garg, C., & Jain, D. K. (2009). An analysis of factors influencing success and failure of IT project. [Noida, India: CDAC.]. *Proceedings of ASCNT, 2009*, 152–158.

Swanson, D., & Mancini, P. (1996). Politics, media and modern democracy – An international study of innovations in electoral campaigning and their consequences. London, UK: Westport.

Taguchi, G. (1986). *Introduction to quality engineering*. White Plains, NY: Asian Productivity Organization.

Taguchi, G. (1993). *Taguchi on robust technology development: Bringing quality engineering upstream*. New York, NY: ASME Press. doi:10.1115/1.800288

Taksa, I., & Spink, A. (2007). Evaluating usability of a long query meta search engine. In *Proceedings of the 40th Annual Hawaii International Conference on System Sciences, 2007*, (p. 82). IEEE Press.

Tam, M. M. C., Chung, W. W. C., Yung, K. L., David, A. K., & Saxena, K. B. C. (1994). Managing organizational DSS development in small manufacturing enterprises. *Information & Management, 26*(1), 33–47. doi:10.1016/0378-7206(94)90005-1

Tapping, D., Luyster, T., & Shuker, T. (2002). *Value stream management: Eight steps to planning, mapping, and sustaining lean improvements.* New York, NY: Productivity Press.

Tariq, A., & Manarvi, I. A. (2011). Defect tend analysis of F-7P aircraft through maintenance history. In *Proceedings of 2011 IEEE Aerospace Conference (AERO 2011)*, (pp. 1-8). Washington, DC: IEEE Computer Society.

TAW3. (2011). *A tool for the analysis of web sites, based on the W3C - Web content accessibility guidelines 1.0 (WCAG 1.0)*. Retrieved from http://www.tawdis.net/ingles.html

Taylor, M. A. (2002). *The 5 reasons why most projects fail and what steps you can take to prevent it.* Retrieved from http://www.idii.com/wp/Tse5reasonswhyprojectsfail.pdf

Taylor, S., & Todd, P. A. (1995). Understanding information technology usage: A test of competing models. *Information Systems Research, 6*(2), 144–176. doi:10.1287/isre.6.2.144

Telford, J. (2007). A brief introduction to design of experiments. *Johns Hopkins APL Technical Digest, 27*(3).

Thite, M. (1999). Leadership: A critical success factor in IT project management. In *Proceedings of Portland International Conference on Management of Engineering and Technology*, (pp. 298-303). PICMET.

Tripathi, P., Pandey, M., & Bharti, D. (2010). Towards the identification of usability metrics for academic web-sites. In *Proceedings of the 2nd International Conference on Computer and Automation Engineering (ICCAE)*, (vol. 2, pp. 393-397). Singapore, Singapore: ICCAE.

Tsang, K. F., Tung, H. Y., & Lam, K. L. (2009). *ZigBee: From basics to designs and applications.* Upper Saddle River, NJ: Prentice Hall.

Tung, H. Y., Tsang, K. F., Lam, K. L., Tung, H. C., Zheng, R. J., Ko, K. T., & Lai, L. L. (2011). A WiMAX-ZigBee energy management system for green education. *Smart Grid and Renewable Energy, 2*(4), 338–348. doi:10.4236/sgre.2011.24039

Turner, R. (1980). The mass media and preparations for natural disaster. In *Disasters and the Mass Media.* Washington, DC: The National Research Council.

United States Department of Defense. (1998). *DoD integrated product and process development handbook.* Washington, DC: DoD.

Van Buren, M., & Safferstone, T. (2009). The quick wins paradox. *Harvard Business Review, 87*(1), 54–61.

van Riel, A. C. R., Liljander, V., & Jurriëns, P. (2001). Exploring consumer evaluations of e-services: A portal site. *International Journal of Service Industry Management, 12*(40), 359–377. doi:10.1108/09564230110405280

Venkatesh, V. (1999). Creation of favourable user perceptions: Exploring the role of intrinsic motivation. *Management Information Systems Quarterly, 23*(2), 239–260. doi:10.2307/249753

Venkatesh, V., & Spier, C. (2000). Computer technology training in the workplace: A longitudinal investigation of the effect of mood. *Organizational Behavior and Human Decision Processes, 79*(1), 1–28. doi:10.1006/obhd.1999.2837

Vijayapriya, T., & Kothari, D. P. (2011). Smart grid: An overview. *Smart Grid and Renewable Energy, 2*(4), 305–311. doi:10.4236/sgre.2011.24035

Vredenburg, K., Isensee, S., & Righi, C. (2001). *User-centered design: An integrated approach.* Upper Saddle River, NJ: Prentice Hall.

Vroom, V. H., & Yetton, P. W. (1973). *Leadership and decision making.* Pittsburgh, PA: University of Pittsburgh Press.

W3C. (2011). *W3C markup validation service.* Retrieved from http://validator.w3.org/

Walkup, G. W., Jr., & Ligon, J. R. (2006). *The good, bad, and ugly of stage-gate project management process as applied in oil and gas industry.* Paper presented at the SPE Annual Technical Conference and Exhibition. San Antonio, TX.

Walsh, J. (2012). *Europe's internet revolt: Protesters see threats in antipiracy treaty.* The Christian Science Monitor.

Wang, G., & Cheng, T. C. E. (2007). Customer order scheduling to minimizing total weighted completion time. *Omega, 35,* 623–626. doi:10.1016/j.omega.2005.09.007

Wang, L., Bretschneider, S., & Gant, J. (2005). Evaluating web-based e-government services with a citizen-centric approach. In *Proceedings of the 38th Annual Hawaii International Conference on System Sciences (HICSS 2005).* Hawaii, HI: IEEE Press.

Wang, W. P. (2009). Toward developing agility evaluation of mass customization systems using 2-tuple linguistic computing. *Expert Systems with Applications, 36*(2), 3439–3447. doi:10.1016/j.eswa.2008.02.015

Warkentin, D. (1998). *Electric power industry in non-technical language.* Tulsa, OK: PennWell.

Warne, L., & Hart, D. (1996). The impact of organizational politics on information systems project failure - A case study. In *Proceedings of the 29th Annual Hawaii International Conference on System Sciences,* (pp. 191-201). IEEE.

Wateridge, J. (1995). IT projects: A basis for success. *International Journal of Project Management, 13*(3), 169–172. doi:10.1016/0263-7863(95)00020-Q

Waterman, R. Jr, Peters, T., & Phillips, J. (1980). Structure is not organization. *Business Horizons, 23*(3), 14–26. doi:10.1016/0007-6813(80)90027-0

Watts, F. (2011). *Engineering documentation control handbook: Configuration management and product life-cycle management.* Waltham, MA: Elsevier.

Webster, J., Trevino, L. K., & Ryan, L. (1993). The dimensionality and correlates of flow in human computer interactions. *Computers in Human Behavior, 9*(4), 411–426. doi:10.1016/0747-5632(93)90032-N

Weick, K., & Quinn, R. (1999). Organizational change and development. *Annual Review of Psychology, 50,* 361–386. doi:10.1146/annurev.psych.50.1.361

Weischedel, A., & Huizingh, E. K. R. E. (2006). Website optimization with web metrics: A case study. In *Proceedings of the 8th International Conference on Electronic Commerce - ICEC 2006,* (pp. 463-470). Fredericton, Canada: ICEC.

Welker. (1966). System effectiveness. In W. Ireson (Ed.), *Reliability Handbook.* New York, NY: McGraw-Hill.

Wellman, B. (1997). The road to utopia and dystopia on the information highway. *Contemporary Sociology, 26*(4), 445–449. doi:10.2307/2655085

Wellman, B., & Gulia, M. (1995). When social networks meet computer networks: The policy implications of virtual communities. New York, NY: American Sociological Association (ASA).

Welsh, J. (2003). *Jack: Straight from the gut.* New York, NY: Grand Central Publishing.

Wenger, D., James, T., & Faupel, C. (1980). *Disaster beliefs and emergency planning.* Newark, NJ: Disaster Research Center.

Wenger, D., & Quarantelli, E. L. (1989). *Local mass media operations, problems and products in disasters report series # 19.* Newark, NJ: Disaster Research Center.

Wessel, G., & Burcher, P. (2004). Six sigma for small and medium-sized enterprises. *The TQM Magazine, 16*(4), 264–272. doi:10.1108/09544780410541918

Wheeler, D. J. (2004). *The six sigma practitioner's guide to data analysis.* New York, NY: SPC Press.

Who.Is. (2011). *Website.* Retrieved from http://whois.is/

Wilson, E. J., & Sherrell, D. L. (1993). Sources effects in communication and persuasion research: A meta-analysis of effect size. *Journal of the Academy of Marketing Science, 21.*

Winklhofer, H. (2001). Organizational change as a contributing factor to IS failure. In *Proceedings of the 34th Hawaii International Conference on System Sciences,* (pp. 1-9). IEEE.

WIPO. (2011). *Article 32.* Retrieved from http://www.wipo.int/pct/en/texts/articles/a32.htm

Wolf, M. (1993). *Teoria delle comunicazioni di massa.* Milan, Italy: Bompiani.

Womack, J., & Jones, D. (1996). *Lean thinking: Banish waste and create wealth in your corporation.* New York, NY: Simon and Schuster.

Womack, J., Jones, D., & Rose, D. (1991). *The machine that changed the world: The story of lean production.* New York, NY: Harper Perennial.

Wong, X., Jingchun, F., & Ming, L. (2009). Research on IT project life cycle. In *Proceedings of the IEEE Second International Conference on Intelligent Computation Technology and Automation,* (pp. 244-247). IEEE Press.

Woorank. (2011). *A SEO website analysis tool.* Retrieved from http://www.woorank.com

World Intellectual Property Organization. (1995). *The first twenty-five years of the patent cooperation treaty (PCT) 1970-1995.* Washington, DC: WIPO Publications.

WTO. (2012). *Article 33.* Retrieved from http://www.wto.org

Wu, H., Liu, Y., Ding, Y., & Liu, J. (2004). Methods to reduce direct maintenance costs for commercial aircraft. *Aircraft Engineering and Aerospace Technology, 1*(1), 15–18. doi:10.1108/00022660410514964

Wylie, R., Orchard, R., Halasz, M., & Dube, F. (1997). IDS: Improving aircraft fleet maintenance. In *Proceedings of 14th National Conference on Artificial Intelligence and Innovative Applications of Artificial Intelligence,* (pp. 1-8). IEEE.

Wysocki, R. (2012). *Effective project management: Traditional, agile, extreme* (6th ed.). Indianapolis, IN: John Wiley & Sons.

Xenu. (2011). *Link sleuth.* Retrieved from http://home.snafu.de/tilman/xenulink.html#Description

Xu, Q., Ning, H., & Chen, W. (2009). Video-based foreign object debris detection. In *Proceedings of IEEE International Workshop on Imaging Systems and Techniques (IST 2009),* (pp.119-122). IEEE Press.

Xue, M., Harker, P. T., & Heim, G. R. (2004). *Incorporating the dual customer roles in e-service design.* Philadelphia, PA: The Wharton Financial Institutions Center.

Yang, J. (1998). *Scheduling with batch objectives.* (PhD Thesis). The Ohio State University. Columbus, OH.

Yang, J. (2005). The complexity of customer order scheduling problems on parallel machines. *Computers & Operations Research, 32,* 1921–1939. doi:10.1016/j.cor.2003.12.008

Yang, J., & Posner, M. E. (2005). Scheduling parallel machines for the customer order problem. *Journal of Scheduling, 8,* 49–74. doi:10.1007/s10951-005-5315-5

Yaobin, L., & Tao, Z. (2007). A research of consumers' initial trust in online stores in China. *Journal of Research and Practice in Information Technology, 39*(3). Retrieved from http://www.jrpit.acs.org.au/jrpit/JRPITVolumes/JRPIT39/JRPIT39.3.167.pdf

Yeo, K. T. (1993). System thinking and project management – Time to routine. *International Journal of Project Management, 11*(2). doi:10.1016/0263-7863(93)90019-J

Yeo, K. T. (2002). Critical failure factors in information system projects. *International Journal of Project Management, 20,* 241–246. doi:10.1016/S0263-7863(01)00075-8

Yin, R. (2008). *Case study research: Design and methods* (4th ed.). Thousand Oaks, CA: Sage Publications, Inc.

Younus, B., & Manarvi, I. A. (2011). Defect trend analysis of airborne fire control radar using maintenance history. In *Proceedings of 2011 IEEE Aerospace Conference,* (pp 1-5). Big Sky, MT: IEEE.

Yourdon, E. (1997). *Death march: The complete software developer's guide to surviving "mission impossible" projects.* Upper Saddle River, NJ: Prentice Hall Publications.

Yousef, D. A. (1998). Predictors of decision-making styles in a non-western country. *Leadership and Organization Development Journal, 19*(7), 366–373. doi:10.1108/01437739810242522

Yuki, G. (1994). *Leadership in organization* (3rd ed.). Englewood Cliffs, NJ: Prentice- Hall.

Zeithaml, V. A., Parasuraman, A., & Malhotra, A. (2000). *A conceptual framework for understanding e-service quality: Implications for future research and managerial practice.* Working paper, report no. 00-115. Marketing Science Institute.

Zhang, Y., Zhu, H., & Greenwood, S. (2004). Web site complexity metrics for measuring navigability. In *Proceedings of the Fourth International Conference on Quality Software,* (pp. 172-179). Braunschweig, Germany: IEEE.

Zhong, J., & Bhattacharya, K. (2002). Towards competitive market for reactive power. *IEEE Transactions on Power Systems, 17*(4), 1206–1215. doi:10.1109/TPWRS.2002.805025

About the Contributors

Saqib Saeed is an Assistant Professor at the Computer Science Department at Bahria University Islamabad, Pakistan. He has a Ph.D. in Information Systems from University of Siegen, Germany, and a Masters degree in Software Technology from Stuttgart University of Applied Sciences, Germany. He is also a Certified Software Quality Engineer from American Society of Quality. His research interests lie in the areas of human-centered computing, computer-supported cooperative work, empirical software engineering, and ICT4D.

Mohammad Ayoub Khan is working with Center for Development of Advanced Computing (Ministry of Communication and IT), Govt. of India, as a Scientist, with interests in CAD VLSI design tools and technique, hardware-software code sign, VLSI (electronic design automation, circuit optimization, timing analysis), placement and routing in network-on-chip (buffer management, interconnects and port design), RFID technologies, etc. He has more than 7 years' experience in his research area. He has published more than 50 papers in reputed journals and international IEEE/Springer conferences. He is contributing to the research communities by various volunteer activities. He has served as conference chair in various reputed international conferences like ICMLC 2010, ICSEM 2010, ICRTBAIP 2010, ICIII 2010, to name a few. He is member of professional bodies of IEEE, ISTE, IACSIT, ACEE, and IAENG. He is also member of editorials/reviewers board for *IEEE Communications Letters*, *IEEE Transaction of Industrial Informatics*, Springer CCSP, Elsevier Computers and Electrical Engineering.

Rizwan Ahmad received B.Sc. Electrical and Electronics Engineering degree from Islamic University of Technology, in 2000, M.Sc. degree in Communication Engineering and Media Technology from University of Stuttgart, Germany, in 2004, and Ph.D. in Electrical Engineering degree from Victoria University, Melbourne, Australia, in 2010. In addition to academic assignments, Dr. Ahmad has worked with several industries, such as Sony, Fujitsu Siemens, and British Telecom. He is currently working as a Postdoctoral Research Fellow at College of Engineering, Qatar University, Qatar. His research interests include MAC protocols, spectrum and energy efficiency, and performance analysis for wireless communication and networks. Dr. Ahmad was a recipient of the prestigious IPRS scholarship from the Australian government. He has published and served as a reviewer for IEEE journals and conferences. He has also served as committee member for various conferences.

* * *

Hisham Abdelsalam holds a Master of Science and a Ph.D. in Mechanical Engineering (Old Dominion University, Norfolk, Virginia, USA). He obtained his Bachelor degree with honors in Mechanical Engineering from Cairo University (Cairo, Egypt). Dr. Abdelsalam is an Associate Professor in the Operations Research and Decision Support Department, Faculty of Computers and Information, Cairo University. In 2009, Dr. Abdelsalam was appointed as the director of the Decision Support and Future Studies Center in Cairo University. During the past four years, Dr. Abdelsalam has led several consultancy and research projects and published fourteen scholarly articles on e-government.

Emad A. Abu-Shanab earned his PhD in Business Administration, majoring in MIS area in 2005 from Southern Illinois University – Carbondale, USA. He earned his MBA from Wilfrid Laurier University in Canada, and his Bachelor in Civil Engineering from Yarmouk University (YU) in Jordan. He is an Associate Professor in MIS, where he teaches courses like Operations Research, e-Commerce, e-Government, introductory and advanced courses in MIS, Production Information Systems, and Legal Issues of Computing. His research interests are in areas like e-government, technology acceptance, and e-learning. He has published many articles in journals and conferences and authored two books in the e-government area. Dr. Emad worked as an Assistant Dean for Students' Affairs, Quality Assurance Officer in Oman, and the Director of Faculty Development Center at YU.

Ashraf Al-Saggar worked in Irbid District Electricity Company (IDECO) since 2003, and now holds a system administrator position. He earned his Bachelor degree in Computer Information System (CIS) from Philadelphia University in 2008 and finished his Master degree in Management Information System (MIS) from Yarmouk University in 2012. He is an expert in system administration and data management, and has interest in IT project management research and IT management in general.

Izzat Alsmadi is an Associate Professor in the Department of Computer Information Systems at Yarmouk University in Jordan. He obtained his Ph.D. degree in Software Engineering from NDSU (USA). His second Master in Software Engineering from NDSU (USA) and his first Master in CIS from University of Phoenix (USA). He has a B.Sc. degree in Telecommunication Engineering from Mutah University in Jordan. He has several published books, journals, and conference articles largely in software engineering fields.

A. Q. Ansari is working with the Department of Electrical Engineering at Jamia Millia Islamia (A Central University by an Act of Parliament), New Delhi. He has also served as Professor and Head, Department of Computer Science, and as Dean, Faculty of Management Studies and Information Technology at Jamia Hamdard (Hamdard University), New Delhi. Professor Ansari did his B. Tech. (Electrical, Low Current) from AMU, Aligarh, and received the M. Tech. (IEC) and Ph. D. degrees from IIT Delhi and JMI, New Delhi, respectively. His studies are in the areas of computer networks, networks-on-chip, image processing, and fuzzy logic. He has published about eighty research papers in international and national journals and proceedings of conferences. Prof. Ansari is a Fellow of IETE, IE (I), and National Telematics Forum (NTF). He is a Senior Member of IEEE (USA) and Computer Society of India and Life Member of ISTE, ISCA, and National Association of Computer Educators and Trainers (NACET). He is also presently the Chairman of the Delhi Chapter of the IEEE-Computational Intelligence Society.

Martin L Bariff is Associate Professor of Information Management in the Stuart School of Business at Illinois Institute of Technology, Chicago, IL, USA. He also is a CPA, CBPP, and CISA. Previously, he was a faculty member of The Wharton School (Penn) and The Weatherhead School (CWRU). Prof. Bariff teaches courses on Information Management and Control, Business Analytics, and Project Management. His research interests include impacts of IT on organization strategy, design, and performance; business agility and change; information overload consequences on productivity; and enterprise risk management. Currently, Prof. Bariff is Treasurer of the Information Overload Research Group and has held leadership positions in a number of academic and practitioner organizations. He has published in a variety of scholarly and professional journals. Prof. Bariff has consulted for both corporate and government organizations.

Eleonora Benecchi is a PhD Student at the Faculty of Communication Sciences at the Universitá della Svizzera Italiana. She collaborated in different research projects with Universitá della Svizzera Italiana focusing on television and radio programming and worked at University of Bologna, where she was Teaching Assistant of Theories and Techniques of Radio-TV Language. She is the author of the book *Anime: Cartoni con l´Anima* (2005). The main research interests are fandom studies, Japan animation, and TV-radio programming. She is also tutor of the radio network Psicoradio.

Edward T. Chen is Professor of Management Information Systems of Operations and Information Systems Department in the Manning School of Business at University of Massachusetts Lowell. Dr. Chen has published over sixty refereed research articles in scholarly journals such as *Information & Management*, *Journal of Computer Information Systems*, *Project Management*, *Comparative Technology Transfer and Society*, *Journal of International Technology and Information Management*, *International Journal of Innovation and Learning*, etc. Dr. Chen has served as vice-president, board director, track chair, and session chair of many professional associations and conferences. Professor Chen has also served as journal editor, editorial reviewer, and ad hoc reviewer for various academic journals. Dr. Chen has received the Irwin Distinguished Paper Award at the Southwestern Federation of Administrative Disciplines conference and the Best Paper Award at the International Conference on Accounting and Information Technology. His main research interests are in the areas of Project Management, Knowledge Management, and Green IT.

Reem Dawoud holds a Master of Business Administration with a concentration in Finance from Maastricht School of Management (MSM). She obtained her Bachelor degree in Business Administration from Cairo University (Cairo, Egypt) in 1992. She attended courses in Liquidity Forecasting at Bank of England, UK, 2008, and a course in US-Monetary Policy Implementation at the New York Federal Reserve, USA, 2009. Reem Dawoud is currently a Financial Consultant in the field of Portfolio Management for two of the top leading Asset Management firms in Egypt; previous to that, Mrs. Dawoud was Head of Money Market Department at Central Bank of Egypt. She had commenced her career in 1996 as a Financial Analyst in a leading brokerage firm in the field of equity markets.

Hatem Elkadi was born in Egypt in 1960. He graduated from the Faculty of Engineering, Cairo University, in 1983. He got his Ph.D. at the University of Lille, France, in 1993. He is currently an Assistant Professor at Cairo University. He is Advisor for Strategic Projects at MSAD (Ministry of State for Administrative Development) and supervises a number of national e-government projects. He was the Director of the Egyptian eGovernment Services Delivery Program, which was ranked 23rd worldwide in 2010, and in 2008, he won the first prize for the All Africa Public Service Innovation Award. He is member of the "National Dispute Settlement Committee for ICT issues," as well as the steering committee for the "National ID Card project." During his career, he managed several successful ICT projects with the government of Egypt, private sector, and NGOs, as well as consulted for National Projects in Yemen and Kuwait.

Maneesha Gupta received the B.E. degree in Electronics and Communication Engineering from Government Engineering College, Jabalpur, in 1981, M.E. in Electronics and Communication Engineering from Government Engineering College, Jabalpur, in 1983, and Ph.D. in Electronics Engineering (Analysis, Synthesis, and Applications of Switched Capacitor Circuits) from Indian Institute of Technology, Delhi, in 1990. She held the positions of Lecturer in Electronics and Communication Engineering Department at Government Engineering College, Jabalpur, from 1981 to 1982, Kota Engineering College, Kota, from 1986 to 1988, YMCA Institute of Engineering, Faridabad, in 1998, and Netaji Subhas Institute of Technology, New Delhi, from 1998 to 2000. She is currently working as Professor in Electronics and Communication Engineering (ECE) Department of the Netaji Subhas Institute of Technology, New Delhi. She has authored and co-authored over 50 research papers in the above areas in various international/ national journals and conferences. She got best paper award for her paper in *IETE Journal of Education* in 2001. Her teaching and research interests are switched capacitors circuits and analog signal processing.

Mudassir Hussain is a graduate of College of Aeronautical Engineering PAF Academy Risalpur, and is currently pursuing his Masters in Engineering Management at Center of Advanced Studies in Engineering.

Assad Iqbal is a Computer Information System Engineer. He graduated from NWFP University of Engineering and Technology, Peshawar, Pakistan, and did his M.Sc. in Engineering Management from the Department of Engineering Management, Center for Advance Studies in Engineering (CASE), Islamabad, Pakistan. Currently, he is pursuing his PhD in Engineering Management.

Seifedine Kadry is an Associate Professor of Applied Mathematics in the Faculty of General Education at American University of the Middle East Kuwait. He received his Masters degree in Modeling and Intensify Calculus (2001) from the Lebanese University – EPFL-INRIA. He did his Doctoral research (2003-2007) in Applied Mathematics from Blaise Pascal University-Clermont Ferrand II, France. He worked as Head of Software Support and Analysis Unit of First National Bank, where he designed and implemented the data warehouse and business intelligence; he has published one book and more than 25 papers on applied math, computer science, and stochastic systems in peer-reviewed journals.

Farrukh Masood Khawaja holds a Master degree in Embedded Systems Engineering from University of Stuttgart, Germany, and Bachelors in Computer Science from International Islamic University Islamabad, Pakistan. He has extensive industrial experience in well-known multinational companies like Ericsson Telecommunication Frankfurt, Altran Technologies Frankfurt, Micronas GmbH Munich, Marconi Communications Stuttgart, and Agilent Technologies Stuttgart. His expertise lies in the area of software verification and quality assurance.

346

Irfan Anjum Manarvi is an Aerospace Engineer. He graduated from PAF College of Aeronautical Engineering PAF Academy Risalpur, and did his M Phil and PhD from Department of Design Manufacture and Engineering Management, University of Strathclyde, Glasgow, United Kingdom. He is currently a Professor at Iqra University Islamabad Campus and is a Visiting Faculty Member at Center of Advanced Studies in Engineering.

Vincenzo De Masi holds a Master in Media Management from the Faculty of Communication Sciences at the University of Lugano, holds a four-year degree in DAMS Cinema at the University of Bologna. He has started a PhD program at University of Zurich and Lugano with a dissertation on Chinese Animation, and he is an assistant with the same Institute and also filmmaker.

Muhammad Asim Qazi was born in 1979. He received Bachelor of Aeronautical Engineering degree in Aerospace Discipline from College of Aeronautical Engineering, Risalpur. Presently, he is doing Masters in Engineering Management from Center for Advanced Studies in Engineering (CASE), Islamabad.

Hammad Ahmed Rafiq was born in 1974 at Rawalpindi, Pakistan. He received his Bachelors of Aeronautical Engineering degree in Avionics Discipline from College of Aeronautical Engineering Risalpur. Presently, he is doing Masters in Engineering Management from Center for Advance Studies in Engineering (CASE), Islamabad.

Mohammad Anwar Rahman is an Assistant Professor at the University of Southern Mississippi, USA. He joined the university in 2008 followed by his PhD in Engineering Science, concentration on Industrial Engineering. Rahman developed several new courses, such as Global Supply Chain Management, Industrial Modeling, Production Technology, in the Department of Industrial Engineering Technology at USM. He published research works in several academic and professional journals in the area of stochastic modeling for supply chain problems and decision-making application under uncertainty where information is little known and/or risk averse. He is partnering with Mississippi Department of Education, Mississippi Department of Transportation through funded projects and affiliated with International Conference on Industrial Engineering and Operations Management Program Committee. He is an executive board member of Center for Logistic Trade and Transportation—a research center for transportation logistics.

Hassanali Rassouli is a Mechanical Engineer and a Member of the American Society of Mechanical Engineers (ASME) with more than twenty-five years of engineering and engineering management practice in R&D and product design and development. He started his engineering practice as a draftsman, while he was a mechanical engineering student at Iran University of Science and Technology. He was acquainted with modular design method when he was working as design engineer for a shock-absorber manufacturing company and continued his engineering practice as a design engineer and later as an engineering and R&D manager mostly for automotive OEM suppliers, where he learned about product development approaches like APQC and standards like QS-9000 and ISO/TS16949. His career development started from component design to product design and engineering system design. He is now working as an engineering and management consultant.

Paolo Renna is an Assistant Professor at Department of Environmental Engineering and Physics in the Engineering Faculty of Basilicata University (Italy). He took Ph.D. degree at Polytechnic of Bari in Advanced Production Systems. His academic researches principally deal with the development of innovative negotiation and production planning in distributed environments and manufacturing scheduling in dynamic environment.

Kashif Saeed is currently working as Reservoir Engineer. He holds a Bachelor of Honour in Geology and Master in Geophysic and a Master degree in Reservoir Engineering. Integration of geo-sciences and engineering, static modeling, dynamic simulation, history matching, and uncertainty quantification are his key areas of interest.

Irene Samanta is working as a Lecturer in International Marketing in the Department of Business Administration TEI of Piraeus in Greece. She obtained her Ph.D. degree from the University of the West of Scotland (UK) in modeling the relationships in B2B firms under e-marketing practices. She obtained M.Sc. in International Marketing from University of Paisley (UK), and B.A. in Business Administration from the Technological Education Institute of Piraeus. She has taught the last eight years in M.Sc. program in International Marketing of the University of the West of Scotland and M.Sc. program in International Business Management in TEI of Piraeus. She has supervised a large number of M.Sc. dissertations concentrated in the area of marketing, electronic relationships. She is author and co-author of several papers in various scientific journals, book chapters, conference proceedings, and several articles in economic journals. She is the co-author of three textbooks. She participates as an active member in academic associations, including the European Association of Operations Management, European Academy of Management, and the Hellenic Institution of Operational Research. In addition, she participates as an active member in professional associations, including the Industrial and Commercial Chamber of Piraeus and Commercial Association of Piraeus.

Kamaljeet Sandhu is a Senior Lecturer of Accounting and Information Systems at the School of Business, Economics, and Public Policy of the University of New England. He earned his Ph.D. in Information Systems from Deakin University, Melbourne. His teaching and research expertise are in Electronic Services and Services Management at Universities, Corporate Governance and Law, Accounting Information Systems, Management Accounting, Asset Management, and E-Learning.

Satish Chandra Tiwari received the B.Sc. and M.Sc. degree in Electronics from University of Delhi and Budenlkhand University, respectively. He received the M.Tech degree in VLSI Design from Indraprastha University, New Delhi, in 2009. He currently is pursuing the Ph.D. degree in Electronics Engineering at the Division of ECE, NSIT, University of Delhi. From March 2009 to July 2009, he was an intern in IC Design Group, Cadence Design Systems, Noida, India, where he worked on generation and validation of test cases for Cadence Encounter. His research interests include low power high-density digital circuits. He has more than seven publications in international journals/conferences. He has two patents pending with IPO.

Muhammad Kashif Yaqoob is specialist in the domain of G&G information and project management systems. Currently, he is working as Specialist Information Management and focuses on designing integrated workflows for G&G applications. He has 13 years of experience in the exploration and production industry with focus on geophysics, project management, and G&G data management.

Georg Ziegler has developed a pre/post processor for the industry simulator Eclipse and works as a Reservoir Engineer. He has 30 years of experience in the exploration and production industry, from geophysics over geology to reservoir engineering.

Index

A

accessibility metrics 280, 286, 299
Advanced Metering Infrastructure (AMI) 209
Advanced Product Quality Planning (APQP) 62-63, 69-70, 81
aircraft ground operations 241
aircraft structure 244
air crew 239, 243, 245
Alternating Current (AC) 204, 208, 218, 258
Anti-Counterfeiting Trade Agreement (ACTA) 98
Area Control Error (ACE) computation 208
Automated Meter Reading (AMR) 209
Automatic Generation Control (AGC) system 208
aviation industry 128, 237
aviation safety 237

B

bottom up 36
brain dominance 226, 228
Bulk Electric System (BES) 204-205, 209
bullwhip effect 84-85, 90-92, 94-95
Burke-Litwin Causal model 29, 31
business process 32-33, 42, 47, 50

C

centralized inventory 87, 89
change management 23-25, 32-46, 63, 73-74, 76, 163
change triggers 40
Collaborative Planning, Forecasting, & Replenishment (CPFR) 94
component replacement 112, 120-122
concurrent engineering 68, 70-71, 76, 79
configuration management 63, 74-76, 81-83
CONstant Work in Process (CONWIP) 1, 3, 5-6, 11, 13-14, 18-20
Continental Aviation Engineering (CAE) 76, 128

continued usage 312, 314-316
copyright 97-99, 172
corporate management 62
corrective action 70, 160, 184-185, 248
critical success factors 188-191, 199-200, 303
customer order scheduling 1-4, 18-20
customer service 75, 85, 91, 95, 211, 218, 278

D

decentralized information system 91-92, 95
decision making styles 219-225, 228-236
Decision Style Inventory (DSI) 222-223, 226-228
decision support 186, 193, 219, 221-222, 229, 234-236
defect categorizing 113
defective components 129
Defect Prevention (DP) 129
Define, Measure, Analyze, Improve, and Control (DMAIC) 49-50, 52, 54-55, 73
descriptive statistics 302
Design for Assembly (DFA) 63, 70, 72, 80
Design for Manufacturability (DFM) 63, 70-72, 80
Design for X (DFX) 71-72
Design of Experiments (DOE) 63, 70, 80
Direct Current (DC) 61, 81-83, 110, 126, 148, 204, 208, 218, 236, 250, 260-261, 299, 321
Direct Maintenance Man Hour/Flying Hour (DMMH/FH) 112
disaster plans 253
discrete event simulation 8, 20
distribution system 208-209
dynamic conwip 1, 5-6, 13-14, 18

E

earthquake 252-257, 259-260
ease of use 277, 310-312, 314-321
Economic Order Quantity (EOQ) 87-88
efficiency measures 90

Electronic Data Interchange (EDI) 94-95
Electronic Services Acceptance Model (E-SAM) 310-311, 314-315, 319-320
energy industry 204, 206, 212, 216
engine intake 244-245
engine reliability 128
engine start up 245
environmental sustainability 48-49
escalation rate 188
e-services 285, 287-288, 302-322
Executive Information Systems (EIS) 221-222, 229, 234, 236
explicit knowledge 145-146

F

Facebook 27, 259, 262, 264, 267-268, 272
factor analysis 313-315, 318, 320, 322
Failure Modes and Effect Analysis (FMEA) 63, 70, 73, 78
Federal Energy Regulatory Commission (FERC) 205, 217
financial metrics 278, 288, 292-293
finished goods 2-4, 6-7, 9-12, 14, 18-19, 21, 86-87, 89
flight hours 129-131, 134
flight safety 128, 237
Foreign Object Damage (FOD) 237-250
Fuel Control Unit (FCU) 130-132
Fuel Distributor (FD) 135, 138
Fukushima 252-255, 258
functional model diagram 207, 218
fuzzy approach 1, 10, 13, 18-19
fuzzy logic 2, 7, 21

G

generators 207
Geometric Dimensioning and Tolerancing (GD&T) 63, 79-80, 82
green energy 205, 207, 218
ground crew 238, 243, 245-246, 249
Group Support Systems (GSS) 221-222, 229, 231-232, 236

H

hazard identification 174, 176, 180-181
Health, Safety, and Environment (HSE) management 151-152, 167-169, 181, 186
human factor 238-239, 243, 245-246, 249

I

Independent Power Producers (IPPs) 204-205
infringement 99-100
Integrated Diagnostic System (IDS) 112, 126
integrated product and process development 62-63, 68, 83
Intellectual Property Rights (IPR) 97-98, 109
international students 302-309, 311-312, 314, 319
Internet Service Providers (ISPs) 310
Investor Owned Utility (IOU) 204-205
IT project management 188, 190-191, 194, 198, 200-202

J

J69-T-25A engine 128
jet engines 128, 238, 250
journalism 148, 253-255, 260
Just In Time (JIT) production 76, 145

K

Kaizen model 33
Knowledge Management (KM) 143, 146-147
Kolb-Frohman Consulting model 31
Kotter 8-Step model 28

L

lead-time 19, 71, 89, 91
lean methodology 143-146
lean product development 62-63, 68, 70-71
lean thinking 62-64, 83, 150
Lewin-Schein Three-Stage model 25
Life Cycle Cost (LCC) 129

M

Management Information Systems (MIS) 95, 214, 221, 229, 232
managerial levels 219, 221-222, 226-228
managing change 36, 45, 47
manufacturing utilization 6-7, 11, 13, 18
marketing metrics 279, 289, 295-296
Mean Time Between Failure (MTBF) 8, 129, 132
Mean Time To Repair (MTTR) 8, 129
measurement systems 49
media 23, 27, 44, 149, 212, 252-255, 257-263, 266-267, 270-273
Meter Data Management Systems (MDMS) 209
MI-17/171 helicopters 111-112, 118, 120-122, 124

MI-172 helicopters 111-112, 114-115, 119-125
microblogging 266

N

National Electric Reliability Council (NERC) 205-206, 210
natural disasters 254, 257, 260-261
navigability metrics 280, 284, 289, 296
new product development 62-64, 66-68, 70, 73
No Fault Found (NFF) 112, 123

O

Occupational Safety and Health Administration (OSHA) 170-171
One Time Inspections (OTIs) 123, 129
online community 264-265
ontology 286, 300
operating hours 124, 129, 131, 134-135, 138, 140
order batching 91-92, 94
order release 1-2, 10-11, 21
organizational change 22-24, 29-31, 34, 36-47, 202
Organizational Decision Support System (ODSS) 221-222, 229, 232
organizational leadership 48
over the wall process 68

P

patentability 97, 100, 102, 104
patents 97-110
performance metrics 31, 277-278, 287-288, 291
Personal Protective Equipment (PPE) 180
postponement 89
Power of Habit model 30, 34
process design 49, 70, 77
process quality metrics 279, 282, 289
product attributes 76, 277
product design process 63, 67-68, 73
product development strategy 63, 65
productivity 22, 64, 66, 77, 82, 140, 144-146, 148-150, 182, 278, 282, 311-312, 315
Product Lifecycle Management (PLM) 63, 74, 76-77
product planning 63, 65, 68, 78
product portfolio management 63, 66, 68
program governance 49
project deliverables 22, 24, 153
project management 22-24, 31, 35, 37, 39-40, 43-46, 70, 76, 148, 151-155, 157-158, 166-167, 186, 188-194, 198-202, 214, 221, 275
project management lifecycle 189

project manager 22-24, 39-41, 44, 155-158, 162-164, 167, 186, 189, 191, 193-195, 199-201, 221, 234, 298
project organization 154-157, 167, 174, 186
project plan 55-56, 152-153, 158-159, 162, 166-167, 191
project quality improvement 160
project scheduling 191
pull-based replenishment 89

Q

Quality Function Deployment (QFD) 63, 70, 73, 77-78
quality management 49, 68, 82, 151-152, 172, 182-186
quality metrics 276, 279, 282, 284, 289, 295, 298

R

Regional Reliability Organizations (RROs) 205
risk assessment 49, 174, 176, 210
risk management 151-152, 155, 161-163, 167, 193, 199, 210, 218

S

safety standards 175, 179
search engine 270, 301
shortage gaming 91-92
simulation environment 1-2, 4, 8, 18
simultaneous engineering 63, 68, 70-71, 76
Six Sigma 48-50, 55, 60-61, 71, 73, 82-83
Smart Grid 203, 209-210, 212-213, 216-218
social media 23, 27, 44, 262, 270-271
social networking 262, 264-267, 269-270, 273
social technographics 267
software development 75, 140, 142-143, 146-149, 193, 195, 201, 275
software metrics 275-276, 286, 298, 300
software process 142-143, 147-148, 192
sources of information 255, 257, 260, 280, 304
stakeholders 23-24, 28-29, 31-37, 39-44, 46, 50-51, 53, 63-64, 75, 129, 147, 153, 157-158, 162, 166-167, 175, 182, 185, 189-193
strategic planning 219
Supervisory Control and Data Acquisition (SCADA) System 210, 212, 218
supply chain management 76, 84-85, 95
Switch model 26, 34-35
synchronization 77, 94-95
system analysis 210

T

tacit knowledge 145
Taguchi method 81
technical manuals 112, 122, 124
technical support 111, 149
technical training 111, 122, 125
Technology Acceptance Model (TAM) 310, 312, 314-317
technology planning 63, 66-67, 82
technology usage 304, 309
television news 252, 256-257
threat assessment 210
time management 164-165
top-down 36
Total Productive Maintenance (TPM) 145
Total Quality Management (TQM) 36, 49, 61, 76
trademark 99, 102
trade promotion 91-94
transfer functions 49
transmission lines 208

troubleshooting process 128
tsunami 252-257, 259-260

U

usage metrics 278, 287, 289, 291
user adoption 304
user experience 281, 302-303, 308, 310, 312, 317-320

V

value stream 63, 150
virtual community 262, 264, 267, 269-270, 272-273
virtual inventory bin 89

W

War of Currents 204, 218
Web 2.0 262-263, 266-267
web content metrics 279, 289
workforce management 203, 211-212, 214, 218
Work in Process (WIP) 2-3, 5-6, 9-11, 13, 18

CPSIA information can be obtained at www.ICGtesting.com
Printed in the USA
BVOW051049250113

311436BV00007B/134/P